LONDON MATHEMATICAL SOCIETY LECTURE NOTE SERIES

Managing Editor: Professor J.W.S. Cassels, Department of Pure Mathematics and Mathematical Statistics, University of Cambridge, 16 Mill Lane, Cambridge CB2 1SB, England

The titles below are available from booksellers, or, in case of difficulty, from Cambridge University Press.

46 p-adic analysis: a short course on recent work, N. KOBLITZ
50 Commutator calculus and groups of homotopy classes, H.J. BAUES
59 Applicable differential geometry, M. CRAMPIN & F.A.E. PIRANI
66 Several complex variables and complex manifolds II, M.J. FIELD
69 Representation theory, I.M. GELFAND *et al*
86 Topological topics, I.M. JAMES (ed)
87 Surveys in set theory, A.R.D. MATHIAS (ed)
88 FPF ring theory, C. FAITH & S. PAGE
89 An F-space sampler, N.J. KALTON, N.T. PECK & J.W. ROBERTS
90 Polytopes and symmetry, S.A. ROBERTSON
92 Representation of rings over skew fields, A.H. SCHOFIELD
93 Aspects of topology, I.M. JAMES & E.H. KRONHEIMER (eds)
96 Diophantine equations over function fields, R.C. MASON
97 Varieties of constructive mathematics, D.S. BRIDGES & F. RICHMAN
98 Localization in Noetherian rings, A.V. JATEGAONKAR
99 Methods of differential geometry in algebraic topology, M. KAROUBI & C. LERUSTE
100 Stopping time techniques for analysts and probabilists, L. EGGHE
104 Elliptic structures on 3-manifolds, C.B. THOMAS
105 A local spectral theory for closed operators, I. ERDELYI & WANG SHENGWANG
107 Compactification of Siegel moduli schemes, C.-L. CHAI
109 Diophantine analysis, J. LOXTON & A. VAN DER POORTEN (eds)
113 Lectures on the asymptotic theory of ideals, D. REES
114 Lectures on Bochner-Riesz means, K.M. DAVIS & Y.-C. CHANG
116 Representations of algebras, P.J. WEBB (ed)
118 Skew linear groups, M. SHIRVANI & B. WEHRFRITZ
119 Triangulated categories in the representation theory of finite-dimensional algebras, D. HAPPEL
121 Proceedings of *Groups - St Andrews 1985*, E. ROBERTSON & C. CAMPBELL (eds)
128 Descriptive set theory and the structure of sets of uniqueness, A.S. KECHRIS & A. LOUVEAU
129 The subgroup structure of the finite classical groups, P.B. KLEIDMAN & M.W. LIEBECK
130 Model theory and modules, M. PREST
131 Algebraic, extremal & metric combinatorics, M.-M. DEZA, P. FRANKL & I.G. ROSENBERG (eds)
132 Whitehead groups of finite groups, ROBERT OLIVER
133 Linear algebraic monoids, MOHAN S. PUTCHA
134 Number theory and dynamical systems, M. DODSON & J. VICKERS (eds)
135 Operator algebras and applications, 1, D. EVANS & M. TAKESAKI (eds)
137 Analysis at Urbana, I, E. BERKSON, T. PECK, & J. UHL (eds)
138 Analysis at Urbana, II, E. BERKSON, T. PECK, & J. UHL (eds)
139 Advances in homotopy theory, S. SALAMON, B. STEER & W. SUTHERLAND (eds)
140 Geometric aspects of Banach spaces, E.M. PEINADOR & A. RODES (eds)
141 Surveys in combinatorics 1989, J. SIEMONS (ed)
144 Introduction to uniform spaces, I.M. JAMES
146 Cohen-Macaulay modules over Cohen-Macaulay rings, Y. YOSHINO
148 Helices and vector bundles, A.N. RUDAKOV *et al*
149 Solitons, nonlinear evolution equations and inverse scattering, M. ABLOWITZ & P. CLARKSON
150 Geometry of low-dimensional manifolds 1, S. DONALDSON & C.B. THOMAS (eds)
151 Geometry of low-dimensional manifolds 2, S. DONALDSON & C.B. THOMAS (eds)
152 Oligomorphic permutation groups, P. CAMERON
153 L-functions and arithmetic, J. COATES & M.J. TAYLOR (eds)
155 Classification theories of polarized varieties, TAKAO FUJITA
156 Twistors in mathematics and physics, T.N. BAILEY & R.J. BASTON (eds)
158 Geometry of Banach spaces, P.F.X. MÜLLER & W. SCHACHERMAYER (eds)
159 Groups St Andrews 1989 volume 1, C.M. CAMPBELL & E.F. ROBERTSON (eds)
160 Groups St Andrews 1989 volume 2, C.M. CAMPBELL & E.F. ROBERTSON (eds)
161 Lectures on block theory, BURKHARD KÜLSHAMMER
162 Harmonic analysis and representation theory, A. FIGA-TALAMANCA & C. NEBBIA
163 Topics in varieties of group representations, S.M. VOVSI
164 Quasi-symmetric designs, M.S. SHRIKANDE & S.S. SANE
166 Surveys in combinatorics, 1991, A.D. KEEDWELL (ed)
168 Representations of algebras, H. TACHIKAWA & S. BRENNER (eds)
169 Boolean function complexity, M.S. PATERSON (ed)
170 Manifolds with singularities and the Adams-Novikov spectral sequence, B. BOTVINNIK
171 Squares, A.R. RAJWADE
172 Algebraic varieties, GEORGE R. KEMPF
173 Discrete groups and geometry, W.J. HARVEY & C. MACLACHLAN (eds)

174 Lectures on mechanics, J.E. MARSDEN
175 Adams memorial symposium on algebraic topology 1, N. RAY & G. WALKER (eds)
176 Adams memorial symposium on algebraic topology 2, N. RAY & G. WALKER (eds)
177 Applications of categories in computer science, M. FOURMAN, P. JOHNSTONE & A. PITTS (eds)
178 Lower K- and L-theory, A. RANICKI
179 Complex projective geometry, G. ELLINGSRUD *et al*
180 Lectures on ergodic theory and Pesin theory on compact manifolds, M. POLLICOTT
181 Geometric group theory I, G.A. NIBLO & M.A. ROLLER (eds)
182 Geometric group theory II, G.A. NIBLO & M.A. ROLLER (eds)
183 Shintani zeta functions, A. YUKIE
184 Arithmetical functions, W. SCHWARZ & J. SPILKER
185 Representations of solvable groups, O. MANZ & T.R. WOLF
186 Complexity: knots, colourings and counting, D.J.A. WELSH
187 Surveys in combinatorics, 1993, K. WALKER (ed)
188 Local analysis for the odd order theorem, H. BENDER & G. GLAUBERMAN
189 Locally presentable and accessible categories, J. ADAMEK & J. ROSICKY
190 Polynomial invariants of finite groups, D.J. BENSON
191 Finite geometry and combinatorics, F. DE CLERCK *et al*
192 Symplectic geometry, D. SALAMON (ed)
193 Computer algebra and differential equations, E. TOURNIER (ed)
194 Independent random variables and rearrangement invariant spaces, M. BRAVERMAN
195 Arithmetic of blowup algebras, WOLMER VASCONCELOS
196 Microlocal analysis for differential operators, A. GRIGIS & J. SJÖSTRAND
197 Two-dimensional homotopy and combinatorial group theory, C. HOG-ANGELONI, W. METZLER & A.J. SIERADSKI (eds)
198 The algebraic characterization of geometric 4-manifolds, J.A. HILLMAN
199 Invariant potential theory in the unit ball of $\mathbf{C}^n$, MANFRED STOLL
200 The Grothendieck theory of dessins d'enfant, L. SCHNEPS (ed)
201 Singularities, JEAN-PAUL BRASSELET (ed)
202 The technique of pseudodifferential operators, H.O. CORDES
203 Hochschild cohomology of von Neumann algebras, A. SINCLAIR & R. SMITH
204 Combinatorial and geometric group theory, A.J. DUNCAN, N.D. GILBERT & J. HOWIE (eds)
205 Ergodic theory and its connections with harmonic analysis, K. PETERSEN & I. SALAMA (eds)
206 An introduction to noncommutative differential geometry and its physical applications, J. MADORE
207 Groups of Lie type and their geometries, W.M. KANTOR & L. DI MARTINO (eds)
208 Vector bundles in algebraic geometry, N.J. HITCHIN, P. NEWSTEAD & W.M. OXBURY (eds)
209 Arithmetic of diagonal hypersurfaces over finite fields, F.Q. GOUVÊA & N. YUI
210 Hilbert C*-modules, E.C. LANCE
211 Groups 93 Galway / St Andrews I, C.M. CAMPBELL *et al*
212 Groups 93 Galway / St Andrews II, C.M. CAMPBELL *et al*
214 Generalised Euler-Jacobi inversion formula and asymptotics beyond all orders, V. KOWALENKO, N.E. FRANKEL, M.L. GLASSER & T. TAUCHER
215 Number theory 1992–93, S. DAVID (ed)
216 Stochastic partial differential equations, A. ETHERIDGE (ed)
217 Quadratic forms with applications to algebraic geometry and topology, A. PFISTER
218 Surveys in combinatorics, 1995, PETER ROWLINSON (ed)
220 Algebraic set theory, A. JOYAL & I. MOERDIJK
221 Harmonic approximation, S.J. GARDINER
222 Advances in linear logic, J.-Y. GIRARD, Y. LAFONT & L. REGNIER (eds)
223 Analytic semigroups and semilinear initial boundary value problems, KAZUAKI TAIRA
224 Computability, enumerability, unsolvability, S.B. COOPER, T.A. SLAMAN & S.S. WAINER (eds)
225 A mathematical introduction to string theory, S. ALBEVERIO, J. JOST, S. PAYCHA, S. SCARLATTI
226 Novikov conjectures, index theorems and rigidity I, S. FERRY, A. RANICKI & J. ROSENBERG (eds)
227 Novikov conjectures, index theorems and rigidity II, S. FERRY, A. RANICKI & J. ROSENBERG (eds)
228 Ergodic theory of $\mathbf{Z}^d$ actions, M. POLLICOTT & K. SCHMIDT (eds)
229 Ergodicity for infinite dimensional systems, G. DA PRATO & J. ZABCZYK
230 Prolegomena to a middlebrow arithmetic of curves of genus 2, J.W.S. CASSELS & E.V. FLYNN
231 Semigroup theory and its applications, K.H. HOFMANN & M.W. MISLOVE (eds)
232 The descriptive set theory of Polish group actions, H. BECKER & A.S. KECHRIS
233 Finite fields and applications, S. COHEN & H. NIEDERREITER (eds)
234 Introduction to subfactors, V. JONES & V.S. SUNDER
235 Number theory 1993–94, S. DAVID (ed)
236 The James forest, H. FETTER & B. GAMBOA DE BUEN
237 Sieve methods, exponential sums, and their applications in number theory, G.R.H. GREAVES, G. HARMAN & M.N. HUXLEY (eds)
238 Representation theory and algebraic geometry, A. MARTSINKOVSKY & G. TODOROV (eds)
239 Clifford algebras and spinors, P. LOUNESTO
240 Stable groups, FRANK O. WAGNER
241 Surveys in combinatorics, 1997, R.A. BAILEY (ed)
242 Geometric Galois actions I, L. SCHNEPS & P. LOCHAK (eds)
243 Geometric Galois actions II, L. SCHNEPS & P. LOCHAK (eds)
244 Model theory of groups and automorphism groups, D. EVANS (ed)
245 Geometry, combinatorial designs and related structures, J. HIRSCHFELD *et al*

London Mathematical Society Lecture Note Series. 241

Surveys in Combinatorics, 1997

Edited by

R. A. Bailey
Queen Mary and Westfield College, University of London

CAMBRIDGE UNIVERSITY PRESS
Cambridge, New York, Melbourne, Madrid, Cape Town, Singapore, São Paulo

Cambridge University Press
The Edinburgh Building, Cambridge CB2 2RU, UK

Published in the United States of America by Cambridge University Press, New York

www.cambridge.org
Information on this title: www.cambridge.org/9780521598408

First published 1997

A catalogue record for this publication is available from the British Library

ISBN-13 978-0-521-59840-8 paperback
ISBN-10 0-521-59840-0 paperback

Transferred to digital printing 2005

Contents

Preface

The 1997 issue of the British Combinatorial Bulletin contains a short history, written by Norman Biggs, of the early years of the British Combinatorial Conference. The first one was held at Oxford in 1969. The sixth conference, held at Royal Holloway College in 1977, was the first at which a volume containing the invited talks was published in time to be available to participants at the conference. Peter Cameron was the pioneering editor of that volume. Such a volume has been produced for every conference thereafter.

The 1977 conference was also the first one that I attended. There I joined the British Combinatorial Committee, which was formally set up at that meeting although it had effectively existed for some years—the previous conferences didn't just organize themselves. As often happens, I found that being on the committee considerably widened my knowledge of the subject. I left the committee in 1981, but have never lost touch with combinatorial activity in Britain.

I was delighted when I was asked to edit the present volume. In spite of the work involved, I am still delighted. I have had a preview of nine magnificent papers, and come to know their subject matter much better than I would otherwise have done.

At the centre of this volume is a long paper by Bruce Reed about the tree width of graphs. This is a new measure of connectivity. It is intimately linked to the concept of a minor of a graph, which is obtained by erasing an edge or coalescing two vertices joined by an edge, or by a sequence of such operations. Although the idea of 'forbidden minors' was made famous by Kuratowski's characterization of planar graphs in 1930, the main theoretical development of graph minors has taken place over the last decade, led by Neil Robertson and Paul Seymour. At the 1985 British Combinatorial Conference in Glasgow, Seymour talked about the early stages of this work, hot off the press. Now Reed, who has himself been one of the contributors to the area, gives this splendid survey of exceptionally deep, interesting and valuable work, some of the most important work ever in graph theory. It shows that tree width and minors give a rich mathematical theory to graphs in a way that the more obvious concepts of k-connectivity and induced subgraphs do not. It is a long paper for a conference proceedings, but it is a magnificent exposition of a major piece of work. It is laid out in such a way that the non-specialist can read it with ease, while at the same time containing proofs of the important results. This should become the definitive paper on the topic.

Alexander Schrijver's paper is a natural accompaniment to Reed's, because it too is concerned in part with graph minors. He describes some new and intriguing graph parameters which are non-decreasing upon taking minors. One of these, introduced by Colin de Verdière in 1990, gives another characterization of planarity but also has wider applications. A related parameter was introduced by Schrijver and co-workers in 1995. Colin de Verdière's invariant was suggested by ideas from differential geometry; most of the related invariants are more obviously combinatorial. Schrijver gives simpler proofs than those in the literature and poses interesting open questions about the relationship between these parameters (and others).

Perhaps the most famous graph parameters are the chromatic number and chromatic index, the smallest number of colours with which the graph can be properly vertex-coloured and edge-coloured respectively. In a proper vertex-colouring, the colours on the the ends of an edge must be different. Natural restrictions on such a colouring are to demand that each unordered pair of colours is used on at most one edge or at least one edge. These give harmonious and complete colourings respectively. Keith Edwards gives a clear and interesting survey of work on the parameters associated with these two types of colouring, some of it extremely recent. The paper will form a very useful background for further research in this area.

One use of graphs is as tool to study partially ordered sets (posets). In a poset, for each pair x and y of distinct elements, either $x < y$ or $y < x$ or x and y are incomparable. The comparability graph of the poset is obtained by joining x to y whenever x and y are comparable. Some properties of posets are constant over all posets with the same comparability graph—comparability invariant.

In applications such as scheduling of tasks or dating archaeologicial finds, the basic objects are intervals on the real line. These can be turned into posets by considering when the whole of one interval is to the left of the whole of another. Thus begins the theory of interval orders. Or they can be turned into graphs—interval graphs—by joining any two intervals whose intersection is not empty. Tom Trotter has contributed extensively to the theory of posets and its links with graph theory. Here he gives an excellent survey of the topics of interval orders and interval graphs.

Although it is also about graphs, Cheryl Praeger's paper has quite a different focus. Her motivation is groups of automorphisms of graphs. She considers highly symmetric graphs, those which are at least vertex-transitive (i.e. which admit groups of automorphisms which are transitive on the vertices.) Often she demands that the graph be 2-arc transitive, which means that its automorphism group is transitive on ordered paths of length two. A well-studied stronger property than transitivity is primitivity: a group acts primitively if it preserves no non-trivial partition.

Praeger defines a quasiprimitive group action as a natural generalization of a primitive group action, and defines a graph to be quasiprimitive if it admits a quasiprimitive group of automorphisms. Surveying work in which she and her collaborators have played a major role, she gives a complete categorization of quasiprimitive groups into eight types, parallel to the categorization given for primitive groups by the celebrated O'Nan–Scott theorem. She summarizes what has been done so far towards the classification of quasiprimitive graphs, especially those that are 2-arc transitive. This is an excellent example of how advances in permutation group theory following the classification of finite simple groups have made feasible such classifications of graphs. This is a very clear account of work still in progress.

Of course, automorphisms are a natural tool for studying all kinds of combinatorial structures, not just graphs. Sometimes we want to find all structures admitting an automorphism group with certain properties, as in Praeger's paper. At the other extreme, sometimes we study one particular example of a combinatorial structure precisely because its automorphisms are so interesting. John Conway's paper takes this point of view. It presents a new construction of the Mathieu group M_{12}, one of the simplest available.

Most constructions of sporadic simple groups such as M_{12} either produce some combinatorial structure (Steiner system, Hadamard matrix, code, ...), show that it is unique and deduce properties of its automorphism group (so that the actual automorphisms are not easy to construct), or else produce the permutations directly in such a way that it is hard to see the combinatorial structure. Conway's new construction takes the latter approach—all the permutations are produced at the outset—but the combinatorial properties are surprisingly easy to verify. Indeed, the exposition is skilfully constructed: once the descriptive parts of the paper have been read, the proofs are almost obvious.

A byproduct of this approach is the result that there exists a sharply 6-transitive set of permutations of 13 objects. They are the permutations of 12 counters and one hole in an analogue of the 15-puzzle played on the 13-point projective plane. This result will be interesting to researchers on extremal combinatorics, specifically, metric properties of sets of permutations.

Clement Lam's use of automorphisms is more conventional. The setting is 2-designs, also known as as balanced incompete-block designs, or just block designs—collections of k-subsets (called blocks) of a v-set with the property that every 2-subset is contained in the same number of blocks. Here is a very useful article about the use of the BDX program (demonstrated at the conference) to search for block designs. The reader is taken through an extended worked example—the search for designs which have a certain set of parameters and which admit an automorphism of order seven with no fixed points or blocks—and sees details of calculations which are not usually provided in published articles on mathematics. The reader is led so gently through the example that (s)he should have no difficulty in trying out BDX alone.

One of the classical sources of block designs is projective geometry: the points are the points of the geometry and the blocks are subspaces or conics, possibly with additional points. Thus we are led to questions in geometry such as: how may points are in the intersection of an object of one sort with an object of another sort? What size can a minimal blocking set be? (A blocking set is a set of points which meets every line.) Tamás Szőnyi's paper is a wide-ranging and detailed survey of such questions. It shows many important applications of two important results—Weil's theorem and Segre's lemma of tangents. It updates the material on blocking sets reported by Aart Blokhuis at the 1993 British Combinatorial Conference at Keele. A variety of results is described. The flavour of the proofs is given, rather than their technicalities, and further probable developments are discussed.

The papers by Edwards, Reed, Schrijver and Trotter all refer to problems in computational complexity. A typical problem is the following: given a graph G with n vertices, can we decide whether G is Hamiltonian? More specifically, is there an algorithm that will decide the answer within time that is linear in n, or polynomial in n, or otherwise bounded by some function of n? During the past decade or so a new type of complexity problem has been considered: not "can we?" but "how many?" and not "exactly" but "approximately". A typical example now is: given a graph G with n vertices, how many forests does G contain as induced subgraphs? Here "how many" means to within an order of magnitude. Once again one wants to know what sort of function of n bounds the running time for an algorithm that answers the question. Dominic Welsh's paper provides a fascinating survey of the very interesting material in this area, and thus rounds off this volume of papers.

In 1977 the task of the editor was to obtain typescripts from the speakers, by a deadline, and to submit camera-ready copy to the publisher. Over the twenty intervening years, what has changed? The deadlines are still there, and so is a publisher, even though Cambridge University Press has replaced Academic Press. Typescripts, with mathematical symbols either hand-written or typed from 'golf-balls', have vanished, being replaced first by word-processors then by mathematical type-setting packages such as TEX. As this happened, camera-ready copy at first became more diverse then converged as mathematicians came to agree on the use of a few systems. These proceedings are, I believe, the first in this series for which the editor has made a serious attempt at uniformity by asking all the authors to use the same system and giving them a style file to encourage them.

The authors have cooperated marvellously, often to tight schedules. Although they had not all used LATEX before, they all submitted their papers in fairly standard LATEX, using a rather simple style file which I had provided. My heartfelt thanks to them all.

None of this would have been possible without TEX, specifically the current version of LATEX, which is designed for people like me who are fussy about what their written mathematics looks like but do not want to program anything much more complicated than 'this is the end of the statement of the theorem'. So thanks to Donald Knuth for giving TEX to the world; to Leslie Lamport, for the original version of LATEX and for the manual which taught me to think in terms of generic mark-up; and to the LATEX3 team, led by Frank Mittelbach and Chris Rowley, for upgrading and maintaining LATEX, for incorporating features that I asked for, and for answering lots of my questions.

The other big change over the twenty years is the arrival of electronic mail. All the papers were submitted by email. Almost all of them were sent on to referees by email, and reports came back by the same route. During the editing phase the myriad queries such as 'Do you mean `\log` or `\ln` there?' were dispatched and answered promptly, all by email. Without email it would simply have been impossible to do the task within the given time-scale. I am grateful to the authors and the referees (as well as various people who helped me with the names of journals and books unfamiliar to me) for dealing with everything as quickly as they did.

In editing these proceedings I have tried to strike a fine balance between respecting the authors' wishes about how their mathematics is presented and providing the uniformity of layout that, by melting into the background, enables the reader to concentrate on the content and so read it more easily. If I have done my job well its effects should be invisible.

Finally, thanks to numerous people in the School of Mathematical Sciences at Queen Mary and Westfield College. Not only have they provided hardware, software, advice and moral support. They have been very patient with me while other jobs have been left on one side as I worked on these proceedings. This is particularly true of the other members of the local organizing team of this British Combinatorial Conference—Peter Cameron, Leonard Soicher and Shirley Wilkinson.

Rosemary A. Bailey
School of Mathematical Sciences
Queen Mary and Westfield College
Mile End Road
London E1 4NS
r.a.bailey@qmw.ac.uk

16 April 1997

M_{13}

J. H. Conway

Summary The group M_{12} has no transitive extension, but the object of the title is the next best thing: a *set* of permutations which is an extension of M_{12}. We give an elementary construction, based on a moving-counter puzzle on the projective plane of order 3, and provide easy proofs of some of its properties.

1 Introduction

Long ago I was intrigued by the fact that M_{12}, É. Mathieu's celebrated quintuply transitive group on 12 letters, shares some structure with $L_3(3)$, which acts doubly transitively on the 13 points of the projective plane PG(2, 3), of which it is the automorphism group.

To be more precise, the point-stabilizer in $L_3(3)$ is a group of structure $3^2{:}2S_4$ that permutes the 12 remaining points imprimitively in four blocks of 3, and there is an isomorphic subgroup of M_{12} that permutes the 12 letters in precisely the same fashion. Again, the line-stabilizer in $L_3(3)$ is a group of this same structure that permutes the 9 points not on that line in a doubly transitive manner, while the stabilizer of a triple in M_{12} is an isomorphic group that permutes the 9 letters not in that triple in just the same manner.

In the heady days when new simple groups were being discovered right and left, this common structure inevitably suggested that there should be a new group that contained both M_{12} and $L_3(3)$, various copies of which would intersect in the subgroups mentioned above. Of course this turned out not to be the case, but some years ago I found an almost equally satisfactory explanation — M_{12} and $L_3(3)$ are indeed both subgroups of the same object, but that object is not a group! I call it M_{13}.

Sections 2–6 will be purely descriptive, and contain numbered assertions, which will be proved in Section 7.

2 Definition of M_{13}

Since

M_{12} is a set of 95040 = 12.11.10.9.8 permutations of 12 letters, (1)

we might expect that

M_{13} is a set of 1235520 = 13.12.11.10.9.8 permutations of 13 letters; (2)

and since

M_{12} is quintuply transitive, (3)

we might hope that

$$M_{13} \text{ is sextuply transitive.} \tag{4}$$

These expectations turn out to be true, but we must be careful about the meanings of the terms.

Unfortunately, the word "permutation" retains two distinct senses — it may refer either to a particular *arrangement* of n objects, or to a particular operation of *rearranging* them, the latter usage being common among group theorists and the former among the public at large.

We shall define M_{13} to be a particular set of permutations of 13 objects in the lay sense, namely certain ways of putting 12 lettered counters and one hole on the points of a projective plane $P = \mathrm{PG}(2,3)$.

We can think of this in terms of a "13-puzzle" analogous to Sam Loyd's famous 15-puzzle, wherein 15 square tiles and one hole are arranged in a 4×4 tray, the object being to proceed from one given arrangement to another by a sequence of moves in each of which the hole is exchanged with one of the adjacent tiles.

In our 13-puzzle, any counter a determines at any time a line that joins its present position to that of the hole. A move of the puzzle is to put the counter into the hole (that is, onto the "holy point", as we shall call it), *and at the same time to interchange the positions of the other two counters* b *and* c *on this line.* We shall refer to this as the move $a|bc$.

To avoid circumlocution, we refer to the point occupied by the counter labelled a as "point a", and the holy point as "point $\circ$".

M_{13} consists of all the arrangements (of counters and hole) that can be obtained from a given one by moves of this type.

In the language of category theory, M_{13} is a *groupoid*, (a category in which all arrows are invertible), whose objects are the 13 positions of the hole, and whose arrows are the rearrangements produced by legal sequences of moves; the initial and terminal objects of an arrow are the positions of the hole before and after the moves are made.

3 The first few moves

Figures 1–4 show the way we shall draw the projective plane P. In Figure 1, the points of the plane are numbered 0–11 and ∞, and to avoid awkward bends some of the points on some lines are indicated by hooks. We have chosen this particular way of drawing the plane so as to emphasize the close relationship with the "MINIMOG" array (see Figure 2).

Figure 3 shows the usual coordinatization of the plane. The 9 points on the right are those of the corresponding "Euclidean" or "affine" plane, the point marked XY being that with affine coordinates (X, Y) or projective coordinates $(X, Y, 1)$. This is extended to the projective plane by adjoining the "line at

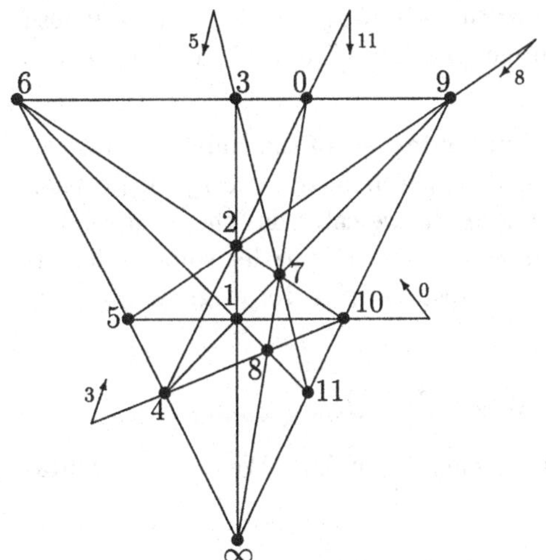

Figure 1: PG(2, 3)

6	3	0	9
5	2	7	10
4	1	8	11

Figure 2: The MINIMOG

infinity" on the left, whose point marked m (for $m = 0, 1, 2$) has projective coordinates $(1, m, 0)$ and lies on the three parallel lines $y = mx + c$, while that marked ∞ has coordinates $(0, 1, 0)$ and lies on the three "vertical" lines $x = c$.

In Figure 4 we display only the y-coordinates of the points other than ∞. There are 4 "vertical" lines (passing through the point ∞), and the 9 others meet these in the points determined by one of the 9 words

0 000, 0 111, 0 222, 1 012, 1 120, 1 201, 2 021, 2 102, 2 210

of the "tetracode" (see [3], [2]). The typical tetracode word is

$$m \;\; c \, c+m \, c+2m$$

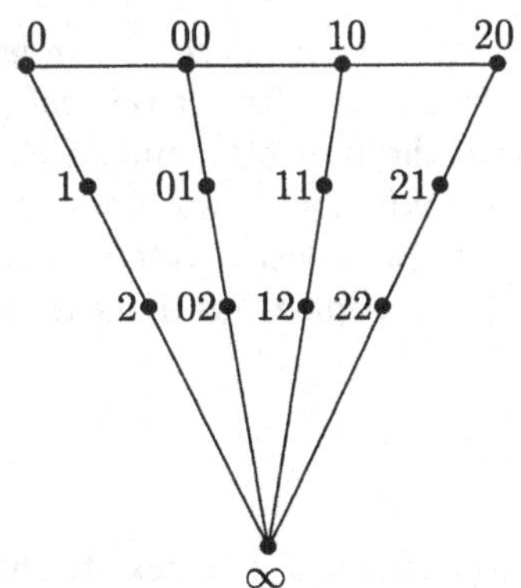

Figure 3: Coordinates

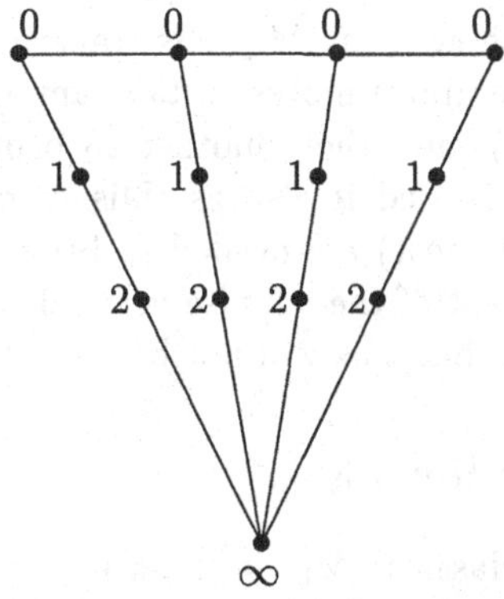

Figure 4: Tetracode coordinates

in which the last three digits form an arithmetic progression (mod 3), whose "slope" m is the first digit. For example, the line $\{5, 3, 7, 11\}$ of Figure 1 is called 1 012 in Figure 4.

It should be obvious that the set of permutations of the counters that can be achieved by move sequences that restore the hole to its original position at ∞ forms a group; anticipating a later result, we call this "the group M_{12}".

Figure 5 shows a few successive moves in the 13-puzzle, whose effect is to move the hole around a triangle and restore it to its original position, interchanging the four pairs

$$(4, 5),\ (3, 0),\ (6, 9),\ (10, 11)$$

of counters as it does so. This gives us an element of M_{12}. We call a permutation of this sort a *triangular permutation.*

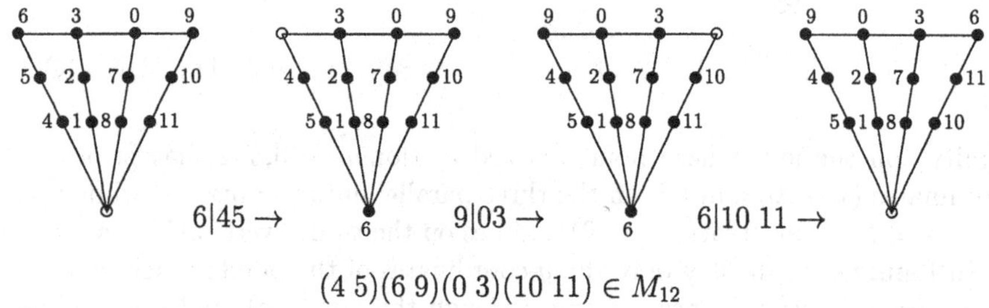

Figure 5: A triangular permutation

Look at the action of this permutation on the middle two of the four "vertical" lines. The points $1, 2, 7, 8$ are fixed, and 3 and 0 are interchanged. By symmetry, we see that M_{12} contains a permutation that swaps any two points lying on distinct verticals, and leaves the other four points on these verticals unchanged.

This shows that M_{12} acts transitively on the 12 counters, since we can use one of the above moves to take any given counter off the leftmost vertical (if necessary), and then another to bring it back to the topmost point of that vertical. Indeed it also establishes double transitivity, since two more such moves (at most) are needed to bring the second of two given counters to the second point of the leftwise vertical. One could prove triple transitivity in the same way, but this will not be necessary.

4 The hexads

The classical M_{12} is known to permute a collection of 132 hexads that form a so-called Steiner system $S(5, 6, 12)$. How do we recognize these in our picture?

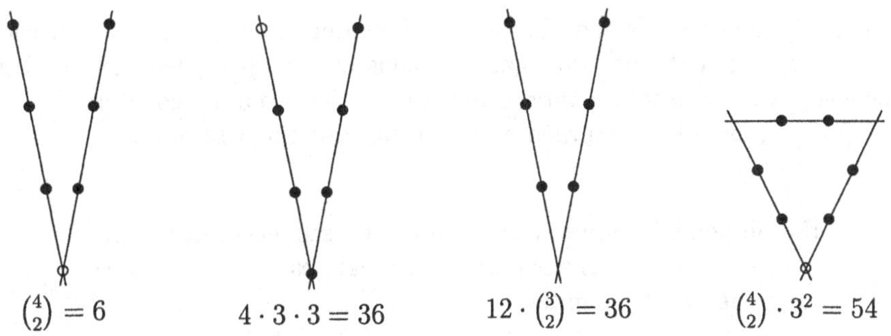

Figure 6: Hexads

Figure 6 shows the answer. At any instant, two lines contain either 6 counters and the hole, in which case these counters form a hexad; or 7 counters, from which we form a hexad by removing the counter at the intersection of the two lines. There is a further type of hexad, consisting of the points (other than the vertices) lying on the lines of a triangle one of whose vertices is the holy point. The Figure shows the four different geometrical appearances that a hexad can assume, and counts the hexads, showing that there are 132 in all.

Note that the triangular permutation shown in Figure 5 fixes the hexad $\{0, 1, 2, 3, 7, 8\}$ and induces the transposition $(0\ 3)$ on it.

One could easily give a case-by-case verification that

> a move of the 13-puzzle takes every hexad to another hexad, although possibly of a different shape. (5)

One could also check that

> no two distinct hexads can contain the same 5 points, and hence the hexads form an $S(5, 6, 12)$, (6)

and that

> M_{12} acts transitively on the hexads. (7)

5 The doublings of M_{13}

The group M_{12} is known to have Schur multiplier of order 2. This reveals itself by the existence of a group $2M_{12}$ that has a homomorphism onto M_{12} with kernel of order 2. This group can be constructed as a group of *monomial permutations* of ± 12 letters: that is to say, it permutes 24 symbols, say

$$+a, -a, +b, -b, \ldots, +k, -k, +l, -l$$

in such a way that we obtain the permutations of M_{12} simply by ignoring the signs. The non-trivial element of the kernel negates all 24 symbols. Is there an analogous $2M_{13}$?

Indeed there is! We can obtain it by labelling the opposite sides of each of our counters with the two signed versions of the appropriate letter. But now when we move some counter into the hole, the two other counters that we interchange must also be turned over. We call this the ± 13-puzzle.

It turns out that

> the monomial permutations realized by sequences of the new moves that return the hole to its original position ∞ do indeed form the usual group $2M_{12}$. (8)

Moreover,

> $2M_{12}$ is *doubly transitive in the monomial sense*; that is, given two pairs a, b and c, d of distinct letters and any signs $\alpha, \beta, \gamma, \delta$, there is an element of $2M_{12}$ which maps αa to γc and βb to δd. (9)

Since we have already shown the double transitivity, it is enough to show that $2M_{12}$ contains an element fixing one counter of the ± 13-puzzle and reversing another. Such an element is easily discovered. For example, the product of the triangular permutations obtained from the triangles $\infty 69$, $\infty 30$, and $\infty 6\ 10$ fixes 4 and negates 3.

The automorphism group of the classical M_{12} is a group $M_{12}.2$ that permutes 24 letters in two sets of 12, with the properties

> each permutation of the original M_{12} extends uniquely to a permutation of the 24 letters, and an outer automorphism interchanges the two sets of 12 letters. (10)

Is there an analogous $M_{13}.2$?

Indeed there is! To obtain it, we enlarge the 13-puzzle to the "26-puzzle" by adjoining an additional hole $\bigcirc$ and 12 new counters

$$A, B, \ldots, K, L$$

that we associate with the lines of P in the same way as the old ones are associated with the points. However, we also demand that the "holy point" must always be incident with the "holy line". This entails that whenever we make a point-move $a|bc$, the four points involved must form the holy line. Dually, we now have "line-moves", say $A|BC$, for which the four lines involved must pass through the holy point.

Any sequence of moves in the 26-puzzle yield a legal sequence of moves in the 13-puzzle if we just ignore the line-counters. It might seem that the incidence condition would restrict our freedom; but in fact an arbitrary sequence of point-moves can still be performed, by interleaving them with the appropriate line-moves. Thus, if we wish to follow a point move on the line L_1 by another on a different line L_2, we need merely interpose the line-move that moves the hole from L_1 to L_2.

In this way, for any permutation of M_{12} we can find a sequence of alternating point and line moves that effects that permutation of the point counters, and also restores both holes to their original positions (say, the point ∞ and the line at infinity). It turns out that

> the resulting permutation of the line-counters is uniquely determined, giving the extension of M_{12} to a group on 24 letters, doubly transitive on both sets of counters. (11)

We therefore define $M_{13}.2$ to be the set of *all* permutations of the 26 counters and holes obtainable from the starting position by moves of the 26-puzzle.

The group $M_{12}.2$ has a double cover $2M_{12}.2$ containing the group $2M_{12}$ that we described earlier. This also has an analogue, $2M_{13}.2$, obtained by using both sides of both sets of counters. Provided that $\circ, a, b, c$ lie on the current holy line $\bigcirc$, we can make a point-move $a|bc$ that puts a in the hole and interchanges and negates both b and c; dually, provided $\bigcirc, A, B, C$ pass through the current holy point $\circ$, we can make the line-move $A|BC$ that puts A in the hole and interchanges and negates B and C. We call it the "$\pm$26-puzzle".

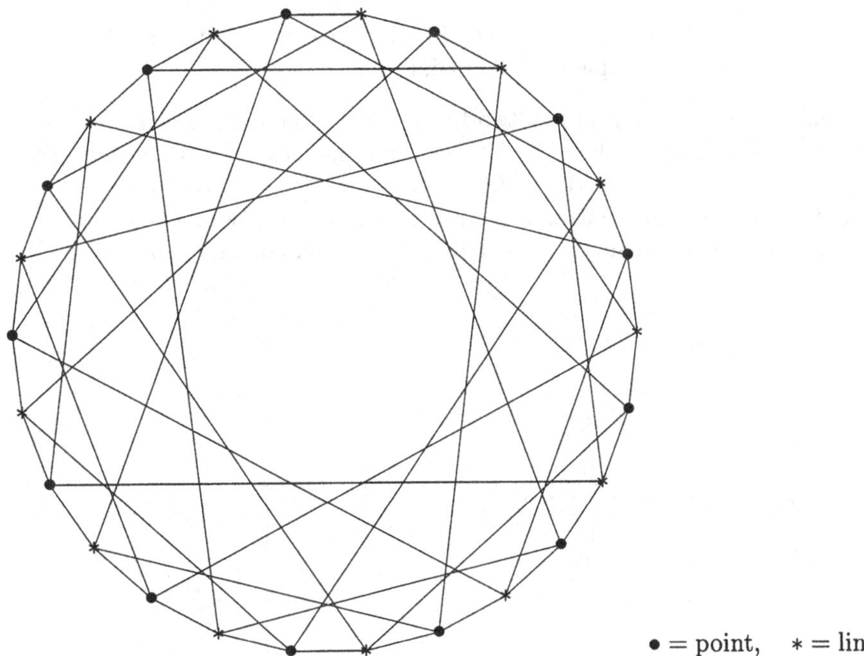

Figure 7: The incidence graph of PG(2, 3)

6 The inner product

We regard the symbols on the counters of the ± 26-puzzle as vectors $\pm u$ (for points) and $\pm V$ (for lines), and introduce an *inner product* between the two sets. We define (u, V) to be -1 just if either u and V are incident with each other, or both are incident with the current holes; otherwise $(u, V) = +1$. In terms of the *incidence graph* of the plane (which has vertices that correspond to the points and lines, with edges corresponding to the incident pairs, see Figure 7), (u, V) is equal to -1 just if either is *accessible* from the other — that is to say, if there is a path from u to V that contains no other counter.

It is quite remarkable that

> this inner product is unchanged by making any legal move of the ± 26-puzzle. (12)

To see this, look at the effect of a point-move $a|bc$ in the incidence graph. Figure 8 shows that the same line-counters N, P, Q, R, S, T are accessible from a before and after the move, while those accessible from either b or c before the move are precisely those *not* accessible from $-b$ or $-c$ after it. (Thus, R, S, T, U, V, W are initially accessible from b, while N, P, Q, X, Y, Z become accessible from $-b$.)

It turns out that

> the hexads that were introduced in an *ad hoc* way earlier can now be given a simple uniform definition. (13)

If V and W are labels from any two line-counters, there are 6 point-counters whose labels u satisfy $(u, V) = (u, W)$, which form a hexad $[V, W]$; the remaining 6 satisfy $(u, V) = -(u, W)$ and form the hexad $[V, -W] = [-V, W]$.

Let H be the Gram matrix of the inner product: its rows and columns are indexed by points and lines, the entry in row u and column V being (u, V).

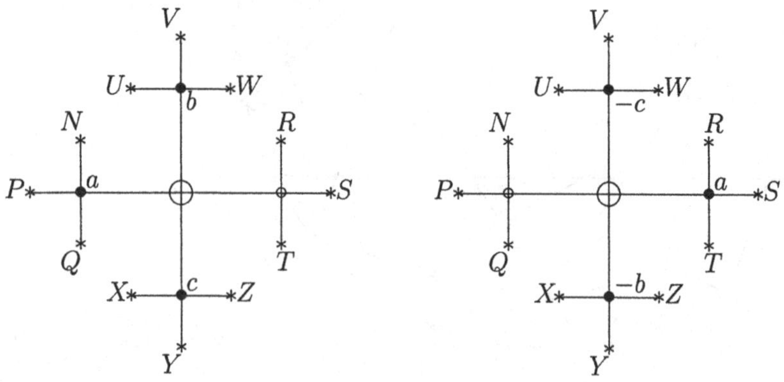

Figure 8: Invariance of the inner product

Then any two columns of H agree and disagree in sets of points forming hexads. So $H^\top H = 12I$, and H is a *Hadamard matrix*.

7 The proofs

It is trivial to check (13): in the new notation, the representative hexads of Figure 6 are as shown in Figure 9. Again, since the hexads are defined in terms of the inner product, (12) immediately implies (5). We can also see that they must be permuted transitively (as claimed in (7)): to move $[V, W]$ to $[X, \pm Y]$ it suffices to move V to X and W to $\pm Y$ (using the dual of (9)).

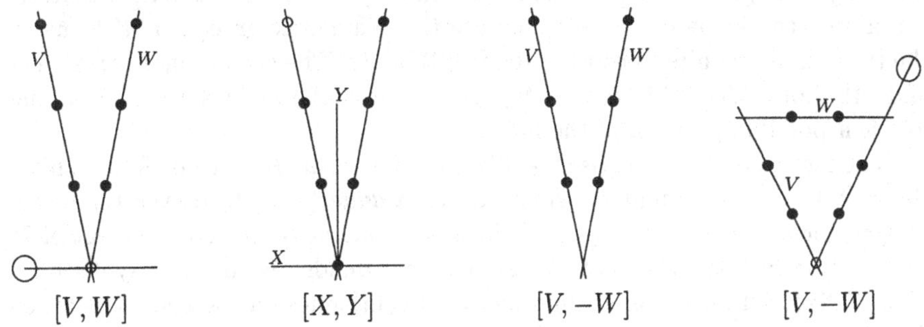

Figure 9: Hexads defined by the inner product

We now prove (10) and (11). Let P and Q be the matrices representing the monomial permutations of points and lines derived from any sequence of moves. Then $P^{-1}HQ = H$, by (12). In particular, $Q = H^{-1}PH$ is uniquely determined by P (since H is invertible). The interchange of point and line counters is effected by a polarity of the plane which swaps the positions of the two holes.

Consider a sequence of point and line moves which leaves every point counter fixed (possibly reversed). Then P is a diagonal matrix with $P^2 = I$. Suppose that $P \neq \pm I$. Since $H^\top PH = 12Q$, we see that P has six entries $+1$ and six entries -1, and the set of six positions where the $+1$s occur meets every hexad in 0, 3 or 6 points. However, no such set can exist. So $P = Q = \pm I$.

The proof of (6) is most easily seen using the Hadamard matrix H, or (for convenience of notation) $H^\top$. By column sign changes, the first three rows of $H^\top$ may be assumed to be

$$
\begin{matrix}
+ & + & + & + & + & + & + & + & + & + & + & + \\
+ & + & + & + & + & + & - & - & - & - & - & - \\
+ & + & + & - & - & - & + & + & + & - & - & -
\end{matrix}
$$

If these rows are V, W, X, then clearly $[V, \pm W] \cap [V, \pm X] = 3$. For any other row Y, if Y has a entries $+$ among the first three, then $[V, \pm W] \cap [X, \pm Y]$

is $2a$ or $6 - 2a$, which is even in either case. So no two hexads can meet in five points. The average number of hexads containing a set of five points is $132.6/\binom{16}{5} = 1$; so any five points lie in a unique hexad.

What is the order of M_{12}? We know by (7) that M_{12} acts transitively on hexads. Moreover, the transpositions induced on the hexad $\{0, 1, 2, 3, 7, 8\}$ by triangular permutations as in Figure 5 generate S_6. Further, the group K fixing a hexad pointwise is trivial (see below). So

$$|M_{12}| = 132.6! = 12.11.10.9.8.$$

Thus, (1) holds. The quintuple transitivity of M_{12} follows: give two 5-tuples, we find $g \in M_{12}$ mapping the hexad $\mathcal{H}$ containing the first tuple to the hexad $\mathcal{H}'$ containing the second; then we may use the symmetric group on $\mathcal{H}'$ to move the tuple to its required position. Thus, (3) holds. The corresponding facts (2) and (4) about M_{13} hold because M_{13} is the union of 13 translates of M_{12}, one for each possible position of the hole.

The fact that the pointwise stabilizer K of a hexad $\mathcal{H}$ is trivial follows from the structure of the Steiner system. Given a *duad* $\{x, y\} \subseteq \mathcal{H}$, the three sets $\mathcal{H}' \setminus \mathcal{H}$, for hexads $\mathcal{H}' \supseteq \mathcal{H} \setminus \{x, y\}$, form a *syntheme* on the complement of $\mathcal{H}$ (a partition into three sets of size 2). Distinct duads give distinct synthemes. Now K fixes all duads on $\mathcal{H}$, and hence all synthemes on the complement; so $K = 1$.

Finally, the identification of M_{12} and the related groups with the "classical" versions follow from properties of the classical versions: for example, there is a unique Steiner system, whose automorphism group is M_{12}; and there is a unique Hadamard matrix whose group of monomial automorphisms is $2M_{12}$ (see [2]).

8 Further comments

In view of the fact that

$$13.132 = 1716 = \binom{13}{6},$$

we might wonder if it is possible to partition all the 6-element subsets of a set of size 13 into disjoint copies of $S(5, 6, 12)$, each omitting one point of the set. In fact this is impossible, even though the analogous partition of 4-element subsets of a 9-element set can be done in two distinct ways [1]. So it is worth seeing how close we get. For each position of the hole, the 132 hexads define a $S(5, 6, 12)$ on the remaining points. However, not every 6-element set occurs as a hexad in one of these Steiner systems, and some occur more than once. Ignoring the hole, the hexads are of three geometric types. The union of two lines with the intersection removed occurs as a hexad for 7 positions of the hole; the three sides of a triangle with the vertices removed, for 3 positions; and the union of two lines with a point on one line removed, just once.

Similar puzzles can be defined on some other configurations, and also define interesting groups. If we use the projective plane of order 2, so that a move consists of a cyclic permutation of the hole and two counters on a line with it, we obtain the alternating group A_6. We can realise its triple cover $3A_6$ by using 3-sided counters (rotated in opposite directions during a move), and a group $A_6.2$ by using line-counters as well as point-counters.

On the Petersen graph, we can play the game using two holes, which must always be adjacent. A move consists of permuting cyclically the three neighbours of a hole (the other hole and two counters). We obtain the group $L_2(7)$ (if the holes return to their original positions) or $L_2(7).2 = \mathrm{PGL}(2,7)$ (if they are allowed to be interchanged by the move sequence).

References

[1] P. J. Cameron & C. E. Praeger, Partitioning into Steiner systems, in *Combinatorics '88, Volume 1* (eds. A. Barlotti *et al.*), Mediterranean Press, Rende (1991), 61–71.

[2] J. H. Conway, R. T. Curtis, S. P. Norton, R. A. Parker & R. A. Wilson, *An ATLAS of Finite Groups*, Oxford University Press, Oxford (1985).

[3] R. T. Curtis, The Steiner system $S(5,6,12)$, the Mathieu group M_{12}, and the "Kitten", in *Computational Group Theory* (ed. M. Atkinson), Academic Press, (1984).

Department of Mathematics
Princeton University
Princeton, NJ 08544
U.S.A.

The Harmonious Chromatic Number and the Achromatic Number

Keith Edwards

Summary The harmonious chromatic number of a graph is the least number of colours in a vertex colouring such that each pair of colours appears on at most one edge. The achromatic number of a graph is the greatest number of colours in a vertex colouring such that each pair of colours appears on at least one edge. This paper is a survey of what is known about these two parameters, in particular we look at upper and lower bounds, special classes of graphs and complexity issues.

1 Introduction

A short survey of harmonious colourings was given by Wilson [80] in 1990. Since then a number of new results have appeared, and the close relationship between harmonious chromatic number and achromatic number has been observed. The purpose of this new survey is to outline what is known about these parameters, and suggest some open problems. A more detailed summary of results on the achromatic number, with a rather different emphasis, can be found in the forthcoming survey by Hughes and MacGillivray [51].

We begin with the definitions of the two parameters.

Definitions A *harmonious colouring* of a graph G is a proper vertex colouring of G such that, for any pair of colours, there is at most one edge of G whose endpoints are coloured with this pair of colours. The *harmonious chromatic number* of G, denoted $h(G)$, is the least number of colours in a harmonious colouring of G.

Definitions A *complete colouring* of a graph G is a proper vertex colouring of G such that, for any pair of colours, there is at least one edge of G whose endpoints are coloured with this pair of colours. The *achromatic number* of G, denoted $\psi(G)$, is the greatest number of colours in a complete colouring of G.

Remark Although these colourings have usually been called *complete colourings*, the terms *irreducible colouring* [64], *full colouring* [32, 46], *correct colouring* [8] and *achromatic colouring* [84] have also been used.

The achromatic number was introduced by Harary, Hedetniemi and Prins [41] in 1967. They considered homomorphisms from a graph G onto a complete graph K_n. A homomorphism from a graph G to a graph G' is a function $\phi: V(G) \to V(G')$ satisfying $u \sim v \Rightarrow u\phi \sim v\phi$. This induces an obvious mapping from $E(G)$ to $E(G')$. It is easy to see that a complete colouring of G with n colours corresponds precisely to a *complete* homomorphism of G

onto K_n, i.e. one whose induced edge mapping maps $E(G)$ *onto* $E(K_n)$. They noted that the smallest n for which such a complete homomorphism exists is just the chromatic number $\chi(G)$ of G. They considered the largest n for which such a homomorphism exists. This was later named the achromatic number $\psi(G)$ by Harary and Hedetniemi [40]. In the first paper [41] it is shown that there is a complete homomorphism from G onto K_n if and only if n satisfies $\chi(G) \leq n \leq \psi(G)$.

In a later paper, Frank, Harary and Plantholt [33] introduced the concept of a *line-distinguishing colouring.* This is similar to a harmonious colouring, for any pair i, j of distinct colours, there is at most one edge with endpoints coloured i, j, but in addition, for each colour i, there may be at most one edge with both endpoints coloured i. Thus a line-distinguishing colouring need not be proper. The *line-distinguishing chromatic number* $\lambda(G)$ is defined in the obvious way as the least number of colours in such a colouring of G. The authors attribute the definition to Pierre Duchet. It was shown by Hopcroft and Krishnamoorthy [48] that determining $\lambda(G)$ is NP-hard, and a short proof of the same result by David S. Johnson is also given in the paper. They refer to λ as the harmonious chromatic number, but this name has subsequently been used for the definition which we have given above.

Although there is an obvious similarity (or perhaps duality) between the definitions of h and ψ given above, this does not appear to have been noticed until recently.

There are several other ways of defining the harmonious chromatic number. A harmonious colouring of G with n colours corresponds to a homomorphism from G to K_n such that the induced map from $E(G)$ to $E(K_n)$ is one-to-one. Then $h(G)$ is just the least n for which such a map exists.

Another way of looking at this is the following: Given a graph G, if we identify two adjacent vertices of G we get a loop, and if we identify two vertices at distance 2, we get a multiple edge. However if we identify two vertices distance 3 or more apart, we get a new simple graph. We can repeat the process until the resulting graph has diameter 2. Then $h(G)$ is just the smallest number of vertices in a diameter 2 graph which can be obtained from G in this way.

We can also look at the process the other way round. A *detachment* of a graph G is a graph H which is obtained from G by splitting each vertex into one or more vertices, and such that $E(H) = E(G)$ (the edges incident with vertex v in G are "shared out" arbitrarily among the vertices into which v is split). A subdetachment is just a subgraph of a detachment. Then G has a harmonious colouring with n colours if and only if G is a subdetachment of K_n. See Nash-Williams [74] for a survey of detachments.

Notation We will usually denote the number of vertices of a graph by n and the number of edges by m. In particular, m will *always* mean the number of edges of the graph in question, and this will usually be assumed without comment. The maximum degree of a graph G is denoted $\Delta(G)$.

2 Basic properties

2.1 Simple bounds

There are two simple but important bounds for h and ψ. The first arises from the observation that, in any harmonious colouring of a graph, any vertex, and all of its neighbours, must all have distinct colours. Hence $h(G) \geq \Delta(G) + 1$. This bound is met for example by a complete graph or a star. More generally, we observe that any two vertices which are adjacent, or have a common neighbour, must have distinct colours. Hence for any graph of diameter 2, the harmonious chromatic number equals the number of vertices, and in general, $h(G)$ is at least the number of vertices in the largest diameter 2 subgraph of G.

The other bound arises from a comparison of the number of edges of G and the number of pairs of colours which are available. If G is harmoniously coloured with k colours, then the number of pairs of colours, $\binom{k}{2}$, must exceed the number of edges of G. Similarly, for a complete colouring, the number of pairs of colours must be at most the number of edges. This motivates the following definitions.

Definitions Let m be a positive integer. Then we define $Q(m)$ to be the least positive integer k such that $\binom{k}{2} \geq m$, and $q(m)$ to be the greatest integer k such that $\binom{k}{2} \leq m$. We also define $R(m)$ to be $\binom{Q(m)}{2} - m$ and $r(m)$ to be $m - \binom{q(m)}{2}$.

Remark It is easily calculated that

$$Q(m) = \left\lceil \frac{1+\sqrt{8m+1}}{2} \right\rceil \qquad \text{and} \qquad q(m) = \left\lfloor \frac{1+\sqrt{8m+1}}{2} \right\rfloor.$$

Also $0 \leq R(m) < Q(m) - 1$ and $0 \leq r(m) < q(m)$.

From the discussion above, we have the following crucial result:

Proposition 2.1 *For any graph G with m edges,*

$$h(G) \geq Q(m) \qquad \textit{and} \qquad \psi(G) \leq q(m).$$

Much of the interest in these parameters is focussed on attempting to find classes of graphs for which these bounds are attained, or nearly attained.

In the special case when $m = \binom{k}{2}$ for some k, there may be a colouring in which each pair of colours appears on precisely one edge. We will call such a colouring an *exact colouring*. An exact colouring is clearly both a harmonious colouring and a complete colouring. It is clear that for a graph with $\binom{k}{2}$

edges, the following are equivalent: i) G has an exact colouring, ii) $h(G) = k$ and iii) $\psi(G) = k$. This observation, made by N. Cairnie, is the starting point for relating the two parameters. Any result for harmonious colouring involving graphs for which $h(G)$ is near to $Q(m)$ is likely to have a counterpart involving graphs for which $\psi(G)$ is near to $q(m)$. The correspondence is not however simple in most cases. Achromatic number counterparts of several harmonious colouring results are proved in [19].

2.2 Partial colourings

Techniques for colouring graphs often involve some sequential colouring of the vertices, so that at some stage we have a partial colouring of the graph which is subsequently extended to a colouring of the whole graph. For harmonious colourings, it is not the case that a colouring of an induced subgraph can always be extended to a colouring of the whole graph, even if extra colours are allowed. This is because we may have an uncoloured vertex adjacent to two vertices with the same colour. For this reason we make the following definition:

Definition A partial harmonious colouring of a graph G is a colouring of a subset of the vertices of G which is a harmonious colouring of the graph induced by the coloured vertices, and is such that no uncoloured vertex has two or more coloured neighbours with the same colour.

We note that if H is a subgraph of G, then any harmonious colouring of G induces one of H, hence $h(H) \leq h(G)$.

If H is an induced subgraph of G, then a complete colouring of H can always be extended to a complete colouring of G. This can be seen as follows: Suppose that we have a complete colouring of H. Colour the remaining vertices sequentially. Suppose v is the next vertex to be coloured. Then if v has neighbours with each of the colours used so far, colour v with a completely new colour, otherwise colour v with any existing colour not used on any of its neighbours. In either case the colouring remains complete. We thus have

Proposition 2.2 *If H is an induced subgraph of G, then*

$$\psi(H) \leq \psi(G).$$

Note however that we can have a (non-induced) subgraph H of G with $\psi(H) > \psi(G)$, for example $\psi(C_4) = 2$ while $\psi(P_4) = 3$.

2.3 Degree sum

Let S_c be the set of vertices which have colour c in a harmonious colouring. Then the neighbours of the vertices in S_c must all have distinct colours, none of which can be c. Thus the sum of the degrees of the vertices in S_c is less than the number of colours. This motivates the following definition:

Definition Suppose that we have a (possibly partial) harmonious colouring of a graph G with C colours, and let S_c be the set of vertices with colour c. Then we define the *degree sum* of colour c to be $\sum_{v \in S_c} d(v)$.

Note that for an exact colouring with C colours, the degree sum of each colour must be exactly $C-1$.

3 General bounds

3.1 Upper bounds for $h(G)$

The first upper bound for general graphs is that due to Lee and Mitchem [57] who showed that

$$h(G) \leq (\Delta^2+1)\left\lceil n^{1/2} \right\rceil .$$

This was improved, independently, both by Zhikang Lu [60] and by McDiarmid and Luo Xinhua [67] who proved, respectively, that

$$h(G) \leq 2\Delta \left\lfloor n^{1/2} \right\rfloor$$

and

$$h(G) \leq 2\Delta(n-1)^{1/2}.$$

Both of these results are obtained by a sequential colouring method, in which the size of each colour class is kept below some limit t. It is then shown that there must always be a spare colour available to allow the next vertex to be coloured.

By a refinement of this method, limiting the degree sum of each colour class rather than its size, it was shown in [29] that

$$h(G) \leq 2(2\Delta m)^{1/2} + (2\Delta - 1)\Delta \tag{1}$$

which improves the previous bounds if G is not regular. In fact a slightly stronger result is proved. For any integer k, the k-core $V_{(k)}$ of a graph G is the set (possibly empty) of vertices which remain after vertices of degree less than k have been repeatly removed until the minimum degree is at least k. Then we have the following:

Theorem 3.1 *Let G be a graph with m edges and maximum degree Δ. Then for any integer $k \geq 2$,*

$$h(G) \leq \max\left\{\left|V_{(k)}\right|, 2(2(k-1)m)^{1/2} + (2k-3)\Delta\right\}.$$

For any graph, $\left|V_{(\Delta+1)}\right| = 0$, which gives inequality (1) above. However for some classes of graphs, much better results can be obtained. For example, for any tree T, $\left|V_{(2)}\right| = 0$, hence

$$h(T) \leq 2(2m)^{1/2} + \Delta.$$

Similarly, for any planar graph $\left|V_{(6)}\right| = 0$, and for a graph of genus $\gamma \geq 1$, $\left|V_{(7)}\right| \leq 12(\gamma - 1)$. Thus we have:

Theorem 3.2 *For any planar graph G with m edges,*

$$h(G) \leq (40m)^{1/2} + 9\Delta.$$

For any graph G of genus $\gamma \geq 1$, with m edges

$$h(G) \leq \max\left\{12(\gamma-1), (48m)^{1/2} + 11\Delta\right\}.$$

By a different technique, Krasikov and Roditty [55] improved the upper bound for general graphs to roughly $\Delta(2n)^{1/2}$. Their result does not however give improved bounds for the special classes above.

3.2 Lower bounds for $h(G)$

The only significant lower bound for $h(G)$, apart from the easy ones given above, is due to Alon, and is given in [55]. He considered the probability of a fixed colouring of n vertices being a valid harmonious colouring for a random graph $G_{n,p}$, for suitable p, and hence proved the following:

Theorem 3.3 *Suppose that Δ satisfies $10^4 \log_2(2n) \leq \Delta \leq (n \log_2(2n))^{1/2}$. Then if n is sufficiently large there exists a graph G with n vertices and maximum degree at most Δ for which*

$$h(G) \geq \frac{\Delta n^{1/2}}{80(\log_2 n)^{1/2}}.$$

This shows that even for graphs with modest vertex degrees, we may have

$$h(G) \geq \frac{\Delta^{1/2}(2m)^{1/2}}{80(\log_2 n)^{1/2}}.$$

Hence in general it is not the case that $h(G)$ is close to $Q(m)$ when m is large, even for quite sparse graphs. The method used by Alon does not however appear to extend to graphs of *bounded* degree.

Zagaglia Salvi [85] showed that for any graph G the line-distinguishing chromatic number $\lambda(G)$ is at least the chromatic index $\chi'(G)$, as conjectured by Harary and Plantholt [42].

3.3 Upper bounds for $\psi(G)$

The simplest upper bound for $\psi(G)$ is due to Harary and Hedetniemi [40] and relates $\psi(G)$ to $\alpha_0(G)$, the size of the minimum vertex cover of G. Suppose that a graph $G = (V, E)$ has a vertex cover X of size k, and G has a complete colouring with at least $k+2$ colours. Then at least two of the colours occur only on $V - X$. But since $V - X$ is an independent set, this is a contradiction. Thus we have proved that for any graph G,

$$\psi(G) \leq \alpha_0(G) + 1.$$

Using the relation $\alpha_0(G) + \beta_0(G) = |V| = n$, where $\beta_0(G)$ is the independence number of G, we obtain the alternative form

$$\psi(G) \leq n - \beta_0(G) + 1.$$

Shaoji Xu [83] gave an extensive list of similar relationships among a number of graph parameters, including the achromatic number.

3.4 Lower bounds for $\psi(G)$

There do not appear to be any general lower bounds for $\psi(G)$ analogous to the upper bounds for h. There are however some important lower bounds on the achromatic number of irreducible graphs.

A *congruence* on a graph G is an equivalence relation on the vertex set $V(G)$ such that each equivalence class is an independent set. If C is a congruence on G we can form the quotient graph G/C with vertex set the congruence classes of C, and two classes adjacent if and only if some vertex in one class is adjacent to some vertex in the other. It is easy to see that $\psi(G/C) \leq \psi(G)$.

Hell and Miller [45] defined the *reducing congruence* R on a graph G as follows: two vertices of G are equivalent if and only if they have the same set of neighbours. Note that two such vertices cannot be adjacent, so R is clearly a congruence. The graph G/R is called the *reduced graph.* A graph is *irreducible* if the equivalence classes of R are singletons, i.e. if distinct vertices have distinct neighbourhood sets. As expected, the reduced graph G/R is always irreducible. Hell and Miller define $v(k)$ to be the maximum number of vertices in an irreducible graph with achromatic number k. It is not at all obvious that $v(k)$ is finite; that this is indeed so, for each positive integer k, is the main result of Hell and Miller [46]. Using an estimate for $v(k)$ they also show that for large enough n, an irreducible graph G on n vertices satisfies

$$\psi(G) > \sqrt{\log \log n}.$$

This lower bound was improved by Máté [64] who showed that for any $\varepsilon > 0$,

$$\psi(G) \geq (\tfrac{1}{2} - \varepsilon) \log n / \log \log n$$

provided n is large enough. He also noted that an example of Erdős shows that there is an irreducible graph G_n with n vertices for which

$$\psi(G_n) \leq \log n / \log 2 + 2.$$

Remark There are arbitrarily large graphs with achromatic number 2, for example the complete bipartite graphs $K_{n,n}$. However, as pointed out in [31], the fact that $v(k)$ is finite means that for any d, k, there are (ignoring isolated vertices) only finitely many graphs with maximum degree at most d and achromatic number at most k.

3.5 The values of $h(G)$

The harmonious chromatic number must take a value between the lower bound $Q(m)$ and the number of vertices n. By considering trees formed from a star attached to the end of a path, Mitchem [73] showed that it can take any such value:

Theorem 3.4 *Let $n = m + 1$ and let t satisfy $Q(m) \leq t \leq n$. Then there is a tree T with n vertices such that $h(T) = t$.*

For general graphs Mitchem proved that we can specify the number of vertices *and* the number of edges:

Theorem 3.5 *Suppose that m satisfies $0 \leq m \leq \binom{n}{2}$, and t satisfies $Q(m) \leq t \leq n$, then there is a graph G with n vertices and m edges such that $h(G) = t$.*

4 Special graphs

As with most graph parameters, there are a number of results giving more or less exact determination of the harmonious chromatic number for various special graphs. These rely mainly on ingenious ad hoc arguments based on the special structure of the graphs.

4.1 Paths and cycles

Naturally among the first graphs to be considered, for all three of the parameters h, λ and ψ, were P_n and C_n, respectively the path and cycle on n vertices. The value of the parameters for these graphs is equivalent to the existence of Eulerian graphs with a certain number of vertices and edges. For example, for $h(C_n)$, it is easy to see that C_n can be harmoniously coloured with t colours if and only if there is an Eulerian (simple) graph on t vertices with n edges. By this argument, Mitchem [73] showed the following:

Theorem 4.1 *If $Q(n)$ is odd and $R(n) \neq 1, 2$ then $h(C_n) = Q(n)$. If $Q(n)$ is even and $R(n) \geq Q(n)/2$ then $h(C_n) = Q(n)$. Otherwise $h(C_n) = Q(n) + 1$.*

The corresponding result for paths, proved by Hopcroft and Krishnamoorthy [48] can be stated as follows:

Theorem 4.2 *Let $m = n - 1$ be the number of edges of P_n. If $Q(m)$ is odd, or if $Q(m)$ is even and $R(m) \geq Q(m)/2 - 1$, then $h(P_n) = Q(m)$. Otherwise $h(P_n) = Q(m) + 1$.*

Note that the harmonious chromatic number can fail to meet the lower bound $Q(m)$ even for paths.

Similar results for the line-distinguishing chromatic number of paths and cycles are given by Frank, Harary and Plantholt [33] and Al-Wahabi et al. [2], and for the achromatic number by Geller and Kronk [36], Bories [11], Bories and Jolivet [13] and Hare et al. [43].

Mitchem also showed that if G is the disjoint union of two cycles C_r and C_s, then $h(G) = h(C_{r+s})$. In the same spirit, Georges [38] determined completely the case of a collection of disjoint paths, generalising Hopcroft and Krishnamoorthy's result:

Theorem 4.3 *Let G consist of r disjoint non-trivial paths with m edges in total. If $Q(m)$ is odd, then $h(G) = Q(m)$. If $Q(m)$ is even, then $h(G) = Q(m)$ if $r + R(m) \geq Q(m)/2$, otherwise $h(G) = Q(m) + 1$.*

This result is also implicit in [14].

By considering the decomposition of various complete and complete bipartite graphs into cycles, Georges also gives a complete solution for collections of 4-cycles:

Theorem 4.4 *Let bC_4 be the graph consisting of b disjoint 4-cycles, and let $m = 4b$, the number of edges of the graph.*

If $Q(m)$ *is even and* $R(m) \geq Q(m)/2$, *or*
$Q(m) \equiv 1 (\text{mod } 8)$, *or*
$Q(m) \equiv 3 (\text{mod } 8)$ *and* $R(m) \geq 3$, *or*
$Q(m) \equiv 5 (\text{mod } 8)$ *and* $R(m) \geq 6$, *or*
$Q(m) \equiv 7 (\text{mod } 8)$ *and* $R(m) \geq 5$,

then $h(bC_4) = Q(m)$. *Otherwise* $h(bC_4) = Q(m) + 1$.

Furthermore, Georges established that $h(bC_{4r}) = h(brC_4)$.

4.2 Complete graphs

For a single complete graph K_n, we of course have $h(K_n) = n$. If a graph G consists of a collection of disjoint complete graphs $K^{(1)}, \ldots, K^{(r)}$, where $K^{(i)}$ has k_i vertices, then it is clear that G can be coloured with colours $1, 2, \ldots, C$ if and only if the set $\{1, 2, \ldots, C\}$ has r subsets $S_1, \ldots, S_r$ with $|S_i| = k_i$, satisfying $|S_i \cap S_j| \leq 1$, $i \neq j$. Although there are various partial results on when this is possible, it does not appear to be completely solved. However Georges proved the following result for the case where the number of complete graphs is relatively small.

Theorem 4.5 *Let G be a graph consisting of disjoint complete graphs $K^{(1)}, \ldots, K^{(r)}$, where $K^{(i)}$ has k_i vertices. If each $k_i \geq r - 1$, then*

$$h(G) = \sum_{i=1}^{r} k_i - \binom{r}{2}.$$

If G is a complete bipartite graph $K_{r,s}$ then it is immediate that $h(G) = r+s$ since G has diameter 2. For a graph consisting of 2 complete bipartite graphs $K_{p,q}$ and $K_{r,s}$, where $p = \max\{p,q,r,s\}$, Georges proved that

$$h(K_{p,q} \cup K_{r,s}) = \begin{cases} p+1 & \text{if } p \geq r+s \\ q+r+s & \text{if } r+s > p\,. \end{cases}$$

4.3 Square grids

The square grid $P_n \times P_n$ has $m = 2n(n-1)$ so that $Q(m) = 2n$. By an ingenious construction, Miller and Pritikin [72] show that $h(P_n \times P_n) = Q(m)$ when n is even, but for n odd they are only able to show that $h(P_n \times P_n) \leq Q(m) + 1$.

4.4 Binary and r-ary trees

A number of authors have considered the problem of determining h for the complete r-ary tree of height H, which we denote $T_{r,H}$. Miller and Pritikin [72] considered the binary trees $T_{2,H}$ and showed that $h(T_{2,H}) = O(2^{H/2})$, and Mitchem [73] gave some improved bounds for $T_{2,H}$. Zhikang Lu [61] produced a more efficient colouring scheme to show that, for $k \geq 3$, $h(T_{2,2k}) \leq Q(m) + 2$, and proved a similar result for ternary trees, showing that, for $k \geq 2$, $h(T_{3,2k+1}) \leq Q(m) + 4$. In [63] he also gave some estimates for $h(T_{4,H})$. Mitchem [73] solved the problem for trees of height 2, showing that $h(T_{r,2}) = \lceil 3(r+1)/2 \rceil$.

It follows from the results in [26] and [19] that for each $r \geq 2$, $h(T_{r,H}) = Q(m)$ and $\psi(T_{r,H}) = q(m)$ if H is sufficiently large. Finally the author has recently verified that with one exception, all complete trees of height at least 3 have harmonious chromatic number $Q(m)$. Thus we can state the following.

Theorem 4.6 *Let $T_{r,H}$ be the complete r-ary tree of height H. Then*

$$\begin{aligned} h(T_{r,1}) &= r+1 \\ h(T_{r,2}) &= \lceil 3(r+1)/2 \rceil \\ h(T_{2,3}) &= 7 \ (= Q(m)+1) \\ h(T_{r,H}) &= Q(m) \ \ \textit{otherwise}. \end{aligned}$$

4.5 Stars

For a single star with m edges $K_{1,m}$, clearly $h(K_{1,m}) = m+1$. The case of a graph which is a forest consisting of a number of stars is solved in [30]:

Theorem 4.7 *Let F be a forest consisting of t stars of sizes $m_1 \geq \cdots \geq m_t$. Let $m = \sum_{i=1}^{t} m_i$, the number of edges of F. Then*

$$h(F) = \max\left(\max\left\{ \left\lceil \frac{1}{k} \sum_{i=1}^{k} (m_i + i) \right\rceil : 1 \leq k < Q(m) \right\}, Q(m) \right).$$

4.6 Random graphs

It was noted in Section 2.1 that for any graph of diameter two, the harmonious chromatic number equals the number of vertices. Since almost all graphs have diameter two, it follows immediately that $h(G) = |V(G)|$ for almost all graphs G.

The achromatic number of a random graph is much less straightforward, but was determined quite precisely by McDiarmid [66], who showed the following:

Theorem 4.8 *The proportion of graphs G with n vertices which satisfy*

$$n/(k+1) < \psi(G) < n/(k-1),$$

where $k = (\log_2 n)^{1/2}$, tends to 1 as n tends to infinity.

The upper bound derives from a fairly straightforward probability argument, while the lower bound is obtained from analysis of the behaviour of a very inefficient colouring algorithm, the "bounded sequential algorithm".

5 Asymptotic results

In their original paper on line-distinguishing chromatic number [33], Frank, Harary and Plantholt derive expressions for $\lambda(P_n)$ and $\lambda(C_n)$ and remark that each of these is asymptotic to $\sqrt{2n}$. Since also $\sqrt{2n} \sim Q(m)$ for a path or cycle, we have

$$\lambda(P_n) \sim \lambda(C_n) \sim Q(m).$$

It is also easy to see from Theorems 4.1 and 4.2 that

$$h(P_n) \sim h(C_n) \sim Q(m).$$

Although in this case the values of $h(P_n)$ and $h(C_n)$ are known precisely, there are many classes of graphs for which the exact determination of harmonious chromatic number appears to be very difficult. In most cases the only available lower bound is $Q(m)$, so determining $h(G)$ exactly is likely to be hard unless $h(G) = Q(m)$, and even if equality holds, proving so requires giving (or at least proving the existence of) a colouring which uses all but a very few of the available colour pairs. Describing such a colouring is likely to be very difficult unless G has a lot of structure. We might however expect to have more success in showing that for certain classes of graphs, $h(G) \sim Q(m)$. Of course, being a weaker statement, it is more likely to be true, but the fact that it is less precise also means that techniques (e.g. probabilistic) which "waste" a small proportion of colour pairs may still succeed.

Notation If Γ is a class of graphs, we will use the shorthand

$$h(G) \sim Q(m) \text{ for } G \in \Gamma$$

to mean: For any $\varepsilon > 0$, there is an integer M such that if G is a graph in the class Γ, and G has $m \geq M$ edges, then $Q(m) \leq h(G) \leq (1+\varepsilon)Q(m)$.

Other asymptotic results follow from the theorem of Wilson [82] on the decomposition of complete graphs into copies of a graph:

Theorem 5.1 *Given a graph G, there exist subgraphs $G_1, \ldots, G_t$ of the complete graph K_n such that each edge of K_n belongs to exactly one of the graphs G_i and such that each G_i is isomorphic to G provided that (i) n is sufficiently large, (ii) the number of edges of G divides $\binom{n}{2}$, and (iii) the greatest common divisor of the degrees of the vertices of G divides $n-1$.*

This theorem can be interpreted as giving sufficient conditions for the graph tG consisting of t disjoint copies of some graph G to have harmonious chromatic number n. As described in [29], it follows from this that for any fixed graph H, if we consider the class of graphs G consisting of disjoint copies of H, i.e. $G = kH$ for some k, we have $h(G) \sim Q(m)$. More generally if k is a positive integer and $\Gamma^{(k)}$ is the class of all graphs whose components have at most k vertices, then $h(G) \sim Q(m)$ for graphs G in $\Gamma^{(k)}$.

In [29] this is extended to a considerably wider class of graphs, as we see in the next section.

5.1 Fragmentable graphs

Definition A class Γ of graphs is said to be *fragmentable* if: For any $\varepsilon > 0$, there are positive integers n_0, c, depending on ε, such that if $G \in \Gamma$ is a graph with $n \geq n_0$ non-isolated vertices, then there is a set S of vertices, with $|S| \leq \varepsilon n$ such that each component of $G - S$ has at most c vertices.

Fragmentable classes of graphs include trees, planar graphs and, more generally, graphs of genus at most γ for some fixed $\gamma \geq 0$. They also include some classes of unbounded genus such as d-dimensional grids for fixed $d \geq 3$.

Suppose Γ is a fragmentable class all of whose elements have maximum degree at most d. Let G be an element of Γ and suppose that G has a set of vertices S of size εn which fragments G into components of size at most $c(\varepsilon)$. It follows from the result above that we can find a harmonious colouring of all of these components using at most $(1+\varepsilon)Q(m)$ colours.

We then use the following lemma, proved in [28]:

Lemma 5.2 *Let d be a fixed integer, and let $\varepsilon > 0$. Suppose that G is a graph with maximum degree at most d, n vertices and m edges, and G has a partial harmonious colouring with k colours such that at least $(1-\varepsilon)n$ of the vertices are coloured. Then $h(G) \leq k + 12d^2\varepsilon^{1/2}n^{1/2}$ provided n is large enough.*

This lemma shows that, as we would expect, we can extend the colouring to the small remaining set of vertices S with only a small number of extra colours (although it may be necessary to recolour some already coloured vertices). Therefore $h(G) \sim Q(m)$ for graphs G in Γ. In particular this shows that bounded degree planar graphs can be coloured near-optimally.

5.2 Bounded degree graphs

Finally in [28] is it shown that for any fixed d, $h(G) \sim Q(m)$ for G in the class Γ_d of graphs of maximum degree at most d. We give a rough outline of the proof below.

First consider the case when G is a d-regular bipartite graph with the vertex set V having parts A and B, and suppose we wish to harmoniously colour G using *disjoint* colour sets for A and B. If G has m edges, and we use a colours for the set A, and b colours for B, then we must have $ab \geq m$. Thus the total number of colours, $a+b$, must be at least $2\sqrt{m}$. We will colour G so that each of a and b is about $\sqrt{m}$.

Choose a colour set C_A of size about $\sqrt{m}$, and for each vertex of A, independently choose a colour uniformly from C_A, and use it to colour the vertex. A minor adjustment of the resulting colouring of A may be needed to ensure that no two vertices with a common neighbour receive the same colour; we omit the details of this. With high probability the degree sum of each colour will be roughly the same, i.e. about $\sqrt{m}$ since the total degree of the vertices in A is m.

We now form a d-uniform hypergraph H. The vertex set $V(H)$ is equal to C_A, the set of colours used on A. The edges of H correspond to vertices of G as follows: for any vertex $v \in B$, the hypergraph H has an edge e_v containing the d colours used on the neighbours of v. The degree $d_H(x)$ of a vertex x in the hypergraph H is determined as follows: x is an element of the colour set C_A, and the degree $d_H(x)$ is the degree sum of the colour x in G, which is about $\sqrt{m}$, as noted above. Hence H is approximately regular of degree about $\sqrt{m}$. For any two vertices x, y of H, the number of edges in H containing x and y, called the *codegree* of x and y, is likely to be small in comparison with $\sqrt{m}$. Thus we have exactly the situation in which to apply the Pippenger-Spencer theorem [75] on the chromatic index of uniform hypergraphs. This gives an edge colouring of the hypergraph with about $\sqrt{m}$ colours. Transferring the colours from the edges e_v to the corresponding vertices v of B completes a colouring c of G. This colouring has disjoint colour sets, of size about $\sqrt{m}$, for each of A and B. Furthermore it is a harmonious colouring, for suppose that (v, w), (v', w') are two distinct edges in G, where $v, v' \in A$, $w, w' \in B$. If $c(w) \neq c(w')$, then the two edges have distinct colour pairs. So suppose $c(w) = c(w')$. Then if $w = w'$, the colours $c(v)$ and $c(v')$ must be distinct, since we ensured that any two vertices in A with a common neighbour received different colours. So suppose $w \neq w'$. Then since $c(w) = c(w')$, we know that

e_w , $e_{w'}$ received the same colour in the colouring of H, hence e_w and $e_{w'}$ must be disjoint. But $c(v) \in e_w$ and $c(v') \in e_{w'}$, so we have $c(v) \neq c(v')$. Hence distinct edges have distinct colour pairs, and the colouring is harmonious.

Now suppose that G, rather than being a bipartite graph, is a d-regular r-partite graph, with the r parts all having the same number of vertices. In this case we can do much the same as above, in r stages. At each stage one of the r parts of the graph is coloured. The first part is coloured randomly, each subsequent stage uses the Pippenger-Spencer theorem to extend the colouring to another part. The total degree of each part is $2m/r$, and we will use about $(2m)^{1/2}/(r-1)$ colours for each part. Each colour thus has degree sum about $(2m)^{1/2}(r-1)/r$. In order to construct the hypergraph at each stage we have to assign, randomly, some temporary colours to the uncoloured vertices; these colours are discarded once the stage is complete. The hypergraph has common degree about $(2m)^{1/2}/r$, so the Pippenger-Spencer theorem allows us to colour it with about $(2m)^{1/2}/(r-1)$ colours as required, allowing the process to continue. In total we use about $(2m)^{1/2}r/(r-1)$ colours. Provided r is large, this is close to $Q(m)$.

Finally suppose we have any d-regular graph G. We choose an integer r, and randomly assign the vertices of G to one of r "parts". Of course, we do not truly obtain an r-partite graph, but it is not hard to see that with high probability nearly all the edges of G will join vertices in distinct parts. Thus the graph can be regarded as nearly r-partite. We therefore colour it as above. This may not give a true harmonious colouring, because of edges whose endpoints are in the same part. However there are few such endpoints, and these can be recoloured using the extension Lemma 5.2. Provided we choose r large enough, this suffices to give a colouring of G with roughly $Q(m)$ colours. The extension to all graphs of maximum degree at most d is straightforward. For details of the proof see [28].

In [19], the equivalent result for achromatic number is given, namely that $\psi(G) \sim q(m)$ for G in Γ_d. (Of course $q(m)$ and $Q(m)$ are asymptotically equivalent.)

6 Trees

For most kinds of graph colouring, questions relating to the colouring of trees are usually answered easily, and in many cases are trivial. However, this is far from being the case for harmonious and complete colourings, for, as we shall see in Section 7, these colourings give rise to NP-hard problems on trees. Nonetheless, the simple structure of trees does allow us to obtain better bounds than for most other classes of graph, and in some cases to obtain exact results.

6.1 General bounds

For achromatic number, the best lower bound for trees is that of Farber, Hahn, Hell and Miller [31]. They show that extending a tree by a fairly small number of edges must increase the achromatic number, and so obtain a recurrence which leads to an upper bound for the number of edges of a tree with maximum degree at most d and achromatic number at most k:

Theorem 6.1 *Let T be a tree with m edges, maximum degree at most d, and satisfying $\psi(T) \leq k$. Then*

$$m \leq \begin{cases} (k-1)d + \binom{k-1}{2} & \text{if } k \leq d, \\ (k-1)k + \binom{d-1}{2} & \text{if } k \geq d. \end{cases}$$

They then prove the following corollary:

Theorem 6.2 *Let T be a tree with m edges and maximum degree Δ. Then*

$$\lfloor \tfrac{3}{2} + (m - \tbinom{\Delta-1}{2} - \tfrac{3}{4})^{1/2} \rfloor \leq \psi(T) \leq q(m).$$

Note that for fixed Δ the lower bound here is asymptotic to $m^{1/2} \sim q(m)/\sqrt{2}$.

For harmonious colouring, McDiarmid and Luo Xinhua [67] noted that for a tree T, $h(T) \leq 2(\Delta n)^{1/2}$. They also noted that for complete r-ary trees it could be shown that $h(T) \leq 2(2n)^{1/2}$, and asked if it could be proved that $h(T) = O(n^{1/2} + \Delta)$ for all trees. As mentioned in Section 3.1, it was proved in [29] that for any tree T with m edges,

$$h(T) \leq 2(2m)^{1/2} + \Delta \leq 2Q(m) + \Delta.$$

Note that this is asymptotic to $2Q(m)$ and so appears to be inferior to the bound for $\psi(T)$ given above.

6.2 Exact results

If we wish to colour a graph G harmoniously with C colours $1, \ldots, C$, we have to colour the vertices of G such that each colour pair $\{i, j\}$ is used at most once. It is therefore natural to think of these colour pairs as being (represented by) the edges of the complete graph on the vertices $1, \ldots, C$. Now consider the situation which arises when we have a partial colouring of a graph G (for example at some stage in a sequential colouring process). Assuming that we are not going to recolour any vertices already coloured, then any coloured vertex whose neighbours are also coloured is, in a sense, "inactive". Thus it is helpful to consider the "active" part of G, namely the subgraph A induced by the uncoloured vertices and their coloured neighbours, together with the partial colouring of A. It is also helpful to think of the colour pairs which do not yet appear on any edge of G forming the edges of a graph on the vertices $1, \ldots, C$. We call this graph the unused colour pairs graph U. As the colouring of the graph progresses, the graphs A and U evolve.

Now consider the special case of an exact graph G where m is exactly $\binom{C}{2}$. (In fact this case is not so special since we can always add $\binom{C}{2} - m$ extra isolated edges to the graph, without affecting the value of $h(G)$.) In this case we wish both A and U to evolve away to nothing. Unless G is very highly structured this is likely to be difficult to achieve, as we have to match up the colour pairs and the edges precisely.

One approach is to try to colour a successively larger portion of G in such a way as to achieve progressively more structure in A and U. Eventually they may become sufficiently well structured that a single technique will complete the colouring all at once.

Two such techniques were used in [26, 27] to prove some exact results on the harmonious chromatic number of bounded degree trees. First consider a tree T of maximum degree at most d having at least εn leaves, where ε is some small positive constant. Also suppose that T has exactly $\binom{C}{2}$ edges, and we wish to colour it with C colours. As observed in Section 5.1, trees are *fragmentable*, so by removing some small set S of the vertices, where $|S| = \eta n$ and η is much smaller than ε, we break up the tree into components all of size less than some constant $c(\eta)$. We then remove the leaves from all these components. Then as in Section 5.1, we colour these components in a highly symmetrical way. Thus at this stage the graph U of unused colour pairs is also highly symmetrical. We then extend the colouring to the set S. This to some extent destroys the symmetry of U, but not too much because $|S|$ is much smaller than the number of edges in U. Note that because the tree without its leaves has only $(1 - \varepsilon)\binom{C}{2}$ edges, then by the asymptotic result of Section 5.1 we are able to do this colouring with only C colours, provided the tree is large enough.

Now the active part of the tree, A, consists of a number of stars, namely the deleted leaves and their neighbours. The centres of the stars are coloured, and the leaves are uncoloured. This is exactly the kind of simple structure for A that is desirable. Let the total degree of the star centres which are coloured i be r_i. It is not hard to see that completing the harmonious colouring by colouring A with the unused colour pairs of U is exactly equivalent to orienting the edges of U so that the outdegree of vertex i is exactly r_i for each colour i. Using Hall's theorem we can derive sufficient conditions for this, and in [26] it is shown that it is always possible provided the tree is large enough. Thus we have the following:

Theorem 6.3 *Let d be a positive integer and let $\varepsilon > 0$. Then there exists an integer $M = M(d, \varepsilon)$ such that if T is any tree with $m \geq M$ edges, maximum* *degree at most d and at least εn leaves, where $n = m + 1$, then T satisfies*

$$h(T) = Q(m).$$

Now consider trees which have very few leaves. Since the average degree of a tree is less than 2, almost all the vertices have degree 2, so the tree consists largely of long paths. Note that there is no hope of proving the above result for these trees, for as noted in Section 4.1, arbitrarily long paths P_n have $h(P_n) = Q(m)+1$. More generally, if T has $\binom{C}{2}$ edges where C is even, then it will be impossible to colour T with C colours unless T has at least C vertices of odd degree. This is because the degree sum of each colour must be $C - 1$ which is odd, thus there must be at least one odd degree vertex of each colour. In what follows we will largely ignore such parity problems, and state simply that in some cases one extra colour will be needed.

To colour T we first colour the vertices of odd degree (plus possibly some even degree vertices). To do this we use Theorem 6.3 above, extended to forests. After this, the active part A consists of a number of paths, with their endpoints coloured, but the rest of the vertices uncoloured. The paths may be of many different lengths. We now partition the colour set into 4 sets W, X, Y, Z, where $|X| = |Y|$ and $|X|$, $|Y|$, $|Z|$ are all even. By colouring parts of the paths, we gradually improve the structure of both A and U, until we have used up all colour pairs except those in $X \times Z$ and $Y \times Z$, so that U is a complete bipartite graph, and the active graph A consists entirely of paths of length 4. Furthermore, there are exactly $\frac{1}{2}|X|\,|Z|$ of these paths, with $|X|\,|Z|$ endpoints, and each colour in X occurs on exactly $|Z|$ of these endpoints (but the colours are arranged in an arbitrary way). We then colour the midpoints of all the paths with a colour from Y, so that each colour from Y occurs the same number (i.e. $\frac{1}{2}|Z|$) of times. Now A consists of $|X|\,|Z|$ paths of length 2, each with an X-colour at one end and a Y-colour at the other, and each colour occurring exactly $|Z|$ times. From this we construct a bipartite multigraph M with vertex set $X \cup Y$, and an edge joining $x \in X$ and $y \in Y$ for each path in A which has endpoints coloured x, y. Now M is regular of degree $|Z|$, and hence can be edge coloured with $|Z|$ colours, i.e. those in the set Z. Transferring the colour of an edge to the midpoint of the corresponding path in A completes the colouring, using for A exactly the colour pairs in $X \times Z$ and $Y \times Z$.

Combining this result with Theorem 6.3 gives:

Theorem 6.4 *Let d be a fixed positive integer. Then there is an integer $N(d)$ such that if F is any forest with maximum degree at most d and with $m \geq N(d)$ edges, then*

$$h(F) = Q(m) \text{ or } Q(m) + 1.$$

Furthermore, if F has a least $Q(m) - 2R(m)$ vertices of odd degree, then $h(F) = Q(m)$.

The achromatic number counterpart of this is the following, proved in [19]:

Theorem 6.5 *Let d be a fixed positive integer. Then there is an integer $N(d)$ such that if F is any forest with maximum degree at most d, and with $m \geq N(d)$ edges, then*

$$\psi(F) = q(m) \text{ or } q(m) - 1.$$

Furthermore, if F has at least $q(m) - 2r(m)$ vertices of odd degree, then $\psi(F) = q(m)$.

6.3 Almost all trees

There are many results of the form: almost all graphs G have property P, meaning that the proportion of (unlabelled) graphs on n vertices which have property P tends to 1 as n tends to infinity. In most cases it is known that there are infinitely many graphs which do not have property P. In contrast there appear to be very few results of the form: almost all (unlabelled) trees T have property P. The only general technique for showing such a result is apparently the "forbidden limb" technique of Schwenk [79]. A limb of a tree at a vertex v is formed as follows: Delete some, but not all of the edges incident to v; the component containing v is a limb at v. Schwenk proved that for any given limb R, almost all trees T contain R as a limb (at some vertex). In fact he showed that if r_n is the number of trees on n vertices *not* containing R as a limb, and t_n is the total number of trees on n vertices, then the growth rate of r_n is strictly less than that of t_n. Corollaries of Schwenk's theorem include the well known result that almost all trees have a non-trivial automorphism, and that almost all trees are cospectral, i.e. share the same characteristic polynomial with some other tree. There are also a number of extensions of the latter result. These are all proved by showing that if a tree contains some particular limb R, then it has the desired property, and then appealing to Schwenk's theorem.

In [26] the following result is proved:

Theorem 6.6 *Almost all trees T satisfy $h(T) = Q(m)$.*

This result was conjectured (in a slightly different form) by Frank, Harary and Plantholt in the first harmonious colouring paper [33]. The method used to prove this result is quite different from the forbidden limb technique above. We first prove a slightly stronger version of Theorem 6.3, which shows that a sufficiently large tree which is "nearly" bounded degree satisfies $h(T) = Q(m)$. Nearly bounded degree means that the maximum degree is not too big ($O(\log n)$), and the total degree of the vertices of degree greater than some d is very small ($< n/2^d$). We then show that almost all trees are nearly bounded degree. Together with the fact that almost all trees have plenty of leaves, this gives the result.

The corresponding result that $\psi(T) = q(m)$ for almost all trees is proved in [19].

Remark It should be noted that the set of trees for which $h(T) \neq Q(m)$ is quite large. If s_n is the number of such trees with n vertices, and t_n is, as above, the number of trees on n vertices, then although $s_n/t_n \to 0$ as $n \to \infty$, the growth rate of s_n is equal to that of t_n. This is easily seen by considering the number of trees with maximum degree greater than $Q(m)$. If follows that Theorem 6.6 cannot be proved by the forbidden limb technique.

6.4 Algorithms

It follows from Theorem 6.4 that there is a simple, though not very practical, algorithm which will determine the harmonious chromatic number of a bounded degree tree to within 1 in linear time. The algorithm simply determines if the number of edges of the tree is at least $N(d)$, if so it declares that $h(T)$ is $Q(m)$ or $Q(m)+1$, otherwise it uses exhaustive search to determine $Q(m)$. In fact, with a little more care, it is possible to determine $h(T)$ exactly in polynomial time. Thus we have

Theorem 6.7 *Let d be a positive integer. Then there is a polynomial time algorithm which will determine the harmonious chromatic number $h(F)$ of a forest F with maximum degree at most d.*

In fact we first deal with the case of an exact forest, i.e. one with exactly $\binom{Q(m)}{2}$ edges. The general problem for harmonious chromatic number reduces to this, since any forest can be made exact by adding $R(m) = \binom{Q(m)}{2} - m$ isolated edges to F, without changing the value of $h(F)$. The achromatic number problem is somewhat more tricky, since in general removing $r(m)$ edges to make the forest exact will reduce the value of ψ. Nevertheless, the corresponding result for achromatic number is given in [19], although the algorithm is considerably more complicated.

Theorem 6.8 *Let d be a positive integer. Then there is a polynomial time algorithm which will determine the achromatic number $\psi(F)$ of a forest F with maximum degree at most d.*

7 Computational complexity

As might be expected, the determination of $h(G)$ and of $\psi(G)$ are both NP-hard problems. Yannakakis and Gavril [84] showed that the ACHROMATIC NUMBER problem

Instance Graph G, integer k.

Question Is $\psi(G) \geq k$?

is NP-complete even for the complements of bipartite graphs. This is one of a number of results which follow from their proof that the edge dominating set problem is NP-complete for bipartite graphs of maximum degree 3. Subsequently, ACHROMATIC NUMBER was shown to be NP-complete for bipartite graphs by Farber, Hahn, Hell and Miller [31], using a reduction from PARTITION.

The line-distinguishing chromatic number $\lambda(G)$ was shown to be NP-hard by Hopcroft and Krishnamoorthy [48], and a very short proof by D. S. Johnson of the same result is also given in the paper. This proof is easily adapted for harmonious chromatic number. Consider the problem HARMONIOUS COLOURING:

> **Instance** Graph G, integer k.
>
> **Question** Is $h(G) \le k$?

The following proof, based on Johnson's, shows that this NP-complete:

> HARMONIOUS COLOURING is obviously in NP. For completeness, reduce from INDEPENDENT SET. Let G, k be an instance of INDEPENDENT SET. Construct an instance G', k' of HARMONIOUS COLOURING as follows: G' consists of a component equal to G with one extra vertex u joined to all the other vertices, and a second component which is a k-clique. Set $k' = |V(G)| + 1$. Note that any colouring of the first component must use k' colours, one for each vertex. Then it is easy to see that a harmonious colouring of G' with k' colours exists if and only if G has an independent set of size k.

However as early as the paper by Frank, Harary and Plantholt [33], it was conjectured that the problem is hard for more limited classes of graphs. In this paper, a directed version of the problem is defined; thus in a harmonious colouring of a directed graph we are allowed, for any pair of colours i, j, both an $i \to j$ edge and a $j \to i$ edge. It is pointed out in [33] that the problem of deciding if a directed graph can be harmoniously coloured with k colours is NP-complete even for graphs in which each component is a star with all edges directed outwards. This is done by a reduction from BIN PACKING, roughly as follows: An instance of BIN PACKING consists of a set of k bins, each of capacity B and a collection of objects u_i, $i = 1, \ldots, n$, each with an integer size $s(u_i)$. A solution to the instance is an arrangement of the objects in the bins so that each is exactly full. BIN PACKING is strongly NP-complete (see for example [34]), thus the size of the integers can be assumed to be in unary. It is also easy to restrict to instances for which $k = B + 1$. Then we reduce to DIRECTED HARMONIOUS COLOURING by constructing a graph consisting of n outwardly directed stars, of sizes $s(u_1), \ldots, s(u_n)$. The colour of the centre of a star corresponds to the choice of bin in which to place the corresponding object.

Frank, Harary and Plantholt conjectured that the undirected version of HARMONIOUS COLOURING is NP-complete when restricted to forests. This is proved in [30] by a more sophisticated bin-packing argument; in fact the paper shows that it is NP-complete to determine if a tree with $\binom{k}{2}$ edges has harmonious chromatic number k. It is also shown that the undirected version can be solved in polynomial time if each component is a star.

A rather similar reduction, this time from 3-PARTITION, is used by Bodlaender [8] to show that ACHROMATIC NUMBER is NP-complete even for graphs which are simultaneously cographs and interval graphs. Since the graphs constructed have $\binom{k}{2}$ edges, it follows immediately that the HARMONIOUS COLOURING problem is NP-complete for these graphs also. Likewise it follows from the NP-completeness of HARMONIOUS COLOURING for trees, mentioned above [30], that ACHROMATIC NUMBER is also NP-complete for trees [19], answering a question originally posed by Hedetniemi, Hedetniemi and Beyer [44] and repeated by a number of other authors [8, 21, 31, 52].

It is not however the case that the two problems are equivalent in complexity for all classes of graphs. We can give examples of classes of graphs for which one problem is in P, while the other is NP-complete.

Example 7.1 Let G be any graph, and let G' be the graph formed from G by adding a single new vertex u and joining u to each vertex of G. Then it is easy to see that $\psi(G') = \psi(G) + 1$, so the problem of determining $\psi(G')$ is the same as that of determining $\psi(G)$. However all such graphs G' have diameter at most 2. Since ACHROMATIC NUMBER is NP-complete, and graphs of diameter 2 are easily recognised, it follows that ACHROMATIC NUMBER is still NP-complete when restricted to graphs of diameter at most 2. On the other hand, for any such graph G, $h(G) = |V(G)|$, so the problem HARMONIOUS COLOURING is trivial, and belongs to P.

Example 7.2 Consider the class of graphs known as split graphs, whose vertex set V can be partitioned into two sets A and B, such that A induces a complete graph and B a null graph. It is easy to recognise split graphs and find the sets A and B by looking at the degrees of the vertices.

Now the achromatic number of any split graph is $|A|+1$ if the vertices of B are collectively adjacent to every vertex of A, and $|A|$ otherwise. Clearly this can be determined in polynomial time. Thus the ACHROMATIC NUMBER problem for split graphs belongs to P. However the HARMONIOUS COLOURING problem for split graphs is NP-complete. This can be seen by the following reduction from CHROMATIC NUMBER. Let $G = (V, E)$ be a graph. We construct a split graph G'. The vertex set of G' is $E \cup V$. The edge set of G' is as follows: for each $(u, v) \in E$, there is an edge in G' joining the vertex $(u, v) \in E$ to the vertices u and v in V. In addition, E induces a complete graph in G', and V induces a null graph in G'. Then it is easy to see that $h(G') \leq |E| + k$ if and only if $\chi(G) \leq k$, and the result follows.

This example also shows that HARMONIOUS COLOURING is NP-complete for graphs of diameter 3, since G' will have diameter 3 provided G has no isolated vertices. (I am indebted to Niall Cairnie for this example.)

Fixed number of colours

It should be noted that both HARMONIOUS COLOURING and ACHROMATIC NUMBER can be solved in polynomial time if the number of colours k is fixed. For HARMONIOUS COLOURING this is obvious, since if the graph has more than $\binom{k}{2}$ edges then it cannot be coloured with k or fewer colours, so the answer is "no". If the graph has $\binom{k}{2}$ or fewer edges then we can do an exhaustive search. For ACHROMATIC NUMBER the result is slightly less straightforward, but was proved by Yannakakis and Gavril [84] as follows. Suppose that a graph G satisfies $\psi(G) \geq k$. Then there is a set of $\binom{k}{2}$ edges in G which uses all the colour pairs. These edges have between them at most $k(k-1)$ endpoints. Thus there is an induced subgraph H of G with at most $k(k-1)$ vertices such that $\psi(H) \geq k$. On the other hand, if such a subgraph H exists, then $\psi(G) \geq k$ also. So to determine if $\psi(G) \geq k$ we simply form each induced subgraph of G on at most $k(k-1)$ vertices and check each to see if its achromatic number is at least k. Since the number of such induced subgraphs is at most $\binom{n}{k(k-1)}$ and k is fixed, the algorithm is polynomial. A more efficient $O(m)$ algorithm is given in [31] and relies on the results of Hell and Miller given in Section 3.4. Even this algorithm, however, is only practicable for very small k, because it uses the value of the function $v(k)$ defined in Section 3.4. Of course given that the ACHROMATIC NUMBER problem is NP-complete when k is unbounded, it is likely that any algorithm will have time complexity exponential in k.

Approximation algorithms

An approximation algorithm is one which delivers an approximate solution to a problem. For a maximisation problem such as the determination of achromatic number, an algorithm has *approximation ratio* α if it always produces a solution whose value is at least $\frac{1}{\alpha}$ of the optimum.

Chaudhary and Vishwanathan [21] give a polynomial time approximation algorithm for the achromatic number with approximation ratio $O(n/\sqrt{\log n})$. For graphs of girth at least 7, they give a simple algorithm with approximation ratio $O(n^{7/20})$.

8 Related topics

8.1 Parameters related to achromatic number

Several parameters similar to the achromatic number have been studied. The most closely related is the *pseudoachromatic number*. A *pseudocomplete* colouring is a colouring of the vertices, such that for any pair of colours, there is at least one edge whose endpoints are coloured with this pair of colours. It differs from a complete colouring in that it need not be proper, i.e. the graph induced by a colour class need not be a null graph. The *pseudoachromatic number* $\psi_s(G)$, is the greatest number of colours in a pseudocomplete colouring of G. The pseudoachromatic number was introduced by Gupta [39] and has also been studied by Bhave [7], Sampathkumar and Bhave [78] and Bollobás, Reed and Thomason [10].

Another related concept is the *ajointed number* defined by Cook and Evans [25]. A simple fold is a homomorphism which identifies a pair of nonadjacent vertices having a common neighbour, and a fold is a sequence of simple folds. The ajointed number $\mathrm{aj}(G)$ of a connected graph G is the largest n for which there is a fold onto K_n. It is easy to see that $\chi(G) \leq \mathrm{aj}(G) \leq \psi(G) \leq \psi_s(G)$ for any connected graph G. Further results on $\mathrm{aj}(G)$ are given in [18].

Bollobás, Catlin and Erdős [9] defined the *contraction clique number* $\mathrm{ccl}(G)$ to be the maximum number of colours in a pseudocomplete colouring of G, such that the graph induced by each colour set is *connected*. Clearly $\mathrm{ccl}(G) \leq \psi_s(G)$ for any graph G. Hadwiger's conjecture is that $\mathrm{ccl}(G) \geq \chi(G)$ for any G. For this reason $\mathrm{ccl}(G)$ has also been called the Hadwiger number of G.

8.2 r-reduction number

In Section 1 we described how the harmonious chromatic number of a graph could be defined as the number of vertices in the smallest graph obtainable from G by repeatedly identifying vertices at distance at least 3. This suggests a natural generalisation.

Definition Let G be a (not necessarily simple) graph. Suppose that u, v are two vertices of G satisfying $d_G(u,v) \geq r$. Form a new graph G' by identifying u and v to form a new vertex uv, so that each edge incident with u or v becomes an edge incident with uv. We will call this operation an *elementary r-reduction.*

Definition An *r-reduction* is a sequence of (zero or more) elementary r-reductions. We will also say that H is an r-reduction of G if H is obtained from G by an r-reduction.

Definition Let G be a (not necessarily simple) graph. The *r-reduction number* $\chi^{(r)}(G)$ is the least integer k such that G has an r-reduction with k vertices.

It follows from the discussion above that $\chi^{(3)}(G)$ is equal to the harmonious chromatic number $h(G)$, and it is also easy to see that $\chi^{(2)}(G)$ is the ordinary chromatic number $\chi(G)$. What can be said about $\chi^{(r)}$, $r \geq 4$?

We first make some simple observations:

1. If G is a simple graph then so is any r-reduction of G, $r \geq 3$, so we can restrict our attention to simple graphs.

2. For any graph G, $\chi^{(4)}(G) \geq h(G) \geq \chi(G)$, and in general we have for each $r \geq 2$,
$$\chi^{(r+1)}(G) \geq \chi^{(r)}(G).$$

3. If G is connected, and $r > \operatorname{diam}(G)$, then $\chi^{(r)}(G) = |V(G)|$. More generally, if G has k components, then for all sufficiently large r,
$$\chi^{(r)}(G) = |V(G)| - k + 1.$$

4. If G has girth at least r, then so does any r-reduction of G.

8.3 The 4-reduction number $\chi^{(4)}$

A special case of observation 4 above is that for any bipartite graph G, any 4-reduction of G also has girth at least 4, i.e. is triangle free. Thus if G has more than $k^2/4$ edges, any 4-reduction of G has at least k vertices, i.e. $\chi^{(4)}(G) \geq k$. Thus we have that for bipartite graphs G,
$$\chi^{(4)}(G) \geq 2\sqrt{m}.$$

First consider paths and cycles. If $m = t^2$ for some even integer t, then it is easy to see that $\chi^{(4)}(P_{m+1}) = \chi^{(4)}(C_m) = 2\sqrt{m}$. For the complete bipartite graph $K_{t,t}$ is Eulerian, and hence has a closed Eulerian trail of length m. Thus we can label the vertices of P_{m+1} or C_m with the names of the vertices of $K_{t,t}$ in the order in which they appear in the Eulerian trail, and then perform a sequence of elementary 4-reductions joining any two vertices with the same label until we obtain $K_{t,t}$. In general it is clear that $\chi^{(r)}(C_m) \leq k$ if and only if there is an Eulerian graph on k vertices with m edges and girth at least r. A similar statement holds for paths. However, little is known about how many edges a graph with a given number of vertices and given girth can have.

We now consider bipartite graphs. In Section 5.2 we sketched a proof that a d-regular bipartite graph with m edges could be harmoniously coloured with roughly $2\sqrt{m}$ colours, using distinct colour sets of size about $\sqrt{m}$ for the two parts of the graph. This proof shows that for such graphs, $\chi^{(4)}$ is approximately $2\sqrt{m}$. The result is easily extended to bipartite graphs of maximum degree at most d, hence we have that for bipartite graphs G of maximum degree at most d, $\chi^{(4)}(G) \sim 2\sqrt{m}$.

NP-hardness of $\chi^{(4)}$

We consider the problem 4-REDUCTION NUMBER:

Instance Graph G, integer r.

Question Is $\chi^{(4)}(G) \leq r$?

We will show that this problem is NP-complete even when restricted to graphs in which each component is a star. As for the DIRECTED HARMONIOUS COLOURING problem considered in Section 7, we reduce from the strongly NP-complete problem BIN PACKING. Recall that an instance of BIN PACKING consists of a set of k bins, each of capacity B and a collection U of objects u_i, $i = 1, \ldots, n$, each with an integer size $s(u_i)$. We can assume that the sum of the sizes $\sum_{i=1}^{k} s(u_i)$ is exactly kB, since otherwise a solution is clearly impossible. We can also assume that each $s(u_i) > k$, for otherwise we can multiply each $s(u_i)$, and B, by $k+1$ without altering the solution to the problem. To form our graph G, we take a star of size $s(u_i)$ for each $i = 1, \ldots, n$, and $B - k$ stars of size B. Take $r = 2B$. Now since G is bipartite and has B^2 edges, it follows from above that $\chi^{(4)}(G) \geq 2B$, and that $\chi^{(4)}(G) = 2B$ if and only if G has a 4-reduction which is a triangle free graph on $2B$ vertices with B^2 edges. It is well known that the only such graph is $K_{B,B}$. Hence $\chi^{(4)}(G) = 2B$ if and only if the stars can be packed edge-disjointly into $K_{B,B}$. It is clear that the centres of all the stars of size B must go on the same side of $K_{B,B}$. Hence having packed these stars, we then have to pack the remainder into $K_{k,B}$. Since each $s(u_i) > k$, the centres of these stars must also go on the same side of $K_{B,B}$, i.e. each centre must be identified with one of the k vertices. This is clearly possible if and only if the instance of BIN PACKING has a solution.

Note that HARMONIOUS COLOURING is solvable in polynomial time for this class of graphs.

8.4 Applications

Although ordinary vertex colourings have many applications, this does not appear to be the case for harmonious or complete colourings. However, Cichelli [24] used a idea similar to harmonious colouring to implement perfect hash functions, and it appears that harmonious colourings could have applications to some forms of data compression. The idea is as follows: Suppose that we wish to store a sparse graph efficiently, while still allowing very fast verification of adjacencies. Suppose we can harmoniously colour the graph efficiently, that is with C colours where C is about $Q(m)$. We can store the graph using a $C \times C$ matrix (actually only half of it need be stored since it will be symmetric) and a lookup table for the colours of the vertices. The (i, j)th entry of the matrix will contain the entry (u, v), the unique edge with endpoints coloured i, j (or a special marker if there is no such edge).

In order to determine if u and v are adjacent in the graph, we look up the colours of u and v, say i and j respectively, and then check whether the (i, j)th entry of the matrix is (u, v). Thus checking the adjacency requires only a very small number of simple steps and is constant time. Also the total storage space used is about $n \log n$ bits for the lookup table and $2m \log n$ for the matrix. This is about the same as for an adjacency list representation, which does not however have constant time adjacency checking. In practice this is most likely to be useful in cases where an input stream consists of pairs u, v, and it has to be checked that the pairs satisfy some fixed relation.

9 Open problems

9.1 General bounds

The upper bounds for $h(G)$ given in Section 3.1 are almost certainly not the best possible. It was suggested by Krasikov and Roditty [55] that the right upper bound for general graphs is around $\Delta\sqrt{n}$.

As mentioned in Section 3.2, the only non-trivial lower bound is that of Alon [55]. It seems that a substantially new idea is needed to obtain further lower bounds.

Apart from the results for irreducible graphs, there are few bounds of note for the achromatic number.

9.2 Special graphs

Open problems on special graphs include the complete determination of h for collections of cycles, and, more generally, for all graphs of maximum degree at most 2, and for all collections of complete graphs. For achromatic number, there appear to be only a few results on special graphs apart from those for paths and cycles, though presumably analogues of all the results in Section 4 could be obtained.

9.3 Trees

Although more is known about the harmonious chromatic number of trees than of other graphs, there are still some notable open problems. The best known upper bound for general trees is surely too big; it should be possible to prove that $h(T)$ is no more than about $\sqrt{2}Q(m)$. In fact, it seems quite likely that $h(T) \leq Q(m) + \Delta$ for all trees T.

9.4 Bounded degree graphs

As above, let Γ_d be the class of graphs with maximum degree at most d. We know that $h(G) \sim Q(m)$ for G in Γ_d, as described in Section 5.2, and that $h(T) \leq Q(m) + 1$ for all sufficiently large trees in Γ_d. Between these results is a large gap in our knowledge of the behaviour of $h(G)$ for bounded degree graphs.

Let c_d be the quantity defined by

$$c_d = \max\left\{(h(G) - Q(m)) : G \in \Gamma_d\right\}.$$

Perhaps the most important question is the following:

Question 9.1 *Is c_d bounded for each positive integer d?*

The example of a d-star shows that c_d cannot be bounded independently of d. Further, if $d = 4r + 2$, $r \geq 1$, then there are infinitely many $G \in \Gamma_d$ such that $h(G) - Q(m) \geq d/2$. For if k is any positive integer, there is a d-regular graph with $(k + \frac{1}{2})((k + \frac{1}{2})d - 1)$ vertices. Then $Q(m) = (k + \frac{1}{2})d$, however since each degree sum must be a multiple of d, it is not hard to see that $h(G) \geq (k + 1)d$. Hence $c_d \geq d/2$. A similar result holds for other values of d.

We can also restrict attention to, say, planar graphs. Let p_d be the quantity defined by

$$p_d = \max\{(h(G) - Q(m)) : G \in \Gamma_d,\ G \text{ planar}\}.$$

Question 9.2 *Is p_d bounded for each positive integer d?*

The example of a d-star above shows that p_d also cannot be bounded independently of d, however it seems possible that there is an absolute constant A, such that for any d, $h(G) \leq Q(m) + A$ for all sufficiently large planar $G \in \Gamma_d$.

Finally, let d be a fixed integer. Then if G is a d regular graph on $k(kd+1)$ vertices, we have $m = \binom{kd+1}{2}$, so $Q(m) = kd + 1$. It seems possible that in this case $h(G) = Q(m)$ provided only that G is sufficiently large, although this would be a very strong result.

9.5 Algorithms and complexity

A major open problem is the complexity of HARMONIOUS COLOURING and ACHROMATIC NUMBER for graphs of bounded degree. It seems likely that both problems can be solved in polynomial time for graphs of degree at most 2, but this has not been established. Conversely, for $k \geq 3$, the problems for graphs of degree at most k seem likely to be NP-complete, although for *trees* of degrees at most k, they can be solved in polynomial time, as described in Section 6.4. However all current NP-completeness proofs for these problems rely on the presence of vertices of large degree.

Another open problem concerns the complexity of the following problem:

Instance Graph G with m edges.

Question Is $h(G) = Q(m)$?

We will call this OPTIMAL HARMONIOUS COLOURING. Clearly this problem is just a special case of HARMONIOUS COLOURING in which we set $k = Q(m)$. It remains NP-complete for general graphs and for trees, but it is not known to be NP-complete for regular graphs for example (HARMONIOUS COLOURING *is* NP-complete for regular graphs).

Acknowledgements

I am grateful to Niall Cairnie and the referee for a number of useful comments on this paper.

Remarks on references

The following is intended to be a comprehensive list of references concerning the achromatic number, harmonious chromatic number and line-distinguishing chromatic number. It does not include papers concerned with edge colouring parameters, such as the achromatic index and the point-distinguishing chromatic number. Also omitted are papers concerned only with hypergraphs.

References

[1] J. Akiyama, F. Harary & P. Ostrand, A graph and its complement with specified properties VI: Chromatic and achromatic numbers, *Pacific Journal of Mathematics*, **104** (1983), 15–27.

[2] K. Al-Wahabi, R. Bari, F. Harary & D. Ullman, The edge-distinguishing chromatic number of paths and cycles, in *Proceedings of Graph Theory in memory of G. A. Dirac (Sandbjerg, 1985)* (eds. L. D. Andersen et al.), *Annals of Discrete Mathematics*, 41, (1989), pp. 17–22.

[3] B. Auerbach & R. Laskar, Some coloring numbers for complete r-partite graphs, *Journal of Combinatorial Theory, Series B*, **21** (1976), 169–170.

[4] D. G. Beane, N. L. Biggs & B. J. Wilson, The growth rate of the harmonious chromatic number, *Journal of Graph Theory*, **13** (1989), 291–299.

[5] V. N. Bhat-Nayak & M. Shanthi, Achromatic number of $\bigcup_{i=1}^{5} K_{n_i}$, in *Combinatorial Mathematics and Applications (Proceedings of the International Conference held in honor of Raj Chandra Bose, Calcutta, 1988)* (eds. J. K. Ghosh et al.), *Sankhyā, Series A*, 54, Special Issue, (1992), pp. 71–75.

[6] V. N. Bhat-Nayak & M. Shanti, Achromatic numbers of a graph and its complement, *Bulletin of the Bombay Mathematical Colloquium*, **6** (1989), 9–14.

[7] V. N. Bhave, On the pseudoachromatic number of a graph, *Fundamenta Mathematicae*, **102** (1979), 159–164.

[8] H. L. Bodlaender, Achromatic number is NP-complete for cographs and interval graphs, *Information Processing Letters*, **31** (1989), 135–138.

[9] B. Bollobás, P. A. Catlin & P. Erdős, Hadwiger's conjecture is true for almost all graphs, *European Journal of Combinatorics*, **1** (1980), 195–199.

[10] B. Bollobás, B. Reed & A. Thomason, An extremal function for the achromatic number, in *Graph Structure Theory (Proceedings of the AMS-IMS-SIAM Joint Summer Research Conference on Graph Minors, Seattle, 1991)* (eds. N. Robertson & P. Seymour), *Contemporary Mathematics*, 147, American Mathematical Society, Providence, Rhode Island (1993), pp. 161–165.

[11] F. Bories, Étude du nombre achromatique de certains graphes, in *Colloque sur la Théorie des Graphes (Paris, 1974), Cahiers du Centre d'Études de Recherche Opérationnelle*, 17, (1975), pp. 155–171.

[12] F. Bories, Sur quelques problèmes de colorations complètes de sommets et d'arêtes de graphes et d'hypergraphes, Thèse de 3ème cycle, Paris, 1975.

[13] F. Bories & J.-L. Jolivet, On complete colorings of graphs, in *Recent Advances in Graph Theory, (Proceedings of Second Czechoslovak Symposium, Prague, 1974)* (ed. M. Fiedler), Academia, Prague (1975), pp. 75–87.

[14] A. Bouchet & R. Lopez-Bracho, Decomposition of a complete graph into trails of given lengths, *Discrete Mathematics*, **42** (1982), 145–152.

[15] R. Brewster, Heuristics for computing the achromatic number of a graph, Manuscript, Victoria, B.C., 1987.

[16] R. Brewster & G. MacGillivray, Homomorphically full graphs, *Discrete Applied Mathematics*, **66** (1996), 23–31.

[17] B. E. Brunton, B. J. Wilson & T. S. Griggs, Graphs which have $n/2$-minimal line-distinguishing colourings, *Discrete Mathematics*, **155** (1996), 19–26.

[18] F. Buckley & L. Superville, The ajointed number and graph homomorphism problems, in *The Theory and Applications of Graphs (Proceedings of Fourth International Conference on the Theory and Applications of Graphs, Kalamazoo, Michigan, 1980)* (eds. G. Chartrand et al.), Wiley, New York (1981), pp. 149–158.

[19] N. Cairnie & K. J. Edwards, On the achromatic number of graphs, Manuscript, Dundee, 1996.

[20] G. Chartrand & J. Mitchem, Graphical theorems of the Nordhaus-Gaddum class, in *Recent trends in Graph Theory (Proceedings of First New York City Graph Theory Conference, New York, 1970)* (eds. M. Capobianco, J. B. Frechen & M. Krolik), *Lecture Notes in Mathematics*, 186, Springer, Berlin (1971), pp. 55–61.

[21] A. Chaudhary & S. Vishwanathan, Approximation algorithms for the achromatic number, in *Proceedings of Eighth Annual ACM-SIAM Symposium on Discrete Algorithms (New Orleans, 1997)*, in press.

[22] N.-P. Chiang & H.-L. Fu, On the achromatic number of the cartesian product $G_1 \times G_2$, *Australasian Journal of Combinatorics*, **6** (1992), 111–117.

[23] C. A. Christen & S. M. Selkow, Some perfect coloring properties of graphs, *Journal of Combinatorial Theory, Series B*, **27** (1979), 49–59.

[24] R. J. Cichelli, Minimal perfect hash functions made simple, *Communications of the Association for Computing Machinery*, **23** (1980), 17–19.

[25] C. R. Cook & A. B. Evans, Graph folding, in *Proceedings of 10th Southeastern Conference on Combinatorics, Graph Theory and Computing, Volume I (Boca Raton, 1979)* (eds. F. Hoffman et al.), *Congressus Numerantium*, 23–24, (1979), pp. 305–314.

[26] K. J. Edwards, The harmonious chromatic number of almost all trees, *Combinatorics, Probability and Computing*, **4** (1995), 31–46.

[27] K. J. Edwards, The harmonious chromatic number of bounded degree trees, *Combinatorics, Probability and Computing*, **5** (1996), 15–28.

[28] K. J. Edwards, The harmonious chromatic number of bounded degree graphs, *Journal of the London Mathematical Society*, in press.

[29] K. J. Edwards & C. J. H. McDiarmid, New upper bounds on harmonious colorings, *Journal of Graph Theory*, **18** (1994), 257–267.

[30] K. J. Edwards & C. J. H. McDiarmid, The complexity of harmonious colouring for trees, *Discrete Applied Mathematics*, **57** (1995), 133–144.

[31] M. Farber, G. Hahn, P. Hell & D. J. Miller, Concerning the achromatic number of graphs, *Journal of Combinatorial Theory, Series B*, **40** (1986), 21–39.

[32] B. Fawcett, On infinite full colourings of graphs, *Canadian Journal of Mathematics*, **30** (1978), 455–457.

[33] O. Frank, F. Harary & M. J. Plantholt, The line-distinguishing chromatic number of a graph, *Ars Combinatoria*, **14** (1982), 241–252.

[34] M. R. Garey & D. S. Johnson, *Computers and Intractability*, W. H. Freeman, New York (1979).

[35] D. P. Geller & S. T. Hedetniemi, A proof technique in graph theory, in *Proof Techniques in Graph Theory (Proceedings of Second Ann Arbor Conference on Graph Theory, Ann Arbor, 1968)* (ed. F. Harary), Academic Press, New York (1969), pp. 49–59.

[36] D. P. Geller & H. V. Kronk, Further results on the achromatic number, *Fundamenta Mathematicae*, **85** (1974), 285–290.

[37] D. P. Geller & S. Stahl, The chromatic number and other functions of the lexicographic product, *Journal of Combinatorial Theory, Series B*, **19** (1975), 87–95.

[38] J. P. Georges, On the harmonious coloring of collections of graphs, *Journal of Graph Theory*, **20** (1995), 241–254.

[39] R. P. Gupta, Bounds on the chromatic and achromatic numbers of complementary graphs, in *Recent Progress in Combinatorics (Proceedings of Third Waterloo Conference on Combinatorics, Waterloo, 1968)* (ed. W. T. Tutte), Academic Press, New York (1969), pp. 229–235.

[40] F. Harary & S. T. Hedetniemi, The achromatic number of a graph, *Journal of Combinatorial Theory*, **8** (1970), 154–161.

[41] F. Harary, S. T. Hedetniemi & G. Prins, An interpolation theorem for graphical homomorphisms, *Portugaliae Mathematica*, **26** (1967), 453–462.

[42] F. Harary & M. J. Plantholt, Graphs with the line-distinguishing chromatic number equal to the usual one, *Utilitas Mathematica*, **23** (1983), 201–207.

[43] W. R. Hare, S. T. Hedetniemi, R. Laskar & J. Pfaff, Complete coloring parameters of graphs, in *Proceedings of 16th Southeastern International Conference on Combinatorics, Graph Theory and Computing (Boca Raton, 1985)* (eds. F. Hoffman et al.), *Congressus Numerantium*, 48, (1985), pp. 171–178.

[44] S. M. Hedetniemi, S. T. Hedetniemi & T. Beyer, A linear algorithm for the Grundy (coloring) number of a tree, in *Proceedings of 13th Southeastern International Conference on Combinatorics, Graph Theory and Computing (Boca Raton, 1982)* (eds. F. Hoffman et al.), *Congressus Numerantium*, 36, (1982), pp. 351–363.

[45] P. Hell & D. J. Miller, On forbidden quotients and the achromatic number, in *Proceedings of Fifth British Combinatorial Conference (Aberdeen, 1975)* (eds. C. St.J. A. Nash-Williams & J. Sheehan), *Congressus Numerantium*, 15, (1976), pp. 283–292.

[46] P. Hell & D. J. Miller, Graph with given achromatic number, *Discrete Mathematics*, **16** (1976), 195–207.

[47] P. Hell & D. J. Miller, Achromatic numbers and graph operations, *Discrete Mathematics*, **108** (1992), 297–305.

[48] J. E. Hopcroft & M. S. Krishnamoorthy, On the harmonious coloring of graphs, *SIAM Journal on Algebraic and Discrete Methods*, **4** (1983), 306–311.

[49] M. Horňák & J. Puntigán, On the achromatic number of $K_m \times K_n$, in *Graphs and other Combinatorial Topics, Proceedings of Third Czechoslovak Symposium on Graph Theory, Prague, 1982)* (ed. M. Fiedler), *Teubner-Texte zur Mathematik*, 59, Teubner, Leipzig (1983), pp. 118–123.

[50] F. Hughes, On the achromatic number of graphs, M.Sc. Thesis, University of Victoria, B.C., 1994.

[51] F. Hughes & G. MacGillivray, The achromatic number of graphs: a survey and some new results, *Bulletin of the Institute of Combinatorics and its Applications*, in press.

[52] T. R. Jensen & B. Toft, *Graph Coloring Problems*, Wiley, New York (1995).

[53] J.-L. Jolivet, Graphes parfaits pour une propriété P, in *Colloque sur la Théorie des Graphes (Paris, 1974), Cahiers du Centre d'Études de Recherche Opérationnelle*, 17, (1975), pp. 253–256.

[54] J. Kelly, Difference systems, graph designs, and coloring problems, *Journal of Combinatorial Theory, Series B*, **30** (1981), 144–165.

[55] I. Krasikov & Y. Roditty, Bounds for the harmonious chromatic number of a graph, *Journal of Graph Theory*, **18** (1994), 205–209.

[56] A. Kundrík, The harmonious chromatic number of a graph, in *Proceedings of Fourth Czechoslovakian Symposium on Combinatorics, Graphs and Complexity (Prachatice, 1990)* (eds. J. Nešetřil & M. Fiedler), *Annals of Discrete Mathematics*, 51, (1992), pp. 167–170.

[57] S.-M. Lee & J. Mitchem, An upper bound for the harmonious chromatic number of a graph, *Journal of Graph Theory*, **11** (1987), 565–567.

[58] R. Lopez-Bracho, Études du nombre achromatique des étoiles, Thesis, Université de Paris-Sud, 1981.

[59] R. Lopez-Bracho, Le nombre achromatique d'une étoile, *Ars Combinatoria*, **18** (1984), 187–194.

[60] Zhikang Lu, On an upper bound for the harmonious chromatic number of a graph, *Journal of Graph Theory*, **15** (1991), 345–347.

[61] Zhikang Lu, The harmonious chromatic number of a complete binary and trinary tree, *Discrete Mathematics*, **118** (1993), 165–172.

[62] Zhikang Lu, Estimates of the harmonious chromatic numbers of some classes of graphs (Chinese), *Journal of Systems Science and Mathematical Sciences*, **13** (1993), 218–223.

[63] Zhikang Lu, The harmonious chromatic number of a complete 4-ary tree, *Journal of Mathematical Research and Exposition*, **15** (1995), 51–56.

[64] A. Máté, A lower estimate for the achromatic number of irreducible graphs, *Discrete Mathematics*, **33** (1981), 171–183.

[65] C. J. H. McDiarmid, Colourings random graphs badly, in *Graph Theory and Combinatorics (Proceedings of a Conference, Milton Keynes, 1978)* (ed. R. J. Wilson), *Pitman Research Notes in Mathematics*, 34, Pitman, San Francisco (1979), pp. 76–86.

[66] C. J. H. McDiarmid, Achromatic numbers of random graphs, *Mathematical Proceedings of the Cambridge Philosophical Society*, **92** (1982), 21–28.

[67] C. J. H. McDiarmid & Luo Xinhua, Upper bounds for harmonious colorings, *Journal of Graph Theory*, **15** (1991), 629–636.

[68] F. Milazzo & V. Vacirca, On the achromatic number of permutation graphs, in *Proceedings of First Catania International Combinatorial Conference on Graphs, Steiner Systems and their Applications, Volume 2 (Catania, 1986)* (ed. M. Gionfriddo), *Ars Combinatoria*, 24B, (1987), pp. 71–76.

[69] F. Milazzo & V. Vacirca, On the achromatic number of $G \times K_m$, in *Proceedings of First Catania International Combinatorial Conference on Graphs, Steiner Systems and their Applications, Volume 2 (Catania, 1986)* (ed. M. Gionfriddo), *Ars Combinatoria*, 24B, (1987), pp. 173–177.

[70] Z. Miller, Extremal regular graphs for the achromatic number, *Discrete Mathematics*, **40** (1982), 235–253.

[71] Z. Miller & D. Pritikin, The harmonious coloring number of a graph, in *Proceedings of 250th Anniversary Conference on Graph Theory (Fort Wayne, Indiana, 1986)* (eds. K. S. Bagga et al.), *Congressus Numerantium*, 63, (1988), pp. 213–228.

[72] Z. Miller & D. Pritikin, The harmonious coloring number of a graph, *Discrete Mathematics*, **93** (1991), 211–228.

[73] J. Mitchem, On the harmonious chromatic number of a graph, *Discrete Mathematics*, **74** (1989), 151–157.

[74] C. St.J. A. Nash-Williams, Detachments of graphs and generalised Euler trails, in *Surveys in Combinatorics 1985 (Invited papers for Tenth British Combinatorial Conference)* (ed. I. Anderson), Cambridge University Press, Cambridge (1985), pp. 137–151.

[75] N. Pippenger & J. Spencer, Asymptotic behavior of the chromatic index for hypergraphs, *Journal of Combinatorial Theory, Series A*, **51** (1989), 24–42.

[76] F. Regonati & N. Zagaglia Salvi, Some constructions of λ-minimal graphs, *Czechoslovak Mathematical Journal*, **44(119)** (1994), 315–323.

[77] F. Regonati & N. Zagaglia Salvi, Minimal semicomplete graphs, in *Proceedings of International Conference on Graphs and Hypergraphs (Varenna, 1991)* (eds. O. D'Antona, M. Gionfriddo & N. Zagaglia Salvi), *Journal of Combinatorics, Information and System Sciences*, 19, Forum for Interdisciplinary Mathematics, Delhi (1994), pp. 63–73.

[78] E. Sampathkumar & V. N. Bhave, Partition graphs and coloring numbers of a graph, *Discrete Mathematics*, **16** (1976), 57–60.

[79] A. J. Schwenk, Almost all trees are cospectral, in *New Directions in the Theory of Graphs (Proceedings of Third Ann Arbor Conference on Graph Theory, Ann Arbor, 1971)* (ed. F. Harary), Academic Press, New York (1973), pp. 275–307.

[80] B. J. Wilson, Line-distinguishing and harmonious colourings, in *Graph Colourings (Proceedings of a Conference on Graph Colourings, Milton Keynes, 1988)* (eds. R. Nelson & R. J. Wilson), *Pitman Research Notes in Mathematics*, 218, Longman Scientific & Technical, Essex (1990), pp. 115–133.

[81] B. J. Wilson, Minimal line distinguishing colourings in graphs, in *Proceedings of Combinatorics '90 (Gaeta, 1990)* (eds. A. Barlotti et al.), *Annals of Discrete Mathematics*, 52, (1992), pp. 549–558.

[82] R. M. Wilson, Decomposition of complete graphs into subgraphs isomorphic to a given graph, in *Proceedings of Fifth British Combinatorial Conference (Aberdeen, 1975)* (eds. C. St.J. A. Nash-Williams & J. Sheehan), *Congressus Numerantium*, 15, (1976), pp. 647–659.

[83] Shaoji Xu, Relations between parameters of a graph, *Discrete Mathematics*, **89** (1991), 65–88.

[84] M. Yannakakis & F. Gavril, Edge dominating sets in graphs, *SIAM Journal on Applied Mathematics*, **38** (1980), 364–372.

[85] N. Zagaglia Salvi, A note on the line-distinguishing chromatic number and the chromatic index of a graph, *Journal of Graph Theory*, **17** (1993), 589–591.

Department of Mathematics and Computer Science
University of Dundee
Dundee
DD1 4HN
U.K.
`kedwards@mcs.dundee.ac.uk`

Computer Construction of Block Designs

Clement Lam

Summary This paper uses an extended example to illustrate how to put together a BDX program to construct block designs fixed by an automorphism, given its orbit matrix. It shows how to specify the parameters and the structural information of the designs. It discusses the symmetry group of the problem and isomorph rejection. It explains how to choose a good order of generation to minimize the size of the search. It also shows how to estimate the size of a search and how to partition the problem into subproblems which can be searched in parallel on several computers.

1 Introduction

In a recent paper that I co-authored [1], we wrote:

> "For each orbit matrix, we used the BDX program to try out all possible circulant matrices with the correct row sum."

Here, I would like to expand on this sentence, not so much to bore you with details, but to use it as an example to explain how to use the BDX program. This paper is intended as a companion to the *BDX reference Guide* [7], which is a dry document outlining the syntax and meaning of each of the BDX commands. In this paper, I shall illustrate how the commands can be put together to solve a real problem.

Let me first state what problem we are trying to solve. We want to find all quasi-symmetric 2-$(28, 12, 11)$ designs with intersection numbers 4 and 6, which are fixed by an automorphism of order 7 without fixed points or fixed blocks.

I assume you know what 2-(v, k, λ) designs are. I will introduce the other terminology as we need it. There are two other parameters associated with a 2-(v, k, λ) design D, namely b, the number of blocks, and r, the number of replications. In fact, a 2-(v, k, λ) design is often referred to as a (v, b, r, k, λ) design. For the 2-$(28, 12, 11)$ design, the corresponding values are $b = 63$ and $r = 27$.

Let $P_1, \ldots, P_v$ be a labelling of the points of D and let $B_1, \ldots, B_b$ be a labelling of the blocks of D. We define a matrix $A = (a_{ij})$ by $a_{ij} = 1$ if P_i is on B_j and $a_{ij} = 0$ otherwise. Thus A is a $v \times b$ matrix with each entry either 1 or 0. We call A an *incidence matrix* for D.

Result 1.1 *If A is an incidence matrix for a 2-(v, k, λ) design, then*

$$AA' = nI_v + \lambda J_v$$

where A' is the transpose of A, I_v is the $v \times v$ identity matrix, J_v is the $v \times v$ matrix all of whose entries are $+1$ and $n = r - \lambda$.

BDX is a computer program that is designed to explore how best to enumerate 2-designs. It is built to be easily adaptable to designs of different parameters. It treats all designs as incidence matrices. As a consequence, we tend to treat the words *rows* and *points*, as well as the words *columns* and *blocks*, as synonyms.

BDX takes a sequence of commands as input, executes them and produces output. For example, if you want to enumerate all 2-$(28, 12, 11)$ designs, you can input:

```
28 rows of 27;
63 cols of 12;
lambda = 11;
try row [1:28, 1:63];
```

The first three lines define a 28×63 $(0, 1)$-matrix with 27 ones in each row, 12 ones in each column and in which the inner product of any pair of distinct rows is 11. This defines an incidence matrix of a 2-$(28, 12, 11)$ design. The last line instructs BDX to find all possible ways of putting 1's and 0's into this matrix subject to the constraints given.

I would not suggest you execute this input. The program would run for an extremely long time and find lots of solutions, but will not tell you what the solutions are. If it ever finishes, it will tell you how many solutions it found; but no one is going to live long enough to see it finish.

In the remainder of this paper, Section 2 shows how to formulate the problem for BDX and Section 3 demonstrates how to solve the problem using BDX.

2 Formulating the problem for BDX

A keen reader may observe that we are not trying to find all 2-$(28, 12, 11)$ designs, but rather a restricted subset. Let us first motivate the definition of the quasi-symmetric property.

While the inner products of pairs of distinct rows of the incidence matrix of a block design are always λ, the inner product of pairs of columns are not necessarily constant. These inner products of distinct columns are the *intersection numbers* of the block design.

A *symmetric* design is one where $v = b$ and, consequently, $r = k$. A symmetric design has a unique intersection number λ [5]. A *quasi-symmetric* design is one with only two intersection numbers [6]. In our case, we are looking for an incidence matrix where the inner products of pairs of distinct columns are either 4 or 6.

How do we specify the quasi-symmetric property to BDX? In the literature, $A'A$ is often referred to as the *block intersection matrix* S [2]. So, in BDX, `S` refers to this block intersection matrix. The BDX command to define a quasi-symmetric design with intersection numbers 4 and 6 is:

```
init S = {4 & 6};
```

It initializes the set of possible values for the off-diagonal entries of S to $\{4,6\}$.

An *automorphism* of a block design is an incidence preserving relabelling of its points and blocks. Let us restate it in terms of an incidence matrix A. Let π be a permutation of its columns and let ρ be a permutation of its rows. The pair of row and column permutations (ρ, π) is an automorphism of A if $a_{i,j} = a_{\rho(i),\pi(j)}$ for all rows i and columns j.

In our problem, we are looking for all 2-$(28, 12, 11)$ designs which are fixed by an automorphism of order 7 with no fixed points or blocks. This automorphism partitions the points into 4 orbits and the blocks into 9 orbits, where all the orbits are of size 7. By reordering the rows and columns, we can have a nice presentation of the structures of both the incidence matrix and the row and column permutations. In particular, we choose the image of row 1 under ρ as row 2, the image of row 2 as row 3, and so on. Because the orbits are of size 7, the image of row 7 is row 1. Thus, any row can be row 8, and we can then continue to use the image of row 8 as row 9, etc. After applying this process to all the rows, we can assume that

$$\rho(i) = \begin{cases} i+1 & \text{if } 7 \not| \, i, \text{ and} \\ i-6 & \text{otherwise.} \end{cases}$$

Similarly, applying the process to all of the columns,

$$\pi(j) = \begin{cases} j+1 & \text{if } 7 \not| \, j, \text{ and} \\ j-6 & \text{otherwise.} \end{cases}$$

In BDX, permutations are input in the image form, which simply means that, for a row permutation, we list all the values of $\rho(i)$, $i = 1, \ldots, v$, and, for a column permutation, we list $\pi(j)$, $j = 1, \ldots, b$. Thus, we specify this assumed automorphism information to BDX by:

```
assume row perm =
  2  3  4  5  6  7  1    9 10 11 12 13 14  8   16 17 18 19 20 21 15
 23 24 25 26 27 28 22;
assume col perm =
  2  3  4  5  6  7  1    9 10 11 12 13 14  8   16 17 18 19 20 21 15
 23 24 25 26 27 28 22   30 31 32 33 34 35 29   37 38 39 40 41 42 36
 44 45 46 47 48 49 43   51 52 53 54 55 56 50   58 59 60 61 62 63 57;
```

BDX uses the assumed automorphism to derive the implications of placing a value into the incidence matrix. For example, if $a_{1,1}$ is 0, then $a_{2,2} = a_{\rho(1),\pi(1)}$ is also 0. More generally, if $a_{i,j}$ is x, then $a_{\rho(i),\pi(j)}$ is also x. Thus, with the assumption of this automorphism, the incidence matrix for the 2-$(28, 12, 11)$ design can be partitioned into 4×9 blocks of 7×7 circulant matrices. For later reference, let us label these circulant matrices as $A_{i,j}$, $i = 1, \ldots, 4$ and $j = 1, \ldots, 9$.

0	3	3	3	3	3	4	4	4
4	1	3	3	3	5	2	2	4
4	3	3	3	5	1	2	4	2
4	5	3	3	1	3	4	2	2

Figure 1: The orbit matrix for case 26

A circulant matrix has constant row and column sums. The collection of row and column sums of all the circulant matrices gives a set of structural information about the incidence matrix. The *orbit matrix* or *tactical decomposition* of A with respect to the assumed automorphism is a 4×9 matrix $C = (c_{i,j})$, where $c_{i,j}$ is the column sum of the (i,j)-th circulant matrix $A_{i,j}$. The entries in an orbit matrix satisfy two relations [3], which in our case translate to

$$\sum_{k=1}^{9} c_{ik}c_{jk} = \begin{cases} 77 & \text{if } i \neq j,\ 1 \leq i,j \leq 4, \\ 93 & \text{if } i = j = 1,\ldots,4, \end{cases} \tag{1}$$

and

$$\sum_{k=1}^{9} c_{ik} = 27 \quad \text{for } i = 1,\ldots,4. \tag{2}$$

As Mathon noted in the 13th British Combinatorics Conference [3]:

> "The search for 2-(v,k,λ) designs proceeds in two stages, each of which requires solving a smaller problem than the original. In the first stage we backtrack for all tactical decompositions ..."

Unfortunately, BDX does not yet have the ability to backtrack and find all feasible orbit matrices. This stage has to be done by a separate program. So, I will just summarize the results. A total of 284 non-isomorphic orbit matrices satisfying Equations (1) and (2) are found. By showing further that

$$\sum_{i=1}^{4} c_{ij}^2 = 36,\ 40,\ 44 \text{ or } 48, \text{ for } j = 1, \ldots, 9,$$

these 284 possibilities are reduced to 9 [1].

Let us use case 26 as an example. Its orbit matrix is given in Figure 1. How do we tell BDX this information? BDX has an **under** directive which allows us to specify a set of acceptable row or column sums in a submatrix. For example, since $c_{1,6} = 3$, the row and column sums are both equal to 3 in the submatrix $A_{1,6}$ formed by the rows 1 to 7 and columns 36 to 42 of the incidence matrix. We can specify the row sum information by:

```
under col [36:42];
 in rows[1:7] type=3;
end under;
```

Since $c_{2,6}$, $c_{3,6}$ and $c_{4,6}$ all refer to row sums restricted to the columns from 36 to 42, all these directives can share the same **under** statement as the one for $c_{1,6}$. Thus, the row sum information of the whole 6-th column of C can be encoded as:

```
under cols [36:42];
 in rows[1:7] type=3;
 in rows[8:14] type=5;
 in rows[15:21] type=1;
 in rows[22:28] type=3;
end under;
```

Even though we can declare the row sum information one column block at a time, we still have a long list of statements. There is a special form of the **under** statement which allows us to declare all the row sum information together:

```
under cols           [ 1: 7| 8:14|15:21|22:28|29:35|36:42|43:49|50:56|57:63];
in row[ 1: 7] type={    0|    3|    3|    3|    3|    3|    4|    4|    4};
in row[ 8:14] type={    4|    1|    3|    3|    3|    5|    2|    2|    4};
in row[15:21] type={    4|    3|    3|    3|    5|    1|    2|    4|    2};
in row[22:28] type={    4|    5|    3|    3|    1|    3|    4|    2|    2};
end under;
```

The column sum information can be similarly declared. Please see the complete declaration at the end of this section for more detail.

Next, we consider the question of isomorph rejection. The property of being a block design does not depend on the labellings of the points and blocks. These labellings are needed to define an incidence matrix. Different labellings of the same design give rise to different incidence matrices, but they are essentially the same. Thus, we call two incidence matrices A and $\hat{A}$ *isomorphic* if there exists a row permutation ρ and a column permutation π, such that $a_{\rho(i),\pi(j)} = \hat{a}_{i,j}$ for all i and j. Two matrices are *non-isomorphic* if one cannot be obtained from another by row and column permutations. Since we are interested in designs that are different, we are interested only in non-isomorphic incidence matrices.

Let us consider the submatrix $A_{2,2}$. Since $c_{2,2} = 1$, the row and column sum of $A_{2,2}$ is 1. Remembering that $A_{2,2}$ is a circulant, its entries are completely determined once the first column is known. This first column has 7 entries, one of which is a 1 and the rest 0's. So, there are 7 possibilities for $A_{2,2}$. Now, the first row of $A_{2,2}$ is in fact part of the 8-th row of A. When we reorder the rows of A to derive the circulant structure of its submatrices, we stated that any row can be row 8. We can use any of the rows 8 to 14 as row 8. Thus, we can assume that $a_{8,8} = 1$, reducing the 7 possibilities of $A_{2,2}$ to just one. In other words, the 7 possibilities of $A_{2,2}$ lead to different but isomorphic solutions. Since we are only interested in non-isomorphic solutions, we only have to consider one such possibility. The process of discarding partial solutions because they are isomorphic to some other partial solutions which are accepted for further extension is called *isomorph rejection.*

However, we have to be careful which permutations we can use and which we cannot use for isomorph rejection. For example, we cannot permute row 1 of A to be row 8, because no entry in row 1 of the orbit matrix C is equal to 1. Also, after fixing $a_{8,8} = 1$, we cannot use row permutation to force $a_{8,1} = 1$, even though there are four 1's in the first column of the submatrix $A_{2,1}$. This is because once row 8 is chosen, the choices for rows 9 to 14 are forced.

The permutations that we can use are the ones that preserve both the orbit matrix and the circulant structure of the submatrices $A_{i,j}$. Such a permutation is called a *symmetry.* The set of symmetries form a group called a *symmetry group.*

There are several obvious types of symmetries for designs constructed via orbit matrices:

1. Any cyclic permutation of the row or columns in an orbit under the action of the assumed automorphism group is a symmetry.
2. Any automorphism of the C matrix can be *expanded* into a symmetry of the A matrix.

The row permutation which allows us to fix $a_{8,8} = 1$ is an example of the first type of symmetry. We can also permute column 22 of A to be column 15, because the third and fourth columns of C are equal. Of course, we have to permute columns 22 to 28 into columns 15 to 21 simultaneously. This is an example of the second type of symmetry.

There is a third type of symmetry, which can best be described first in terms of a circulant $X = (x_{i,j})$ of order p and whose indices run from 0 to $p-1$. Let t be co-prime to p. Then the operation of taking an index i to the index $ti \bmod p$ is a permutation on the set of indices. Moreover, the row and column permutations defined by taking $x_{i,j}$ to $x_{ti \bmod p, tj \bmod p}$ preserve the circulant property. If the corresponding row and column permutations are applied to all the circulant submatrices $A_{i,j}$ of A, then both the orbit matrix and the circulant structure of the submatrices are preserved.

In our example, $p = 7$, and if $t = 3$ the row permutation for this symmetry is

```
 1  4  7  3  6  2  5    8 11 14 10 13  9 12    15 18 21 17 20 16 19
22 25 28 24 27 23 26,
```

and its column permutation is

```
 1  4  7  3  6  2  5    8 11 14 10 13  9 12    15 18 21 17 20 16 19
22 25 28 24 27 23 26   29 32 35 31 34 30 33    36 39 42 38 41 37 40
43 46 49 45 48 44 47   50 53 56 52 55 51 54    57 60 63 59 62 58 61.
```

By default, BDX assumes that the symmetry group is one which allows arbitrary row and column permutations. Since we cannot use this default, the symmetry group has to be read in. If the name of the file containing the symmetry group is case2_gp, then the BDX command to read in the group is

```
read_sym_group "case2_gp";
```

Given an orbit matrix, the preparation of the BDX input which specifies its associated incidence matrix and symmetry group is quite involved. So, I have written another program to prepare this input. When fed the orbit matrix, the program produces the following output:

```
28 rows of 27;
63 cols of 12;
lambda = 11;
init S={4&6};
read_sym_group "case26";
assume row perm =
  2  3  4  5  6    7  1  9 10 11    12 13 14  8 16    17 18 19 20 21
 15 23 24 25 26   27 28 22 ;
assume col perm =
  2  3  4  5  6    7  1  9 10 11    12 13 14  8 16    17 18 19 20 21
 15 23 24 25 26   27 28 22 30 31    32 33 34 35 29    37 38 39 40 41
 42 36 44 45 46   47 48 49 43 51    52 53 54 55 56    50 58 59 60 61
 62 63 57 ;
under rows           [ 1: 7| 8:14|15:21|22:28];
in col[ 1: 7] type={     0|    4|    4|    4};
in col[ 8:14] type={     3|    1|    3|    5};
in col[15:21] type={     3|    3|    3|    3};
in col[22:28] type={     3|    3|    3|    3};
in col[29:35] type={     3|    3|    5|    1};
in col[36:42] type={     3|    5|    1|    3};
in col[43:49] type={     4|    2|    2|    4};
in col[50:56] type={     4|    2|    4|    2};
in col[57:63] type={     4|    4|    2|    2};
end under;
under cols           [ 1: 7| 8:14|15:21|22:28|29:35|36:42|43:49|50:56|57:63];
in row[ 1: 7] type={     0|    3|    3|    3|    3|    3|    4|    4|    4};
in row[ 8:14] type={     4|    1|    3|    3|    3|    5|    2|    2|    4};
in row[15:21] type={     4|    3|    3|    3|    5|    1|    2|    4|    2};
in row[22:28] type={     4|    5|    3|    3|    1|    3|    4|    2|    2};
end under;

% generators of the automorphism group of the orbit matrix
%  2: row =  1  3  4  2
%     col =  1  6  3  4  2  5  9  7  8
%  4: row =  1  4  2  3
%     col =  1  5  3  4  6  2  8  9  7
%  8: row =  1  2  3  4
%     col =  1  2  4  3  5  6  7  8  9
% Symmetry Group Order = (1)(6)(7^11)(14)(21)
```

Note that the program lists the generators of the automorphism group of the orbit matrix as part of the comments. These generators are useful in deciding how to arrange the order of generation of the entries of A, so as to use these inherited symmetries efficiently. We shall look into this aspect early in the next section.

3 Towards a search

The basic technique used by BDX is an entry-by-entry *backtrack search*. It first tries a 1 and if there is no contradiction, it continues to the next entry. If there is a contradiction or when it backtracks from the next entry, it tries a 0 and again checks whether it should continue to the next entry. After trying both 1 and 0, it then backtracks to the previous entry.

The efficiency of a backtrack search is greatly influenced by the effectiveness of the contradiction test, which is in turn influenced by the order of generation of the entries. Isomorph rejection is used to further reduce the size of the search. However, it is a time consuming process and is best done when the number of cases to be tested is small. For this reason, isomorph rejection is typically performed in the early levels of the backtrack search, and it is most effective when the order of generation follows the orbit structure of the symmetry group. All these considerations require great flexibility in the order of generation of the entries in A. BDX gives us this flexibility.

Let us continue the example of finding all non-isomorphic solutions with `case26` as the orbit matrix. For ease of explanation, we shall refer to the row of submatrices $A_{i,j}$, $j = 1, \ldots, 9$ as the i-th row block, and the column of submatrices $A_{i,j}$, $i = 1, \ldots, 4$ as the j-th column block. Isomorph rejection involving symmetries which arise from the automorphisms of the orbit matrix require isomorphism testing and should be used up first. The automorphisms of C imply that the second to fourth row blocks can be cyclically permuted. Since all automorphisms fix the first column of C, the symmetry group maps the entries in the first column block amongst themselves. Thus, the first column block is a natural candidate to be generated first, because it allows us to use the permutation of row blocks.

The entries in the submatrix $A_{1,1}$ are all 0's. BDX allows us to set the values of entries in A. The corresponding command is:

```
a[1:7,1:7] = 0;
```

An alternative method is to use the `try` command, which asks BDX to generate all possible ways to fill in the specific entries in A. The `try row` command instructs BDX to backtrack on a row by row basis; and the `try col` command instructs it to backtrack column by column. Thus, to fill in the submatrix $A_{1,1}$, we can do:

```
try col [1:7,1:7];
```

Since the entries must be 0's, there is only one choice and BDX can accomplish this task very quickly.

Next, we consider $A_{2,1}$. The column sum is 4. By using the cyclic row permutation, we can force $a_{8,1}$ to be either 1 or 0. Let us choose $a_{8,1} = 0$ and insert a comment reminding us why this choice is done without loss of generality. Comments begin with a % and are important to remind us why something is done. While the whole BDX program can be regarded as a "proof", the embedded comments serve as explanations of the "proof". In our case, the command is:

```
a[8,1] = 0; % by cyclic row permutation within a row block
```

Let us fill in the rest of the submatrix $A_{2,1}$, print out the first column block, and try the program out. The commands are:

```
try col [8:14, 1:7];
printrow [1:28, 1:7];
exit
```

The command exit terminates the commands to BDX, and it is the only command which does not end with a semicolon. The printrow command prints the first column block in a normal row by column format. In this case, it will have 28 rows and 7 columns. We can also print its transpose by using the printcol command. In our case, the printcol command will use less paper. It may surprise you, but the ability to print out a partially filled incidence matrix nicely is the original motivation for BDX. We were tired of making photocopies of hand drawn incidence matrices and of the tedious process of writing in 1's and 0's. The printrow and printcol commands have many controls to allow us to highlight some structural aspects of the matrix. For example, one can highlight the submatrix structure by placing a blank row between each row block. Thus, to modify the printout so that we print the transpose of the first column block, and to put in the blank rows, we do:

```
printcol [1:7$&8:14$&15:21$&22:28, 1:7];
```

Running this input through BDX produces 15 partial solutions. The first one is:

```
print request from line 47
                11111 1111122 2222222
     1234567 8901234 5678901 2345678

   1 0000000 0111100 ....... .......
   2 0000000 0011110 ....... .......
   3 0000000 0001111 ....... .......
   4 0000000 1000111 ....... .......
   5 0000000 1100011 ....... .......
   6 0000000 1110001 ....... .......
   7 0000000 1111000 ....... .......
```

BDX echoes all the input, which we are not reproducing here. BDX also gives all commands a line number for later reference. For example, in the above output, it states that the printing command is at line 47. One can then go back and find out that this is the `printcol` command. We then can determine that we are printing the transpose. In `printcol`, the row numbers are on top and the column numbers are on the side. Besides printing the known entries, the unknown entries are denoted by a ".".

On exit, BDX also gives a summary of its execution, in the form of a final `printcount`. In our case, the summary is:

```
Final printcount
Cmd  44 (        try)      IN:       1/1.0e+00 OUT:          1/1.0e+00.
 COL        ones          fills     COL        ones           fills
  1       0/0.0e+00     1/1.0e+00    5       0/0.0e+00      1/1.0e+00
  2       0/0.0e+00     1/1.0e+00    6       0/0.0e+00      1/1.0e+00
  3       0/0.0e+00     1/1.0e+00    7       0/0.0e+00      1/1.0e+00
  4       0/0.0e+00     1/1.0e+00
Cmd  45 (       setA)      IN:       1/1.0e+00 OUT:          1/1.0e+00.
Cmd  46 (        try)      IN:       1/1.0e+00 OUT:         15/1.5e+01.
 COL        ones          fills     COL        ones           fills
  1      14/1.4e+01    15/1.5e+01    5      10/1.0e+01     15/1.5e+01
  2      10/1.0e+01    15/1.5e+01    6      10/1.0e+01     15/1.5e+01
  3      10/1.0e+01    15/1.5e+01    7      10/1.0e+01     15/1.5e+01
  4      10/1.0e+01    15/1.5e+01
Cmd  48 (       exit)      IN:      15/1.5e+01 OUT:          0/0.0e+00.
 High water mark of stack used 540 out of 300000 bytes
Execute time was     0.040 Seconds
```

Command 44 is `try col [1:7, 1:7]`. It is entered once and it exited once, meaning that it found only one possibility to fill this submatrix. The next 5 lines give the backtracking details while executing this command. Since it is a `try col`, the submatrix is filled column by column, starting from column 1. Under the heading `ones`, BDX lists the number of times ones have to be placed in the corresponding column, while executing the backtracking for this command. For command 44, all the counts for `ones` are 0, because a one is never placed. Under the heading `fills`, BDX gives the number of times the particular column is completed. Command 45 is `a[8, 1] = 0` and command 46 is another backtracking step, `try col [8:14, 1:7]`. In this case, it finds 15 completions. The run takes 0.040 seconds.

The last completion is:

```
                11111 1111122 2222222
        1234567 8901234 5678901 2345678

      1 0000000 0001111 ....... .......
      2 0000000 1000111 ....... .......
      3 0000000 1100011 ....... .......
      4 0000000 1110001 ....... .......
      5 0000000 1111000 ....... .......
      6 0000000 0111100 ....... .......
      7 0000000 0011110 ....... .......
```

By permuting rows, it can be made to be the same as the first completion. and can be pruned by isomorph rejection. To instruct BDX to do so, we add the command:

```
test isom;
```

With this additional command, BDX reports:

```
Cmd  47 (test isom)       IN:      15/1.5e+01 OUT:       2/2.0e+00.
Context date=46( 2), 2 certs and 13 non-certs, using 0.460 seconds
```

So, the 15 completions were reduced to 2 by isomorph rejection. The term `certs` is a short form for `certificates`, which refers to those completions which survives the isomorph rejection test. The term `non-certs` refers to those that are rejected.

The above isomorph rejection is one of generating all the completions before testing them. Another way is to perform more group theory calculations to avoid generating isomorphic cases. For example, after BDX backtracks and changes the value of $a_{9,1}$ from 1 to 0, it computes the orbit of the entry $a_{9,1}$, under the action of the subgroup of the symmetry group fixing the known part of the incidence matrix up to this point, i.e. the submatrix $A_{1,1}$ and the entry $a_{8,1}$. It then places a 0 in all the entries of this orbit, because if any of these entries is 1, then there exists a symmetry permutation which takes this 1 to the $a_{9,1}$ position. In BDX, we can add the keyword `isom` to the `try` command for this look-ahead type of isomorph rejection. For example:

```
try col isom [8:14, 1:7];
```

In this case, only two non-isomorphic completions are generated. However, the amount of time spent in the extra group theory calculation is more than the saving in the isomorphism testing time. The cost-benefit conclusion would have been reversed if we had not eliminated some of the isomorphic completions by the line

```
a[8,1] = 0;
```

So, it may be worthwhile trying the look ahead version out if we expect a lot of non-certificates. Doing isomorph rejection by hand is a very error prone process. To minimize errors, it is advisable to let the computer do more of the work.

Now, let us finish the first column block by:

```
a[15,1] = 0;
try col [15:21, 1:7];
test isom;
a[22, 1] = 0;
try col [22:28, 1:7];
test isom;
```

We use the cyclic row permutations to force $a_{15,1}$ and $a_{22,1}$ to be 0's. Isomorph rejection is performed after the completion of each circulant submatrix to reduce the number of continuations. With this input, BDX gives 4 non-isomorphic completions of the first column block.

Even though we have started by completing the incidence matrix columnwise, we can continue in any order. In fact, it is better to next complete a row block before returning to complete the matrix columnwise. The reason is that there remain many symmetries which are cyclic column permutations of the columns within a column block, as well as the symmetry which interchanges the third and fourth column block. By completing a row block now, we can use up these symmetries before the number of cases explodes.

Which row block to do next? We want to choose the one with the fewest continuations. One indication of the number of continuations is the number of possibilities without taking into account the symmetry and the fact that any distinct pair of rows must intersect in 11 places. For the first row block, there are $\binom{7}{3}$ ways to fill in $A_{1,2}$, $\binom{7}{3}$ ways to fill in $A_{1,3}$ and so on. Thus, the number of possibilities is $\binom{7}{3}^5\binom{7}{4}^3$. The next three row blocks are essentially identical. Since the entry for the first column block is known at this point, the number of possibilities for each of these row blocks is $\binom{7}{1}\binom{7}{2}^2\binom{7}{3}^3\binom{7}{4}\binom{7}{5}$, which is about 200 times smaller than the number for the first row block. So, we shall continue with the second row block.

In order to use the symmetry that interchanges the third and fourth column block, let us generate $A_{1,3}$ and $A_{1,4}$ first. The commands, as echoed by BDX with line numbers inserted, are:

```
a[8, 15] = 1;
try col [8:14, 15:21];
test isom;
a[8, 22] = 1;
try col [8:14, 22:28];
test isom;
```

Let us select a few lines out of the summary statistics:

```
Cmd  57 (      setA)       IN:       4/4.0e+00 OUT:         4/4.0e+00.
Cmd  58 (       try)       IN:       4/4.0e+00 OUT:        60/6.0e+01.
Cmd  59 (test isom)        IN:      60/6.0e+01 OUT:        14/1.4e+01.
Cmd  60 (      setA)       IN:      14/1.4e+01 OUT:        14/1.4e+01.
Cmd  61 (       try)       IN:      14/1.4e+01 OUT:       210/2.1e+02.
Cmd  62 (test isom)        IN:     210/2.1e+02 OUT:        36/3.6e+01.
Execute time was    5.490 Seconds
```

The 4 completions of the first column block expand to 36 non-isomorphic completions. The amount of time taken is still small.

Are you curious why BDX seems to print every value twice, for example as in `36/3.6e+01`? In fact, the two values have different meanings! The first value `36` gives the actual number of certificates that BDX finds. The second value `3.6e+01` reports the estimated number of certificates. We can instruct BDX to sample selected branches of the search tree and to report the estimated number of completions based on the sampling. Since we have been extending every branch, there is no difference between the actual count and the estimated value. Estimation gives us a way of deciding whether a problem is solvable with a given amount of computing resource. It also helps us determine whether a particular idea of improving the search actually works in practice. The simplest sampling command is the `pass every` directive. It causes BDX to systematically accept one out of a number of cases. For example,

```
pass every 2 times;
```

will accept the first, third, fifth cases and so forth. BDX will assume that the cases skipped behave the same as the accepted cases and it will adjust the estimated completions accordingly.

Let us continue part way on the second row block and use `pass every` to obtain an estimation. Again, we list the version with line numbers inserted. Note also that since the column sum of $A_{2,2}$ is 1, there is only one choice for this circulant, once $a_{8,8}$ is forced to be a 1.

```
63 a[8, 8] = 1;
64 try col [8:14, 8:14];
65 pass every 2 times;
66 a[8, 36] = 0;
67 try col [8:14, 36:42];
68 test isom;
69 pass every 2 times;
70 a[8, 43] = 1;
71 try col [8:14, 43:49];
72 test isom;
73 pass every 4 times;
74 a[8, 50] = 1;
75 try col [8:14, 50:56];
```

```
76 test isom;
```

A selected subset of the summary statistics shows:

```
Cmd  63 (      setA)    IN:     36/3.6e+01 OUT:     36/3.6e+01.
Cmd  64 (       try)    IN:     36/3.6e+01 OUT:     36/3.6e+01.
Cmd  65 (passevery)     IN:     36/3.6e+01 OUT:     18/3.5e+01.
Cmd  66 (      setA)    IN:     18/3.5e+01 OUT:     18/3.5e+01.
Cmd  67 (       try)    IN:     18/3.5e+01 OUT:    108/2.1e+02.
Cmd  68 (test isom)     IN:    108/2.1e+02 OUT:     48/9.3e+01.
Cmd  69 (passevery)     IN:     48/9.3e+01 OUT:     24/9.1e+01.
Cmd  70 (      setA)    IN:     24/9.1e+01 OUT:     24/9.1e+01.
Cmd  71 (       try)    IN:     24/9.1e+01 OUT:    142/5.4e+02.
Cmd  72 (test isom)     IN:    142/5.4e+02 OUT:     71/2.7e+02.
Cmd  73 (passevery)     IN:     71/2.7e+02 OUT:     18/2.6e+02.
Cmd  74 (      setA)    IN:     18/2.6e+02 OUT:     18/2.6e+02.
Cmd  75 (       try)    IN:     18/2.6e+02 OUT:    104/1.6e+03.
Cmd  76 (test isom)     IN:    104/1.6e+03 OUT:     52/7.8e+02.
```

The first `pass every` command is at line 65. It picks 18 of the 36 possibilities to continue. The estimated number of cases is 35. Note that the current version of BDX has some truncation errors and it tends to slightly underestimate. Command 68 reports generating 48 non-isomorphic completions up to $A_{2,6}$ and gives an estimate of 93 non-isomorphic completions. Command 69 chooses 24 out of the 48 cases. Command 72 finds 71 non-isomorphic continuations to $A_{2,7}$, with an estimate of 270. Command 76 estimates 780 non-isomorphic completions up to $A_{2,8}$. These estimates are quite reasonable, because the actual number of non-isomorphic completions are 96 up to $A_{2,6}$, 286 up to $A_{2,7}$ and 842 up to $A_{2,8}$.

I think at this point, you know enough about BDX to complete the generation of the second row block. Maybe you can even complete the rest of the incidence matrix, one column block at a time. Just remember that we still have one class of symmetry left: the cyclic row permutation of the rows in the first row block.

Let me turn to another question: how to save the partial or complete results for later analysis? You may have generated all 69 non-isomorphic solutions for case 26. Printing all the solutions out is one way of keeping a record. But, it is difficult to read it back in for some future work, for example, finding the size of the automorphism group for each solution. BDX provides the `dump` and `restore` directives which allow the user to save the current environment, and to restore it at a later time. To be more precise, the command

```
dump;
```

saves the current state of the incidence matrix A to a file. Since the `try` commands imply a backtracking process, the `dump` command is typically encountered many times, each with a different A. Each of these A's is written to the file as a separate `image`. BDX also saves all the commands that are used to generate these images.

These images can be restored using the `selectimage` command. During a restore operation, BDX will also read in all the commands that generated the images. Thus, one can continue the search as if there is no interruption.

For example, suppose we have `dumped` the 69 solutions for case 26, and we want to find the size of the automorphism group for the first five solutions. The whole BDX input consists of the following 4 lines:

```
selectimage [1:5];
option print_grp_size on;
test isom;
exit
```

The `dump` and `restore` combination is also useful in breaking up a long running job into many smaller jobs. These jobs can then be run in parallel on separate machines. The advantage is that the whole job can be completed in less real world time. The management of the jobs running on many computers can be overwhelming. I would strongly suggest you use B. McKay's AUTOSON package [4] for job management.

4 Conclusion

In this paper, I have illustrated how to use some of the most common commands in BDX. I have also shown you how to reduce the size of the search by using isomorph rejection and by choosing a good order of generating the entries. There are still many other commands and options that I have not touched. However, I hope I have given you enough of a head start that you can continue on your own. There is nothing better than actually trying the package out and you can start today at the conference.

Acknowledgements

Research of the author is supported by the Natural Sciences and Engineering Research Council of Canada and the Fonds pour la Formation de Chercheurs et l'Aide à la Recherche.

References

[1] Y. Ding, S. Houghten, C. Lam, S. Smith, L. Thiel & V. D. Tonchev, Quasi-symmetric 2-(28, 12, 11) designs with an automorphism of order 7, in press.

[2] M. Hall Jr., R. Roth, G. H. J. van Rees & S. A. Vanstone, On designs (22, 33, 12, 8, 4), *Journal of Combinatorial Theory, Series A*, **47** (1988), 157–175.

[3] R. Mathon, Computational methods in design theory, in *Surveys in Combinatorics, 1991* (ed. A. D. Keedwell), *London Mathematical Society Lecture Note Series*, 166, Cambridge University Press, Cambridge (1991), pp. 101–118.

[4] B. D. McKay, Autoson manual, Technical report TR-CS-96-03, Australian National University, 1996.

[5] H. J. Ryser, A note on a combinatorial problem, *Proceedings of the American Mathematical Society*, **1** (1950), 422–424.

[6] M. S. Shrikhande & S. S. Sane, *Quasi-Symmetric Designs*, *London Mathematical Society Lecture Note Series*, 164, Cambridge University Press, Cambridge (1991).

[7] L. Thiel & N. Fink, BDX reference Guide, Internal Report, Concordia University, 1994.

Department of Computer Science
Concordia University
Montreal
Canada
`lam@cs.concordia.ca`

Finite Quasiprimitive Graphs

Cheryl E. Praeger

Summary A permutation group on a set Ω is said to be quasiprimitive on Ω if each of its nontrivial normal subgroups is transitive on Ω. For certain families of finite arc-transitive graphs, those members possessing subgroups of automorphisms which are quasiprimitive on vertices play a key role. The manner in which the quasiprimitive examples arise, together with their structure, is described.

1 Introduction

A permutation group on a set Ω is said to be *quasiprimitive* on Ω if each of its nontrivial normal subgroups is transitive on Ω. This is an essay about families of finite arc-transitive graphs which have group-theoretic defining properties. By a *quasiprimitive graph* in such a family we shall mean a graph which admits a subgroup of automorphisms which not only is quasiprimitive on vertices, but also has the defining property of the family. For example, in the family of all arc-transitive graphs, a quasiprimitive graph is one with a subgroup of automorphisms which is both quasiprimitive on vertices and transitive on arcs. (An *arc* of a graph Γ is an ordered pair of adjacent vertices.)

First we shall describe an approach to studying several families of finite arc-transitive graphs whereby quasiprimitive graphs arise naturally. The concept of quasiprimitivity is a weaker notion than that of primitivity for permutation groups, and we shall see that finite quasiprimitive permutation groups may be described in a manner analogous to the description of finite primitive permutation groups provided by the famous theorem of M. E. O'Nan and L. L. Scott [17, 30]. There are several distinct *types* of finite quasiprimitive permutation groups, and several corresponding distinct types of finite quasiprimitive graphs. Using the family of finite 2-arc transitive graphs as a model, we illustrate the role played by the quasiprimitive members of the family, and also the effectiveness of identifying the type of a quasiprimitive graph in understanding the structure of such graphs. Finally we give a critique of this approach which considers in particular finite quasiprimitive graphs for which the full automorphism group may not be quasiprimitive.

2 Partitions and quotients of permutation groups and graphs

Let G be a transitive permutation group on a set Ω. A *G-invariant partition* $\mathcal{P}$ of Ω is one such that $p^g \in \mathcal{P}$ for all $p \in \mathcal{P}$ and $g \in G$. The *trivial partitions* $\{\Omega\}$ and $\{\{\omega\} : \omega \in \Omega\}$ are G-invariant for all transitive groups G, and a transitive permutation group G on Ω is said to be *primitive* on Ω if these are the only G-invariant partitions of Ω. If G is transitive, but not primitive, on Ω, then G is said to be *imprimitive* on Ω.

One way of seeing that quasiprimitivity is a natural weakening of the concept of primitivity for permutation groups is provided by the notion of a normal partition. For a transitive permutation group G on Ω, we say that a G-invariant partition of Ω is *G-normal* (or simply *normal* if the group G is clear from the context) if it is the set $\mathcal{P}_N$ of orbits in Ω of some normal subgroup N of G. Certainly all G-normal partitions are G-invariant, and both of the trivial partitions are G-normal for every transitive group G since $\{\Omega\}$ and $\{\{\omega\} : \omega \in \Omega\}$ are the orbit sets for G and $\{1_G\}$ respectively. It is immediate from the definition that G is quasiprimitive on Ω if and only if the only G-normal partitions of Ω are the trivial ones.

For a G-invariant partition $\mathcal{P}$, the setwise stabilizer G_p in G of a part $p \in \mathcal{P}$ is a subgroup of G containing the point stabilizer G_α where $\alpha \in p$. Conversely, if $G_\alpha \leq H \leq G$ then the H-orbit $p := \alpha^H$ containing α determines a G-invariant partition, namely $\mathcal{P} := \{\, p^g : g \in G\}$. In fact the lattice of all G-invariant partitions of Ω is isomorphic to the lattice of subgroups of G containing G_α, and hence G is primitive on Ω if and only if G_α is a maximal subgroup of G.

For a transitive permutation group G on Ω and a G-invariant partition $\mathcal{P}$, there correspond two (usually smaller) transitive permutation groups, namely the group $G^{\mathcal{P}}$ of permutations of $\mathcal{P}$ induced by G, and the permutation group G_p^p induced on p by G_p, where $p \in \mathcal{P}$. Note that G_p^p is independent of p up to permutational equivalence since, for $p, p' \in \mathcal{P}$, each element $x \in G$ such that $p^x = p'$ induces a bijection $\bar{x}$ from p to p' (by restriction) and an isomorphism $\hat{x}$ from G_p^p to $G_{p'}^{p'}$ (by conjugation), and the pair $(\bar{x}, \hat{x})$ is easily seen to be a permutational equivalence. As in [23, p. 32] by a *permutational equivalence* between two permutation groups G, H on Ω and Σ respectively we mean a pair (φ, f) such that $f\colon G \to H$ is a group isomorphism, $\varphi\colon \Omega \to \Sigma$ is a bijection and for all $g \in G$, $\alpha \in \Omega$, we have $(\alpha^g)\varphi = (\alpha\varphi)^{(g)f}$.

As abstract groups $G^{\mathcal{P}}$ is a quotient group of G, and we view the permutation group $G^{\mathcal{P}}$ on $\mathcal{P}$ as a quotient of the the permutation group G on Ω. If we choose $\mathcal{P}$ such that G_p is maximal in G for some, and hence all, $p \in \mathcal{P}$, we see that $G^{\mathcal{P}}$ must be primitive on $\mathcal{P}$. Thus every transitive permutation group has at least one primitive quotient. Similarly if $\mathcal{P}$ is a G-normal partition then the permutation group $G^{\mathcal{P}}$ will be called a *normal quotient* of G. By choosing a normal subgroup N to be maximal by inclusion such that N is intransitive on Ω, we can ensure that the normal quotient $G^{\mathcal{P}_N}$ is quasiprimitive. Thus every transitive permutation group has at least one quasiprimitive normal quotient.

One reason for making the latter definition is that there is a useful analogous concept for graphs. For a graph $\Gamma = (V, E)$ with vertex set V and edge set E, if $\mathcal{P}$ is a partition of V then the *quotient graph* of Γ relative to $\mathcal{P}$ is defined to be the graph $\Gamma_{\mathcal{P}} = (\mathcal{P}, E_{\mathcal{P}})$ where $\{p, p'\} \in E_{\mathcal{P}}$ if and only if there exist $x \in p$ and $x' \in p'$ such that $\{x, x'\} \in E$. Quotient graphs inherit some of the properties of the original graphs. For example, quotients of connected graphs are connected. Moreover if the partition $\mathcal{P}$ is invariant under some subgroup G of the automorphism group $\mathrm{Aut}(\Gamma)$, then $\Gamma_{\mathcal{P}}$ admits $G^{\mathcal{P}}$ as a subgroup of automorphisms. In particular, if $G \leq \mathrm{Aut}(\Gamma)$ and G is arc-transitive on Γ, then for each G-invariant partition $\mathcal{P}$ of V, $G^{\mathcal{P}}$ is arc-transitive on $\Gamma_{\mathcal{P}}$. If $\mathcal{P}$ is a G-normal partition of V, say $\mathcal{P}$ is the set of N-orbits in V with N a normal subgroup of G, then we say that $\Gamma_{\mathcal{P}}$ is a *G-normal quotient* of Γ, and write $\Gamma_N = \Gamma_{\mathcal{P}}$.

Note that, if Γ is connected and G is arc-transitive, then for any nontrivial G-invariant partition $\mathcal{P}$, each edge of Γ joins vertices in distinct parts of $\mathcal{P}$. Thus each $p \in \mathcal{P}$ is an independent subset of V and the subgraph induced on p is an empty graph. Moreover, since in this case G is transitive on unordered pairs of adjacent parts $\{p, p'\}$ in $\Gamma_{\mathcal{P}}$, the subgraph of Γ induced on $p \cup p'$ is independent of the pair $\{p, p'\} \in E_{\mathcal{P}}$. In the special case where the edges of this subgraph form a complete matching between p and p', that is where each vertex of p is joined to exactly one vertex of p', we say that Γ is a *cover* of $\Gamma_{\mathcal{P}}$. (This terminology comes from topology with the graph Γ playing the role of a covering space and the parts of the partition $\mathcal{P}$ corresponding to the fibres.)

3 Primitive graph quotients

If G is imprimitive on the vertex set V of Γ then, as explained above, we may choose a G-invariant partition $\mathcal{P}$ of V such that $G^{\mathcal{P}}$ is primitive on the quotient graph $\Gamma_{\mathcal{P}}$, and a G-normal partition $\mathcal{P}_N$ such that $G^{\mathcal{P}_N}$ is quasiprimitive on the normal quotient Γ_N. Of course if Γ is bipartite then the primitive or quasiprimitive quotient may be simply the complete graph K_2 on two vertices. However for non-bipartite graphs Γ, depending on the combinatorial relationships between Γ and its quotients, or normal quotients, a great deal may be learned about Γ from the set of its primitive quotients, or quasiprimitive normal quotients respectively.

In this and the following sections we shall explore these relationships further for several families of finite arc-transitive graphs which have group theoretic defining properties. Whether the primitive or the quasiprimitive quotients are the appropriate ones to study tends to depend on whether the defining property of the family is essentially a global property or a local property. By a local property we mean a condition on the neighbourhood of a vertex (or sometimes on the vertices up to some fixed distance from a given vertex). Families with a global group theoretic defining property are sometimes closed under the formation of nearly all graph quotients relative to partitions invariant under the relevant group, and for these families it is useful to study the primitive quotients of graphs in the family. On the other hand, some families of arc-transitive graphs with a local group theoretic defining property are closed under taking normal quotients, but are not closed under taking arbitrary quotients relative to partitions invariant under the given group. For these families it is the quasiprimitive normal quotients (that is the quasiprimitive graphs in the family) which give better information about typical graphs in the family. We give some examples below.

For a graph $\Gamma = (V, E)$, the *distance* $d(\alpha, \beta)$ between two vertices α and β is the length of the shortest path between them. In particular $d(\alpha, \alpha) = 0$ for each $\alpha \in V$. If Γ is finite and connected then the maximum $d(\Gamma)$ of the distances between pairs of vertices of Γ is called the *diameter* of Γ. For example, if $|V| = n > 1$, then $d(\Gamma) = 1$ if and only if Γ is the *complete graph* K_n. For $i = 0, \ldots, d(\Gamma)$, we set $\Gamma_i := \{(\alpha, \beta) : d(\alpha, \beta) = i\}$, and $\Gamma_i(\alpha) := \{\beta : d(\alpha, \beta) = i\}$. Thus $\Gamma_1(\alpha)$ is the set of neighbours of α in Γ, and we sometimes write this set simply as $\Gamma(\alpha)$.

Example 3.1 A graph Γ is said to be *distance transitive* (or *G-distance transitive* where $G \leq \mathrm{Aut}(\Gamma)$) if, for each $i = 0, \ldots, d(\Gamma)$, $\mathrm{Aut}(\Gamma)$ (respectively G) is transitive on Γ_i. If $\Gamma = (V, E)$ is a G-distance transitive graph, and $\mathcal{P}$ is a nontrivial G-invariant partition of V, then it was shown by D. H. Smith [31] (or see [5, Theorem 4.2.1, and the following remarks]) that either Γ is bipartite, $\mathcal{P}$ is the bipartition of V, and $\Gamma_{\mathcal{P}} = K_2$, or $\Gamma_{\mathcal{P}}$ is $G^{\mathcal{P}}$-distance transitive and Γ is a cover of $\Gamma_{\mathcal{P}}$. Thus non-bipartite finite distance transitive graphs are covers of vertex-primitive distance transitive graphs. In the case of a finite bipartite G-distance transitive graph Γ with bipartition $\mathcal{P}$, Smith also showed that the distance-two graphs defined on the two parts of $\mathcal{P}$ are isomorphic distance transitive graphs, and are not themselves bipartite provided Γ has valency at least 3. (For $p \in \mathcal{P}$, the *distance-two graph* on p is the graph with vertex set p and edges $\{\{\alpha, \beta\} : \alpha, \beta \in p, d(\alpha, \beta) = 2\}$.)

Thus, in a strong sense, understanding the finite primitive distance transitive graphs is the key to understanding the whole family of finite distance transitive graphs. There are many infinite series of finite primitive distance transitive graphs. One which is easy to describe is the family of *Hamming graphs* $H(d, m)$ where $d \geq 2$, $m \geq 3$. The vertex set of $H(d, m)$ is $\mathbb{Z}_m^d$ and vertices $x = (x_1, \ldots, x_d)$ and $y = (y_1, \ldots, y_d)$ are joined by an edge if and only if $x - y$ has *weight* 1, that is $x_i = y_i$ for all except one value of $i = 1, \ldots, d$. The diameter of $H(d, m)$ is d and its automorphism group is $S_m \,\mathrm{wr}\, S_d$. It turns out that the complement $\overline{H(2, m)}$ of $H(2, m)$ is also distance transitive for all $m \geq 3$.

In [28] a study was begun of finite primitive distance transitive graphs guided by the O'Nan–Scott Theorem [17]. It was shown there that, if $\Gamma = (V, E)$ is a G-distance transitive graph with G primitive on vertices, then either Γ is $H(d, m)$ or $\overline{H(2, m)}$ for some $d \geq 2$, $m \geq 3$, or G is a primitive group on V of primitive type HA or AS. These types of primitive permutation groups are defined as follows. A primitive permutation group G of *type* HA on Ω is a group of affine transformations of a finite vector space V, and is of the form $G = NH$ where N is the group of translations of V, H is an irreducible group of nonsingular linear transformations of V, and $\Omega = V$ with the natural action. A primitive permutation group G on Ω has *type AS* if G is an *almost simple group*, that is if $T \leq G \leq \mathrm{Aut}(T)$ for some finite nonabelian simple group T.

Much progress has been made on completing the classification of finite primitive distance transitive graphs, and a good account of this can be found in [12]. Completing this classification is feasible, and it will rely on the finite simple group classification. However it is not yet complete.

Problem 3.2 *Complete the classification of the finite primitive distance transitive graphs.*

Example 3.3 Let s be a positive integer. An *s-arc* in a graph $\Gamma = (V, E)$ is a sequence $\alpha = (\alpha_0, \alpha_1, \ldots, \alpha_s)$ of vertices such that $\{\alpha_{i-1}, \alpha_i\} \in E$ for $i = 1, \ldots, s$. So in particular a 1-arc is simply an arc of Γ. The graph Γ is said to be (G, s)-arc transitive if $G \leq \mathrm{Aut}(\Gamma)$ and G is transitive on the set of s-arcs of Γ. If such a subgroup G exists, we also say simply that Γ is s-arc transitive. For a vertex-transitive graph, the property of s-arc transitivity involves the vertices at distance up to $\lceil s/2 \rceil$ from a given vertex and so may be regarded as a kind of generalised group theoretic local condition on Γ. (To see this consider all the s-arcs α with a given vertex as $\alpha_{\lfloor s/2 \rfloor}$.) Thus it is the cases $s = 1, 2$ which provide genuinely local group theoretic conditions, that is conditions on the neighbourhood of a vertex. To be precise, if G is vertex-transitive on Γ then, for $s = 1, 2$, Γ is (G, s)-arc transitive if and only if G_α is s-transitive on $\Gamma(\alpha)$. (A permutation group is *s-transitive* if it is transitive on ordered s-tuples of distinct points.)

We have already observed that for $s = 1$, if Γ is (G, s)-arc transitive and if $\mathcal{P}$ is a nontrivial G-invariant partition of V, then $\Gamma_{\mathcal{P}}$ is $(G^{\mathcal{P}}, s)$-arc transitive. However this implication is not true if $s = 2$. The situation may be illustrated effectively in the case of $(G, 2)$-arc transitive graphs with G a Suzuki simple group $\mathrm{Sz}(q)$, with $q = 2^{2m+1} \geq 8$. This case may be typical of many families of $(G, 2)$-arc transitive graphs with almost simple automorphism groups G. Let $q = r^2/2$, and let t be a divisor of $2m + 1$, $t > 1$. Then 5 divides $q + \delta r + 1$, where δ is 1 or -1. There are $(\mathrm{Sz}(q), 2)$-arc transitive graphs $\Gamma(5)$ and $\Gamma(2^t)$ of valencies 5 and 2^t respectively. These graphs have $\mathrm{Sz}(q)$-vertex primitive quotients $\Gamma(5)_{\mathcal{P}}$ and $\Gamma(2^t)_{\mathcal{P}}$ such that $\mathcal{P}$ has parts of size $(q + \delta r + 1)/5$ and $q^2(q-1)/\left(2^t(2^t - 1)\right)$ respectively. Also between each pair of adjacent parts of $\mathcal{P}$ there is a single edge or $(q-1)/(2^t - 1)$ edges respectively, so that $\Gamma(5)$ and $\Gamma(2^t)$ are far from being covers of these quotients. Moreover $\Gamma(5)_{\mathcal{P}}$ has valency $q + \delta r + 1$ and is not 2-arc transitive, while $\Gamma(2^t)_{\mathcal{P}}$ is the complete graph K_{q^2+1} on which (by "chance") $\mathrm{Sz}(q)$ acts 2-arc transitively. (These are some of the 2-arc transitive graphs constructed by Xin Gui Fang in his PhD thesis [7], see also [8].)

Thus the property of being 2-arc transitive is not in general inherited by primitive quotients, so a study of finite primitive 2-arc transitive graphs will not yield significant information about typical finite 2-arc transitive graphs. In contrast to this we shall show in the next section that each finite non-bipartite 2-arc transitive graph Γ is a cover of a quasiprimitive 2-arc transitive graph.

4 Normal quotients of locally primitive and locally quasiprimitive graphs

In this section we study several families of vertex transitive graphs which are subfamilies of the family $\mathcal{F}_{1\text{-arc}}$ of finite 1-arc transitive graphs and which contain the family $\mathcal{F}_{2\text{-arc}}$ of finite 2-arc transitive, vertex transitive graphs. For a group theoretic property $\mathcal{Q}$, we shall say that a graph Γ is G-*locally* $\mathcal{Q}$ (or simply *locally* $\mathcal{Q}$) if the action induced by the vertex stabilizer G_α on $\Gamma(\alpha)$ has property $\mathcal{Q}$. Thus, within the class of finite vertex transitive graphs, $\mathcal{F}_{2\text{-arc}}$ is the family of locally 2-transitive graphs, and $\mathcal{F}_{1\text{-arc}}$ is the family of locally transitive graphs. Since all 2-transitive permutation groups are primitive, $\mathcal{F}_{2\text{-arc}}$ is contained in the family $\mathcal{F}_{\text{l-prim}}$ of finite vertex transitive, locally primitive graphs. Similarly, since all primitive permutation groups are quasiprimitive, $\mathcal{F}_{\text{l-prim}}$ is contained in the family $\mathcal{F}_{\text{l-qprim}}$ of finite vertex transitive, locally quasiprimitive graphs, and of course $\mathcal{F}_{1\text{-arc}}$ contains $\mathcal{F}_{\text{l-qprim}}$. Thus

$$\mathcal{F}_{2\text{-arc}} \subset \mathcal{F}_{\text{l-prim}} \subset \mathcal{F}_{\text{l-qprim}} \subset \mathcal{F}_{1\text{-arc}}. \tag{1}$$

We shall investigate normal quotients of graphs in these families. To describe the structure of these normal quotients we need some more notation. A permutation group G on a set Ω is said to be *semiregular* on Ω if the only element of G which fixes a point of Ω is the identity element 1_G. Let $\mathcal{P}$ be a partition of the vertex set of a graph $\Gamma = (V, E)$. Then Γ will be called a *multicover* of $\Gamma_{\mathcal{P}}$ if, for each $\{p, p'\} \in E_{\mathcal{P}}$, each vertex of p is joined to at least one vertex of p'. The multicovers we describe below are all uniform in the sense that, $|\Gamma(\alpha) \cap p'|$ is independent of the choice of the vertex α in p. Multicovers were called pseudocovers in [24], but the term multicover seems a better name for them. The theorem below was proved in [24, Section 1].

Theorem 4.1 *Let $\Gamma = (V, E)$ be a connected $(G, 1)$-arc transitive graph of valency v, and let N be a normal subgroup of G. Then one of the following holds.*

(a) N is transitive on V; or

(b) Γ is bipartite and the N-orbits in V are the two parts of the bipartition; or

(c) $\Gamma_N = (\mathcal{P}_N, E_N)$ is a connected $(G^{\mathcal{P}_N}, 1)$-arc transitive graph of valency v/k where, for each $\{p, p'\} \in E_N$ and each $\alpha \in p$, $|\Gamma(\alpha) \cap p'| = k$, and Γ is a multicover of Γ_N. Moreover,

(i) if in addition Γ is G-locally quasiprimitive then N is semiregular on V and Γ_N is $G^{\mathcal{P}_N}$-locally quasiprimitive (and the permutation group $G_\alpha^{\Gamma(\alpha)}$ acts faithfully on the partition $\mathcal{P}(\alpha)$ defined by $\mathcal{P}(\alpha) := \{\Gamma(\alpha) \cap p' : \{p, p'\} \in E_N\}$);

(ii) if Γ is G-locally primitive then Γ is a cover of Γ_N (that is $k = 1$) and Γ_N is $G^{\mathcal{P}_N}$-locally primitive (and the permutation groups $G_\alpha^{\Gamma(\alpha)}$ and $G_p^{\Gamma_N(p)}$ are permutationally equivalent);

(iii) in particular if Γ is $(G, 2)$-arc transitive then Γ is a cover of Γ_N and Γ_N is $(G^{\mathcal{P}_N}, 2)$-arc transitive.

Thus, if we restrict ourselves to the non-bipartite members Γ of one of the families in (1) we see that all of the normal quotients of Γ lie in the same family as Γ does, and each non-bipartite graph in each of these families has at least one quasiprimitive normal quotient. Moreover each non-bipartite graph in $\mathcal{F}_{\text{1-prim}}$ or $\mathcal{F}_{\text{2-arc}}$ is a cover of a quasiprimitive graph in $\mathcal{F}_{\text{1-prim}}$ or $\mathcal{F}_{\text{2-arc}}$ respectively. Thus the quasiprimitive graphs in each family in (1) are important for understanding the structure of typical non-bipartite graphs in the family. However to test the power of this observation, and ultimately to gain some more concrete knowledge of these families of graphs, we need some detailed information about the structure of finite quasiprimitive permutation groups, similar to the information about finite primitive permutation groups given by the O'Nan–Scott Theorem. At the time when Theorem 4.1 was proved no information of this nature was available in the literature. Thus we first needed to analyse the structure of finite quasiprimitive permutation groups. This analysis was undertaken in [26] and will be discussed in the next section. Then, in Section 6 we will illustrate, with the family $\mathcal{F}_{\text{2-arc}}$, how this structure theory may be applied to help understand the structure of quasiprimitive graphs in certain families of graphs.

It should be emphasised that, of the four families in (1), it is only the family $\mathcal{F}_{\text{1-arc}}$ that is closed under taking arbitrary primitive quotients. The 5-valent $(\mathrm{Sz}(q), 2)$-arc transitive graph $\Gamma(5)$ described in Example 3.3 exhibits properties which may be typical of many other $(G, 2)$-arc transitive graphs with G almost simple. The primitive $(\mathrm{Sz}(q), 1)$-arc transitive quotient $\Gamma(5)_{\mathcal{P}}$ of valency $q + \delta r + 1$ described there is not even $\mathrm{Sz}(q)$-locally quasiprimitive if $q > 8$.

Appropriate means for understanding the bipartite examples in these families are more difficult to find. In the case of bipartite finite distance transitive graphs the distance-two graph induced on a part of the bipartition turns out to be again distance transitive. However each of the families in (1) has a local group theoretic defining property, and consequently the situation is rather different from that for finite distance transitive graphs. There is no natural way to obtain a graph in the relevant family induced on a part of the bipartition of a bipartite graph in one of the families in (1). An approach to studying the structure of finite bipartite 2-arc transitive graphs as covers of smaller graphs in $\mathcal{F}_{2\text{-arc}}$ was made in [27]. The relevant quotients in this case are those for which every nontrivial normal subgroup of the group in question is either vertex-transitive or has as vertex orbits the two parts of the bipartition. The action induced on a bipartite half of such a quotient is not necessarily quasiprimitive. Nevertheless the structure theorem for quasiprimitive groups described in the next section is crucial in analysing the possible structures of such quotients: the results are not as nice as those for the non-bipartite case.

5 Finite quasiprimitive permutation groups

The class of finite quasiprimitive permutation groups may be described in a fashion very similar to the description given by the O'Nan–Scott Theorem [17, 30] for finite primitive permutation groups. This provides essentially an identification of several types of finite quasiprimitive groups such that for each type we have additional information about either the abstract group theoretical structure, or the nature of the action, or both. The quasiprimitive types in most cases are similar to the primitive types from the O'Nan–Scott Theorem. The ordering and the presentation of the various types, and the amount of subdivision of the types preferred by those who wish to use this classification vary according to the requirements of the different applications. For some requirements the nature of the socle or the minimal normal subgroups is most important, while (for example) for others it is the nature of the suborbits, or the existence of regular subgroups that is needed. (The *socle* $\mathrm{soc}(G)$ of a group G is the product of the minimal normal subgroups of G. A permutation group G on Ω is said to be *regular* on Ω if G is both transitive and semiregular on Ω.)

For this exposition I have chosen a subdivision into types which was suggested by Laci Kovacs in 1985 for the types of primitive groups in the O'Nan–Scott Theorem. It is a little finer than the subdivision used in the original paper on the structure of quasiprimitive groups [26], which mirrored the subdivision given in [17] for primitive groups. The subdivision given here was used in [25] for primitive groups and again in [3] for quasiprimitive groups; and moreover in the latter paper a slight refinement of it was introduced.

We shall define eight types of finite quasiprimitive permutation groups. The main theorem in [26] states that every finite quasiprimitive permutation group belongs to exactly one of these types. Let G be a finite quasiprimitive permutation group on Ω and let $\alpha \in \Omega$. Then G has at most two minimal normal subgroups (see the first portion of [26, Section 3]), and if there are two minimal normal subgroups then they are isomorphic and each is equal to the centraliser of the other. In particular if G has an abelian minimal normal subgroup N then N is the unique minimal normal subgroup of G, that is N is the socle of G. Further N is elementary abelian and is regular on Ω, and G is the semidirect product $G = NG_\alpha$, for $\alpha \in \Omega$. Thus we may identify Ω with

$N = \mathbb{Z}_p^d$, which may be viewed as a d-dimensional vector space over a field of prime order p, and choosing α as the zero vector we then have that G_α is an irreducible subgroup of nonsingular linear transformations of N. This is our first quasiprimitive type, and is named HA since G is contained in the Holomorph of the Abelian group N. The *holomorph* $\mathrm{Hol}(N)$ of a group N is the semidirect product $N.\mathrm{Aut}(N)$, where $\mathrm{Aut}(N)$ acts naturally on the normal subgroup N.

HA $\Omega = \mathbb{Z}_p^d$ for a prime p and positive integer d and G is the semidirect product $G = N.G_o$, a subgroup of the affine group $\mathrm{AGL}(d,p)$ on Ω, where N is the group of translations and G_o is an irreducible subgroup of $\mathrm{GL}(d,p)$.

Note that, for this quasiprimitive type, the requirements for quasiprimitivity and primitivity coincide so that a quasiprimitive permutation group of type HA is primitive.

For all other types each minimal normal subgroup N of G is nonabelian, and hence $N = T_1 \times \cdots \times T_k$ for some positive integer k where each $T_i \cong T$, a nonabelian simple group. If there is a second minimal normal subgroup $M \neq N$, then $M \cong N$, both M and N are regular on Ω, and $\mathrm{soc}(G) = N \times M$. In this case G is contained in the holomorph of N, and the second minimal normal subgroup $M = \{\varphi_x^{-1}x : x \in N\}$, where φ_x is the inner automorphism of N induced by x. This case is subdivided into two quasiprimitive types, namely the cases where $k = 1$ and $k > 1$. In the former case the type is named HS since G is contained in the Holomorph of the Simple group T. The case where $k > 1$ is named HC since G is contained in the Holomorph of a Compound group N. (The word "compound" is used here because G is also a subgroup of a group of compound diagonal type, a type which will be described later.) Again for these types HS and HC the conditions for quasiprimitivity and primitivity coincide and each quasiprimitive permutation group of type HS or HC is primitive. For a group N, and for $x \in N$ and $\sigma \in \mathrm{Aut}(N)$, we denote the image of x under σ by x^σ. Also we set $\mathrm{Inn}(N) := \{\varphi_x : x \in N\}$, the *inner automorphism group* of N.

HS $\Omega = T$, a finite nonabelian simple group, and $T.\mathrm{Inn}(T) \le G \le \mathrm{Hol}(T) = T.\mathrm{Aut}(T)$, where for $x \in \Omega$, $t \in T$ and $\sigma \in \mathrm{Aut}(T)$, $t\sigma\colon x \mapsto x^\sigma t^\sigma$. If $\alpha = 1_T \in \Omega$, then $\mathrm{Inn}(T) \le G_\alpha \le \mathrm{Aut}(T)$.

HC $\Omega = T^k = N$, where $k > 1$ and T is a finite nonabelian simple group, and $N.\mathrm{Inn}(N) \le G \le \mathrm{Hol}(N) = N.\mathrm{Aut}(N)$. As for the type HS, for $x, n \in N$ and $\sigma \in \mathrm{Aut}(N)$, $n\sigma\colon x \mapsto x^\sigma n^\sigma$. If $\alpha = 1_N \in \Omega$, then $\mathrm{Inn}(N) \le G_\alpha \le \mathrm{Aut}(N)$ and G_α acts transitively by conjugation on the simple direct factors of N.

There are five further quasiprimitive types and for each type there is a corresponding type of primitive group. However, unlike the types HA, HS, and HC, each of these five types contains imprimitive quasiprimitive permutation groups. Moreover, for each of these quasiprimitive types $N = \mathrm{soc}(G)$ is the unique minimal normal subgroup of G, and for the first four of them the parallel with the corresponding primitive type is very close. These four types are the type AS where G is Almost Simple, type SD where G has a Simple Diagonal action on Ω, type CD where G has a Compound Diagonal action on Ω, and type TW where G is a Twisted Wreath product with regular socle.

AS $T \leq G \leq \mathrm{Aut}(T)$, where T is a finite nonabelian simple group, and $G = TG_\alpha$ with $T \not\leq G_\alpha$. In this type we may have $G_\alpha = \{1_G\}$, that is $N = T$ may be regular on Ω.

For G to be primitive of type AS the stabilizer G_α must be maximal in G, and in particular $G_\alpha \neq \{1_G\}$. The next type is the simple diagonal type SD. A quasiprimitive permutation group of type SD is a subgroup of the group

$$W = \{(a_1, \ldots, a_k)\pi : a_i \in \mathrm{Aut}(T),\ \pi \in S_k,\ a_i \equiv a_j \ (\mathrm{mod}\ \mathrm{Inn}(T)) \text{ for all } i, j\}$$

where $\pi^{-1}(a_1, \ldots, a_k)\pi = (a_{1\pi^{-1}}, \ldots, a_{k\pi^{-1}})$ and $k > 1$. The socle of W is $\mathrm{soc}(W) = \{(t_1, \ldots, t_k) : t_i \in \mathrm{Inn}(T)\}$, and a primitive action of W on T^{k-1} (which we identify with $\mathrm{Inn}(T)^{k-1}$) is defined by

$$(a_1, \ldots, a_k) \colon (t_1, \ldots, t_{k-1}) \mapsto (a_k^{-1} t_1 a_1, \ldots, a_k^{-1} t_{k-1} a_{k-1}), \quad \text{and}$$

$$\pi \colon (t_1, \ldots, t_{k-1}) \mapsto (t_{k\pi^{-1}}^{-1} t_{1\pi^{-1}}, \ldots, t_{k\pi^{-1}}^{-1} t_{(k-1)\pi^{-1}}),$$

for $(a_1, \ldots, a_k)\pi \in W$ and $(t_1, \ldots, t_{k-1}) \in T^{k-1}$, where $t_k = 1_T$. Thus for $\alpha = (1_T, \ldots, 1_T) \in T^{k-1}$, $W_\alpha = A \times S_k$ where $A = \{(a, \ldots, a) : a \in \mathrm{Aut}(T)\}$.

SD $\Omega = T^{k-1}$, where $k > 1$ and T is a finite nonabelian simple group, $N = \mathrm{soc}(W) \leq G \leq W$ with the action defined above, and G acts transitively by conjugation on the k simple direct factors of N.

The name Simple Diagonal comes from the fact that N_α is the full diagonal subgroup $\{(t, \ldots, t) : t \in \mathrm{Inn}(T)\}$ of N. For the corresponding primitive type SD, G_α must act primitively on the simple direct factors of N. For the compound diagonal type CD the group G preserves a product structure on Ω, that is $\Omega = \Delta^l$ for some $l \geq 2$ and $G \leq \mathrm{Sym}(\Delta) \,\mathrm{wr}\, S_l$ where, for $\delta = (\delta_1, \ldots, \delta_l) \in \Omega$ and $(a_1, \ldots, a_l)\pi \in \mathrm{Sym}(\Delta)$ wr S_l,

$$(a_1, \ldots, a_l) \colon \delta \mapsto (\delta_1^{a_1}, \ldots, \delta_l^{a_l}), \quad \text{and} \quad \pi \colon \delta \mapsto (\delta_{1\pi^{-1}}, \ldots, \delta_{l\pi^{-1}}).$$

Moreover the subgroup of $\mathrm{Sym}(\Delta)$ involved is a quasiprimitive group of type SD, hence the name Compound Diagonal.

CD $\Omega = \Delta^l$ and $N = T^k \leq G \leq H \,\mathrm{wr}\, S_l \leq \mathrm{Sym}(\Delta) \,\mathrm{wr}\, S_l$, for some divisor l of k, where $l \geq 2$ and $k/l \geq 2$, T is a finite nonabelian simple group, $H \leq \mathrm{Sym}(\Delta)$, $\mathrm{soc}(H) = T^{k/l}$, and H is quasiprimitive on Δ of type SD; also G acts transitively by conjugation on the simple direct factors of N.

For the primitive type CD the subgroup H induced on Δ must be primitive of type SD. The next type is the twisted wreath type TW. The original definition of a twisted wreath product was given by B. H. Neumann in [22]. The exposition here follows that in Suzuki's book [32, p. 269] and is the same as in [17]. Its use was suggested by Laci Kovacs. The *core* of a subgroup H of a group G is $\mathrm{core}_G(H) := \bigcap_{g\in G} H^g$. The *twisted wreath product* $T\,\mathrm{twr}_\varphi\, P$ of groups T and P relative to φ may be defined as follows. Let P have a transitive action on $\{1,\ldots,k\}$ and let Q be the stabilizer of the point 1 in this action. Suppose that there is a homomorphism $\varphi\colon Q \to \mathrm{Aut}(T)$ such that $\mathrm{core}_P(\varphi^{-1}(\mathrm{Inn}(T))) = \{1_P\}$. Define

$$B := \{f\colon P \to T : f(pq) = f(p)^{\varphi(q)} \text{ for all } p \in P,\ q \in Q\}.$$

Then B is a group under pointwise multiplication and $B \cong T^k$. Let P act on B by

$$f^p(x) := f(px) \text{ for all } p, x \in P.$$

We define $T\,\mathrm{twr}_\varphi\, P$ to be the semidirect product of B by P relative to this conjugation action of P. Such a twisted wreath product $T\,\mathrm{twr}_\varphi\, P$ has a transitive action on B such that B acts by right multiplication and for $f \in B$ and $p \in P$, $p\colon f \mapsto f^p$.

TW $\Omega = T^k = B$, where $k > 1$ and T is a finite nonabelian simple group, G is a twisted wreath product $T\,\mathrm{twr}_\varphi\, P$ defined as above, and G acts on Ω with the action defined above.

The differences between the conditions on P and Q for primitivity and quasiprimitivity of $T\,\mathrm{twr}_\varphi\, P$ are rather subtle. A discussion can be found in [26, Remark 2.1], and in [1, Section 5]. We note in particular that for the quasiprimitive type TW we do not require that the image of φ contains $\mathrm{Inn}(T)$.

For the final quasiprimitive type PA, G preserves a product structure Δ^k on a G-invariant partition of Ω and the subgroup of $\mathrm{Sym}(\Delta)$ involved is quasiprimitive of type AS with socle T. Thus a quasiprimitive group G of type PA induces a faithful product action on this partition of Ω. For a group R and positive integer $k \geq 2$, a *subdirect product* of R^k is a subgroup H of R^k such that, for each $i = 1,\ldots,k$, the projection map $\pi_i\colon R^k \to R$ onto the ith direct factor is surjective when restricted to H, that is, $(H)\pi_i = R$.

PA $N = T^k \leq G \leq H \,\mathrm{wr}\, S_k \leq \mathrm{Sym}(\Delta) \,\mathrm{wr}\, S_k$, where $k > 1$ and T is a finite nonabelian simple group, H is a quasiprimitive permutation group on Δ of type AS with non-regular socle T, and G acts transitively by conjugation on the simple direct factors of N. Choose $\delta \in \Delta$ and set $R := T_\delta$. Then $1 < R < T$. There is a G-invariant partition Ω' of Ω (which is possibly trivial in that the parts of Ω' may have size 1) such that for some $\omega \in \Omega'$, $N_\omega = R^k$ and for $\alpha \in \omega$, N_α is a subdirect product of R^k.

Of all the quasiprimitive types, this last type PA differs the most from its corresponding primitive type. For a primitive group G of type PA, the partition Ω' is trivial and so may be identified with Ω, and the group H is primitive of type AS. It is clear from the descriptions of the eight types that a finite quasiprimitive permutation group belongs to at most one of these types. The main theorem in [26] shows that the converse is also true.

Theorem 5.1 ([26, Theorem 1]) *Each finite quasiprimitive permutation group is permutationally equivalent to a quasiprimitive group in exactly one of the quasiprimitive types HA, HS, HC, AS, SD, CD, TW, PA.*

6 Finite quasiprimitive 2-arc transitive graphs

In order to test the effectiveness or power of our observations about the role of quasiprimitive graphs in the families $\mathcal{F}_{\text{2-arc}}$, $\mathcal{F}_{\text{l-prim}}$ and $\mathcal{F}_{\text{l-qprim}}$, we might try to apply the structure result, Theorem 5.1, to study the possible structures of the quasiprimitive graphs in these families. Here we report on the results of following this strategy for the family $\mathcal{F}_{\text{2-arc}}$ of finite vertex-transitive, 2-arc transitive graphs. The first analysis of quasiprimitive graphs in this family using Theorem 5.1 showed that only half of the quasiprimitive types could occur as quasiprimitive 2-arc transitive subgroups of automorphisms.

Theorem 6.1 ([26, Lemmas 5.2 and 5.3]) *If $\Gamma = (V, E)$ is a finite $(G, 2)$-arc transitive graph such that G is quasiprimitive on V, then G has quasiprimitive type HA, AS, TW, or PA.*

Further, it was observed in [26, Section 6] that, for each of the types HA, AS, TW, or PA, there are examples of $(G, 2)$-arc transitive graphs with G quasiprimitive of the given type. The next step in the study of quasiprimitive 2-arc transitive graphs was to investigate further the nature of the graphs corresponding to each of these quasiprimitive types. It was in fact possible to complete the classification of those of type HA. This was done in [13] in joint work with Sasha Ivanov. Recall that all quasiprimitive permutation groups of type HA are primitive.

Theorem 6.2 ([13]) *If $\Gamma = (V, E)$ is a finite $(G, 2)$-arc transitive graph of valency n such that G is (quasi)primitive on V of type HA, then $|V| = 2^d$, $G = \mathbb{Z}_2^d.G_o \leq \mathbb{Z}_2^d.(\mathrm{Aut}(\Gamma) \cap \mathrm{GL}(d, 2))$ with G_o irreducible, and Γ, d, n, and $\mathrm{Aut}(\Gamma) \cap \mathrm{GL}(d, 2)$ are as in one of the lines of Table 1.*

If $\Gamma = (V, E)$ is a finite primitive $(G, 2)$-arc transitive graph of type HA and of valency n, then it was shown in [13] that G must be of the form $G = N.G_o$, where the normal subgroup $N = \mathbb{Z}_2^d$ acts regularly on vertices, and the subgroup G_o (which is the stabilizer of the identity element of N) has a faithful 2-transitive action of degree n. Furthermore it was shown that N may be

Γ	d	n	$\mathrm{Aut}(\Gamma) \cap \mathrm{GL}(d,2)$	Comments
K_{n+1}	$\log_2(n+1)$	n	$\mathrm{GL}(d,2)$	–
$\frac{1}{2} \cdot Q_n$	$n-1$	n odd	S_n	–
$P_m(a)$	m^a	$\frac{2^{am}-1}{2^a-1}$	$\mathrm{P\Gamma L}(m, 2^a)$	$m \geq 3$
$U(q)$	$\geq q^2 - q + 1$	q^3+1	$\mathrm{P\Gamma U}(3,q)$	$q \equiv 3 \pmod 4$; see below
$\Gamma(C_{23})$	11	23	M_{23}	–
$\Gamma(C_{22})$	10	22	$M_{22}.2$	–

Table 1: Quasiprimitive 2-arc transitive graphs of type HA.

identified with a quotient of the natural $\mathbb{Z}_2G_o$-permutation module $\mathbb{Z}_2^n$ for this 2-transitive action of G_o by some maximal submodule W such that G_o is faithful on $\mathbb{Z}_2^n/W$, and moreover that N and hence W determine Γ. In fact we have $V = N = \mathbb{Z}_2^n/W$ and $\{x+W, y+W\}$ is an edge if and only if $x + y \equiv e \pmod W$ for some weight 1 vector $e \in \mathbb{Z}_2^n$. Thus, for a 2-transitive permutation group G_o of degree n, the finite primitive $(G,2)$-arc transitive graphs of type HA, with point stabilizer G_o, are in one-to-one correspondence with the maximal $\mathbb{Z}_2G_o$-submodules W of the $\mathbb{Z}_2G_o$-permutation module $\mathbb{Z}_2^n$ such that G_o is faithful on $\mathbb{Z}_2^n/W$, or alternatively on the minimal faithful $\mathbb{Z}_2G_o$-submodules of $\mathbb{Z}_2^n$. Most of the examples in Theorem 6.2 are therefore defined in terms of a maximal submodule of the $\mathbb{Z}_2G_o$-permutation module for the 2-transitive group G_o. Occasionally the examples turn out to be isomorphic to some well-known graphs, and in such cases we use their more familiar descriptions. This is the case for the complete graphs in line 1 of the table, and is also the case for the graphs in line 2. The graph $\frac{1}{2} \cdot Q_n$ is the *folded cube*, the antipodal quotient of the n-dimensional cube; it may be defined as the graph with vertex set $V = \mathbb{Z}_2^{n-1}$ with $\{x, y\}$ an edge if and only if $x - y$ has weight either 1 or $n-1$.

The graph $P_m(a)$ ($m \geq 3$, $a \geq 1$) is defined in terms of the $\mathbb{Z}_2G_o$-permutation module $\mathbb{Z}_2^n$ for G_o where $\mathrm{PSL}(m, 2^a) \leq G_o \leq \mathrm{P\Gamma L}(m, 2^a)$ with G_o acting on the $n = (2^{ma}-1)/(2^a-1)$ points of the projective geometry $\mathrm{PG}_{m-1}(a)$. Let W be the $\mathbb{Z}_2G_o$-submodule generated by the characteristic functions of all the complements of hyperplanes of $\mathrm{PG}_{m-1}(a)$. Then the vertex set of $P_m(a)$ is the quotient module $V := \mathbb{Z}_2^n/W^{\perp}$ with edges as described above.

The examples in line 4 probably do not exist (as quasiprimitive examples). When $q \equiv 3 \pmod 4$, the submodule structure of the $\mathbb{Z}_2G_o$-permutation module $V = \mathbb{Z}_2^n$ for the 2-transitive unitary groups G_o, where $\mathrm{PSU}(3,q) \leq G_o \leq \mathrm{P\Gamma U}(3,q)$ and $n = q^3+1$, has not been completely determined. It is believed, but has not yet been proved, that V has no faithful minimal G_o-submodule and therefore provides no quasiprimitive examples in Theorem 6.2. This has been proved by Jane McCorkindale [21] in her D. Phil. thesis in the case where $q \equiv 3 \pmod 8$ and $q+1 \not\equiv 0 \pmod 3$, and has also been verified computationally when $q = 3$, 7, and 11 by Andreas Brouwer (private communication).

The graphs $\Gamma(C_{23})$, $\Gamma(C_{22})$ are constructed in a similar manner from the $\mathbb{Z}_2$-permutation modules for the 2-transitive groups M_{23} and $M_{22}.2$ respectively. Descriptions of these graphs, including distance diagrams, may be found in [5, Theorems 11.3.4 and 11.3.5], see also [13, Description 1.8].

We next turn to the classification of $(G,2)$-arc transitive graphs where G is quasiprimitive of type AS, say $T \leq G \leq \mathrm{Aut}(T)$ for a finite nonabelian simple group T. On the one hand, for some classes of Lie type simple groups T of low Lie rank it should be possible to classify all $(G,2)$-arc transitive graphs. This has been achieved for the classes $\mathrm{PSL}(2,q)$, $\mathrm{Sz}(q)$ and $\mathrm{Ree}(q)$ in [7, 8, 9, 11]. Each of these classifications involved the constructions of several new infinite families of examples, and gave new insights into the nature of such graphs and in particular of their primitive quotients. On the other hand it can be demonstrated that such a complete and explicit classification is not possible for the family of finite alternating groups $T = A_n$. A study was made in [29] of primitive permutation representations of $G = A_n$ or S_n with a 2-transitive suborbit. All examples were classified except those for which a point stabilizer G_α is a primitive permutation group of degree n of primitive type AS and has a faithful 2-transitive permutation representation. Although all possibilities for the 2-transitive group G_α are known (using the classification of the finite simple groups, see [6]) we still do not know all faithful primitive representations of all such groups. Thus it is not feasible at present to complete the classification in [29], and consequently it is not feasible to classify even the $(G,2)$-arc transitive graphs with $G = A_n$ or S_n and G primitive on vertices.

Robert Baddeley [1] has made a study of finite $(G,2)$-arc transitive graphs with G quasiprimitive or primitive on vertices of type TW. There are many examples and his paper gives a general approach to their construction. The situation for finite $(G,2)$-arc transitive graphs with G quasiprimitive of type PA is not quite so clear. Examples may be constructed as follows. However we do not know what "typical" quasiprimitive graphs of this type are like.

Example 6.3 Let $\Delta = \{1,\dots,n\}$ and let $H = \mathrm{Sym}(\Delta)$. Then the transposition $h = (12)$ is such that $H_{1,2} = C_H(h) \cap H_1$. Let $L := (H \times H).\langle \pi \rangle$, where $\pi^2 = 1$ and $(h_1,h_2)^\pi = (h_2,h_1)$ for all $h_i \in H$. Then $L \cong H \mathrm{\,wr\,} \mathbb{Z}_2$. Also let $K = \{(x,x) : x \in H_1\}$ and $g = (1_H, h)\pi$. Define Γ to be the graph with vertex set $V = \{Ky : y \in L\}$ such that $\{Ky, Ku\}$ is an edge if and only if $yu^{-1} \in KgK$. Now $K \cap K^g = \{(x,x) : x \in C_H(h) \cap H_1 = H_{1,2}\}$ and $G := \langle K, g\rangle$ has index 2 in L. It follows from [8, Theorem 2.1] that the connected component Γ' of Γ containing the vertex K is a connected $(G,2)$-arc transitive graph and G is quasiprimitive of type PA on the vertices of Γ'.

7 Full automorphism groups of quasiprimitive graphs

It is immediate from the definition of primitivity that if G is a primitive permutation group on Ω and $G \leq H \leq \mathrm{Sym}(\Omega)$, then H also is primitive on Ω. However this implication is not true if G is an imprimitive quasiprimitive permutation group on Ω. Indeed if $\mathcal{P}$ is a nontrivial G-invariant partition of Ω and if H is the subgroup of all permutations of Ω preserving $\mathcal{P}$, then $G \leq H \leq \mathrm{Sym}(\Omega)$, H is a wreath product $\mathrm{Sym}(p)$ wr $\mathrm{Sym}(\mathcal{P})$ (where $p \in \mathcal{P}$) and the base group $\mathrm{Sym}(p)^{|\mathcal{P}|}$ of H is intransitive on Ω. Thus an overgroup in $\mathrm{Sym}(\Omega)$ of a quasiprimitive group may not be quasiprimitive. Recall that, for a family $\mathcal{F}$ of arc-transitive graphs with a group theoretic defining property $\mathcal{Q}$, a quasiprimitive graph in $\mathcal{F}$ is a graph Γ for which there exists $G \leq \mathrm{Aut}(\Gamma)$ such that G is quasiprimitive on vertices and also G has property $\mathcal{Q}$. It may be the case that Γ is G-quasiprimitive, but $\mathrm{Aut}(\Gamma)$ is not quasiprimitive on vertices.

For most of the properties we considered in Section 4, if $G \leq H \leq \mathrm{Aut}(\Gamma)$ and G has property $\mathcal{Q}$, then H also has property $\mathcal{Q}$, that is the property $\mathcal{Q}$ is inherited by overgroups in $\mathrm{Aut}(\Gamma)$. (This is true for $\mathcal{F}_{\text{1-arc}}$, $\mathcal{F}_{\text{2-arc}}$, and $\mathcal{F}_{\text{1-prim}}$, but not for $\mathcal{F}_{\text{1-qprim}}$.) These observations suggest a number of interrelated questions about overgroups in $\mathrm{Aut}(\Gamma)$ of a given quasiprimitive subgroup G. We pose them in terms of the full automorphism group $\mathrm{Aut}(\Gamma)$, but they are equally interesting with $\mathrm{Aut}(\Gamma)$ replaced by an arbitrary overgroup of G in $\mathrm{Aut}(\Gamma)$.

Questions 7.1 *Suppose that $\mathcal{F}_\mathcal{Q}$ is the family of finite arc-transitive graphs Γ with a certain group theoretic defining property $\mathcal{Q}$ which is inherited by overgroups in* $\mathrm{Aut}(\Gamma)$. *Let $\Gamma = (V,E)$ be a G-quasiprimitive graph in $\mathcal{F}_\mathcal{Q}$, that is, $G \leq \mathrm{Aut}(\Gamma)$, G is quasiprimitive on V, and G has property $\mathcal{Q}$.*

(a) Under what conditions can we be certain that $\mathrm{Aut}(\Gamma)$ *is quasiprimitive on V?*

(b) If $\mathrm{Aut}(\Gamma)$ *is quasiprimitive on V, is it possible that G and* $\mathrm{Aut}(\Gamma)$ *have different quasiprimitive types, and if so what are the possible pairs of quasiprimitive types for G,* $\mathrm{Aut}(\Gamma)$*?*

(c) If G and $\mathrm{Aut}(\Gamma)$ *are quasiprimitive with the same quasiprimitive type, is it possible that* $\mathrm{soc}(G) \neq \mathrm{soc}(\mathrm{Aut}(\Gamma))$, *and if so what are the possibilities for these socles?*

We suspect that, in many of the cases in which $\mathrm{Aut}(\Gamma)$ *is not quasiprimitive on* V*, the centralizer of* G*, or of* $\mathrm{soc}(G)$*, in* $\mathrm{Aut}(\Gamma)$ *may be a nontrivial intransitive normal subgroup. To be more precise suppose that* G *is maximal by inclusion in* $\mathrm{Aut}(\Gamma)$ *such that* G *is quasiprimitive on* V*, and that* $G \neq \mathrm{Aut}(\Gamma)$*. Let* $G < H \leq \mathrm{Aut}(\Gamma)$ *with* G *a maximal subgroup of* H*. Then* H *has an intransitive minimal normal subgroup* M *such that* $M \cap G = 1$ *and* $H = GM$*.*

(d) Under what conditions is it true that $H = G \times M$*?*

If this is the case, then the approach described in this paper to analysing the structure of Γ *would suggest that we study the normal quotient* Γ_M *since often* Γ *is a multicover, or even a cover, of* Γ_M*, and* Γ_M *is* (G/M)*-quasiprimitive. On the other hand if* G *acts nontrivially on* M *(by conjugation), it is often possible to prove that* M *is elementary abelian and can be identified with a faithful irreducible* G*-module.*

(e) What can we say about Γ*,* G *and* M *if* M *is a faithful irreducible* G*-module?*

Many of these questions arose in the study of finite 2-arc transitive graphs. As we have already discussed, the study of finite 2-arc transitive graphs highlighted the importance of finite quasiprimitive permutation groups and motivated the first study of their structure in [26]. They also prompted the questions above. Again we were in a situation where there was insufficient knowledge available about the structure of quasiprimitive permutation groups on a finite set Ω and their overgroups in $\mathrm{Sym}(\Omega)$, even to begin to answer Questions 7.1 (a)–(c). The information was however available for finite primitive permutation groups. The results of [16] give a classification of the overgroups in $\mathrm{Sym}(\Omega)$ of primitive groups of type AS, while the paper [25] completes the classification of all inclusions $G < H \leq \mathrm{Sym}(\Omega)$ with G primitive on Ω and either G, H of different primitive types, or G, H having different socles. A study applying these results to the automorphism groups of vertex-primitive graphs is in progress [20]. It describes the pairs of groups G, H such that $G < H \leq \mathrm{Aut}(\Gamma)$, G is vertex-primitive, and G, H have different primitive types or different socles.

A start has been made in [3] to classifying permutation groups G, H such that $G < H \leq \mathrm{Sym}(\Omega)$, G is quasiprimitive and imprimitive on Ω, and H is primitive on Ω. A summary statement of the results of [3, 16, 25] may be made as follows.

Theorem 7.2 ([3, 16, 25]) *Suppose that* Ω *is a finite set and that* $G < H \leq \mathrm{Sym}(\Omega)$ *with* G *quasiprimitive on* Ω *and* H *primitive on* Ω*, such that either* G *is imprimitive, or* G*,* H *are primitive of different primitive types, or* G*,* H *are primitive of the same type but with different socles. Then either*

(a) the pair (G, H) *belongs to an explicit list of families of examples; or*

(b) *H is of type AS and G is imprimitive; or*

(c) *H is of type PA and G is imprimitive of quasiprimitive type TW or PA.*

The results in [3] required the solution of several classification problems of special kinds of factorisations of almost simple groups (completed in [2]) and a special investigation (in [4]) of the structure of quasiprimitive groups obtained from the blow-up construction introduced by Laci Kovacs [14].

To extend the classification in Theorem 7.2 it is necessary to solve several more classification problems for almost simple groups (see [3, Section 6]). To explain this, suppose that $G < H \leq \mathrm{Sym}(\Omega)$ with G quasiprimitive and imprimitive, and H primitive. Since all primitive inclusions $K < H$, with K, H of different primitive types, or of the same type but with different socles, have been classified in [16, 25], we may assume that each subgroup K such that $G < K \leq H$ is imprimitive. Then by Theorem 7.2 we may assume that H has primitive type AS or PA. Much is known about the case where H has type PA from [4], but some delicate questions remain.

From now on let us suppose that H has type AS, with socle T. We are interested only in the case where $\mathrm{soc}(G) \neq T$, as otherwise G would be quasiprimitive of the same type, and with the same socle as H. Let K be a subgroup of H containing G which is maximal such that $K \not\geq T$. Then, for $\alpha \in \Omega$, since G is transitive on Ω, $H = GH_\alpha$, and hence $H = KH_\alpha$ is a (so-called max$^+$ or max$^-$) factorisation of the almost simple group H. All possibilities for such triples (H, K, H_α) are classified explicitly in [18, 19]. This classification is not of course sufficient to solve our problem as we need to find those triples for which the subgroup K contains a quasiprimitive subgroup G, and to classify all possibilities for the quasiprimitive types of such subgroups G. As a first step let us suppose that G is maximal by inclusion among the quasiprimitive subgroups of K. If K itself is quasiprimitive then of course we will have (H, G, H_α) in the subfamily of triples (H, L, H_α) with $L < H$ and L quasiprimitive. Suppose now that K is not quasiprimitive and let M be a minimal normal subgroup of K which is intransitive on Ω. Since M is intransitive and G is quasiprimitive, we have both $H \neq MH_\alpha$ and $M \cap G = 1$. In particular $|H : H_\alpha| = |\Omega|$ divides $|G|$, which divides $|K : M|$. Thus the following are some of the preliminary problems whose solution will help to extend the classification in Theorem 7.2.

Problem 7.3 *Let H be an almost simple group with socle T. Classify all factorisations $H = AB$ such that B is a maximal subgroup of H and $B \not\geq T$, and*

(a) *$A = S^d$ for some nonabelian simple group S and positive integer d, and $A \cap B = 1$; or*

(b) *A is maximal in H subject to $A \not\geq T$, A has a unique minimal normal subgroup M and $H = MB$; or*

(c) *A is maximal in H subject to $A \not\geq T$, A has a minimal normal subgroup M such that $H \neq MB$ and $|H : B|$ divides $|A : M|$.*

Recall that a quasiprimitive group has either a unique minimal normal subgroup, or two regular nonabelian minimal normal subgroups. Thus we could relatively easily find all the triples (H, L, H_α), with $L < H$ and L quasiprimitive, from solutions to Problem 7.3 (a) and (b). Problem 7.3 (c) should yield sufficient information for us to locate most of the triples (H, L, H_α), with $L < H$ and L not quasiprimitive but such that L has a quasiprimitive subgroup G. Of course even these classifications would not lead to a complete solution of the problem of determining all inclusions of imprimitive quasiprimitive permutation groups in primitive groups, but they seem a reasonable strategy to pursue. There remains of course the problem of classifying the imprimitive quasiprimitive inclusions.

Problem 7.4 *For a finite set Ω, classify all inclusions $G < H \leq \mathrm{Sym}(\Omega)$ such that G, H are both quasiprimitive on Ω and of different quasiprimitive types.*

Let us return to Questions 7.1 on automorphism groups of quasiprimitive graphs. The group theoretic analysis just described should be applied to give some answers to Questions 7.1 (a)–(c). With regard to Question 7.1 (d), it is certainly possible for a $(G, 2)$-arc transitive graph Γ, with G quasiprimitive, to have $\mathrm{Aut}(\Gamma) = G \times C$ with C some small nontrivial cyclic subgroup. An example of such a graph with G quasiprimitive of type TW was found by Robert Baddeley [1, Section 6], and he commented there that such examples seem difficult to find. Recently Cai Heng Li [15] has found an infinite family of examples with G quasiprimitive of type AS. On the other hand, if $\Gamma = (V, E)$ is G-locally primitive with G quasiprimitive of type AS, with $T := \mathrm{soc}(G)$ not regular on V and $C_{\mathrm{Aut}(\Gamma)}(T) = 1$, then it has been shown in [10] that either $\mathrm{Aut}(\Gamma) \leq \mathrm{Aut}\,T$, or $G < Y \leq \mathrm{Aut}(\Gamma)$ with Y almost simple having $\mathrm{soc}(Y) \neq T$, or T is a member of a restricted family of Lie type simple groups over a field of characteristic p and $\mathrm{Aut}(\Gamma)$ contains a semidirect product $N.G$ with $N = \mathbb{Z}_p^d$ a specific faithful absolutely irreducible $\mathbb{Z}_pG$-module.

These observations and results about automorphism groups of graphs Γ are simply sample, or preliminary answers to the questions posed above. Much more work remains to be done.

References

[1] Robert W. Baddeley, Two-arc transitive graphs and twisted wreath products, *Journal of Algebraic Combinatorics*, **2** (1993), 215–237.

[2] Robert W. Baddeley & Cheryl E. Praeger, On classifying all full factorisations and multiple-factorisations of the finite almost simple groups, Technical Report No. 1996/13, University of Leicester, 1995.

[3] Robert W. Baddeley & Cheryl E. Praeger, On primitive overgroups of quasiprimitive permutation groups, Technical Report No. 1996/14, University of Leicester, 1996.

[4] Robert W. Baddeley & Cheryl E. Praeger, Expansion and blow-up decompositions of quasiprimitive permutation groups, in preparation, 1996.

[5] A. E. Brouwer, A. M. Cohen & A. Neumaier, *Distance-Regular Graphs*, Springer-Verlag, Berlin, Heidelberg (1989).

[6] Peter J. Cameron, Finite permutation groups and finite simple groups, *Bulletin of the London Mathematical Society*, **13** (1981), 1–22.

[7] Xin Gui Fang, *Construction and classification of some families of almost simple 2-arc transitive graphs*, Ph. D. Thesis, University of Western Australia, 1995.

[8] Xin Gui Fang & Cheryl E. Praeger, Finite two-arc transitive graphs admitting a Suzuki simple group, Research Report, University of Western Australia, 1996.

[9] Xin Gui Fang & Cheryl E. Praeger, Finite two-arc transitive graphs admitting a Ree simple group, Research Report, University of Western Australia, 1996.

[10] Xin Gui Fang & Cheryl E. Praeger, On graphs admitting arc-transitive actions of almost simple groups, Research Report, University of Western Australia, 1996.

[11] Akbar Hassani, Luz Nochefranca & Cheryl E. Praeger, Finite two-arc transitive graphs admitting a two-dimensional projective linear group, Research Report, University of Western Australia, 1995.

[12] A. A. Ivanov, Distance-transitive graphs and their classification, in *Investigations in the Algebraic Theory of Combinatorial Objects* (eds. I. A. Faradzev et al.), Kluwer, Dordrecht (1994), pp. 283-378.

[13] A. A. Ivanov & Cheryl E. Praeger, On finite affine 2-arc transitive graphs, *European Journal of Combinatorics*, **14** (1993), 421–444.

[14] L. G. Kovacs, Primitive subgroups of wreath products in product action, *Proceedings of the London Mathematical Society*, **58** (1989), 306–322.

[15] Cai Heng Li, A family of 2-arc transitive graphs, in preparation.

[16] M. W. Liebeck, Cheryl E. Praeger & Jan Saxl, A classification of the maximal subgroups of the finite alternating and symmetric groups, *Journal of Algebra*, **111** (1987), 365–383.

[17] M. W. Liebeck, Cheryl E. Praeger & Jan Saxl, On the O'Nan-Scott Theorem for finite primitive permutation groups, *Journal of the Australian Mathematical Society, Series A*, **44** (1988), 389–396.

[18] M. W. Liebeck, Cheryl E. Praeger & Jan Saxl, *The Maximal Factorisations of the Finite Simple Groups and their Automorphism Groups*, *Memoirs of the American Mathematical Society*, 86, No. 432, American Mathematical Society, Providence, Rhode Island (1990).

[19] M. W. Liebeck, Cheryl E. Praeger & Jan Saxl, On factorisations of almost simple groups, *Journal of Algebra*, in press.

[20] M. W. Liebeck, Cheryl E. Praeger & Jan Saxl, On the automorphism groups of finite vertex-primitive, edge-transitive graphs and directed graphs, in preparation.

[21] Jane McCorkindale, The 2-modular representation theory of $PSU_3(q)$, $q \equiv 3 \pmod 4$, D. Phil. Thesis, University of Oxford, 1990.

[22] B. H. Neumann, Twisted wreath products of groups, *Archiv der Mathematik*, **14** (1963), 1–6.

[23] Peter M. Neumann, Gabrielle A. Stoy & Edward C. Thompson, *Groups and Geometry*, Oxford University Press, Oxford (1994).

[24] Cheryl E. Praeger, Imprimitive symmetric graphs, *Ars Combinatoria*, **19A** (1985), 149–163.

[25] Cheryl E. Praeger, The inclusion problem for finite primitive permutation groups, *Proceedings of the London Mathematical Society, Series 3*, **60** (1990), 68–88.

[26] Cheryl E. Praeger, An O'Nan-Scott Theorem for finite quasiprimitive permutation groups and an application to 2-arc transitive graphs, *Journal of the London Mathematical Society, Series 2*, **47** (1993), 227–239.

[27] Cheryl E. Praeger, On a reduction theorem for finite, bipartite, 2-arc transitive graphs, *Australasian Journal of Combinatorics*, **7** (1993), 21–36.

[28] Cheryl E. Praeger, Jan Saxl & Kasuhiro Yokoyama, Distance transitive graphs and finite simple groups, *Proceedings of the London Mathematical Society, Series 3*, **55** (1987), 1–21.

[29] Cheryl E. Praeger & Jie Wang, On primitive representations of finite alternating and symmetric groups with a 2-transitive subconstituent, *Journal of Algebra*, **180** (1996), 808–833.

[30] L. L. Scott, Representations in characteristic p, in *Santa Cruz Conference on Finite Groups, Proceedings of Symposia in Pure Mathematics*, 37, American Mathematical Society, Providence, Rhode Island (1980), pp. 318–331.

[31] D. H. Smith, Primitive and imprimitive graphs, *Quarterly Journal of Mathematics, Oxford, Series 2*, **22** (1971), 551–557.

[32] M. Suzuki, *Group Theory I*, Springer-Verlag, Berlin (1982).

Department of Mathematics
University of Western Australia
Nedlands WA 6907
Australia
praeger@maths.uwa.edu.au

Tree Width and Tangles: A New Connectivity Measure And Some Applications

B. A. Reed

Summary We discuss tree width, a new connectivity invariant of graphs defined by Robertson and Seymour. We present a duality result and a canonical decomposition theorem tied to this invariant. We also discuss a number of applications of these results, including Robertson and Seymour's Graph Minors Project.

1 Introduction

1.1 A taste of things to come

A graph is a set of vertices and an adjacency relation which indicates which pairs of vertices are joined by an edge. Thus, graph theory is essentially the study of connectivity. How then does one measure the connectivity of a graph?

Measuring the connectivity between two vertices is straightforward. Two vertices are said to be *k-connected* if there are k internally vertex disjoint paths between them. A classical theorem of Menger [30] states that vertices a and b are k-connected in a graph G precisely if there is no set X of fewer than k vertices such that a and b lie in different components of $G - X$. Standard alternating paths techniques, see e.g. [21], allow us to find either k internally vertex disjoint a-b paths or such a set X efficiently.

An appropriate definition of a highly connected graph, or of a highly connected piece of a graph is more difficult. The classical approach is to call a graph k-connected if every pair of its vertices is k-connected. This definition, although natural, does not capture the kind of connectivity that will concern us. It focuses on local properties rather than global ones. To illustrate what we mean, we consider the hexagonal lattice $\mathcal{L}$ (see Figure 1). $\mathcal{L}$ has maximum degree three and hence contains no 4-connected subgraph. On the other hand, if X is a set of at most three vertices of $\mathcal{L}$ then $\mathcal{L} - X$ is either connected or has two components, one of which is a vertex. Similarly, if X is a set of at most k vertices then the largest component of $\mathcal{L} - X$ contains all but at most k^2 of the vertices of $\mathcal{L}$. Thus, no cutset of size k "globally disconnects" $\mathcal{L}$. The same is true for sufficiently large and robust finite subgraphs of $\mathcal{L}$.

One of the goals of this paper is to introduce its readers to a connectivity invariant, the bramble number, which measures global connectivity.

Having done so, we will examine those graphs for which this invariant is bounded by a given k. If the bramble number of G is at most two then G is a forest. It turns out that for any k, the graphs with bramble number at most k have a "tree-like" structure. How tree-like depends on the value of k. We will present a precise duality theorem after we have developed the requisite definitions. Now, such tree-like structures are easy to deal with. For example,

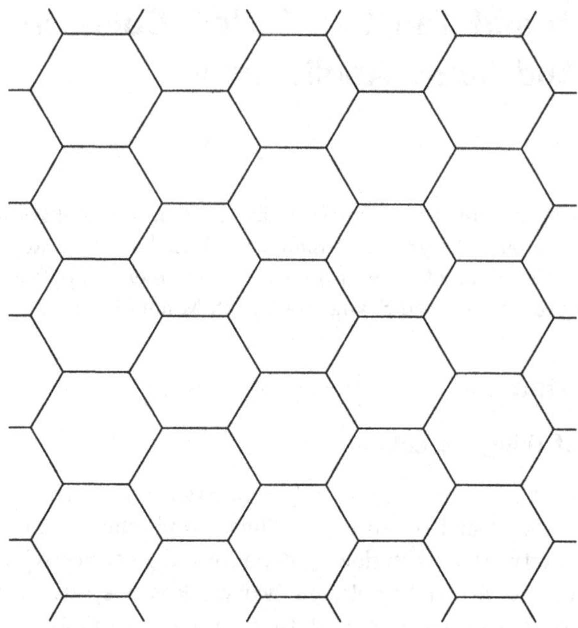

Figure 1: The hexagonal lattice

we will develop efficient algorithms for resolving seemingly difficult (i.e. NP-complete) problems on graphs whose bramble number is bounded. We will also see that many difficult conjectures can be resolved when restricted to such graphs.

We will also examine the structure of graphs with large bramble number. It turns out that if a graph has large bramble number then it must contain a subdivision of a large piece of the hexagonal lattice. This fact allows us to resolve many difficult questions on such graphs.

Combining the two complementary types of results discussed above has proven very fruitful. We will present some of the many results obtained using this technique. We will also discuss an assortment of other results which elucidate the dual notions of brambles and tree-structures. The most important of these is a theorem which states that every graph can be decomposed into a tree structure of pieces each of which corresponds to a maximally connected part of the graph.

We flesh out this skeletal overview in the next four subsections, developing the necessary definitions en route. The theory of brambles and tree decompositions (to use their proper name) was developed mainly by Robertson and Seymour ([36, 37, 38, 41]). The first four sections of this paper survey the development of the theory. We discuss applications of the theory in Sections 5 and 6. Section 5 presents the applications that Robertson and Seymour had in mind when they developed the theory. Section 6 discusses some more recent applications of the theory to packing and covering problems. These results demonstrate that brambles and tree decompositions are not just specialized pieces of mathematical machinery but rather a new central theory which is ripe for further development.

In what follows, we may confound a graph with its vertex or edge set for notational convenience. We will often use $X + y$ to denote $X \cup \{y\}$. Also, we will often use n for $|V(G)|$ particularily when measuring computational complexity.

1.2 Measuring connectivity

In this section, we present a number of invariants which measure the global connectivity of a graph. We shall take as our archetypes of globally connected graphs, subdivisions of certain finite subgraphs of the hexagonal lattice. Specifically, we shall consider *walls*.

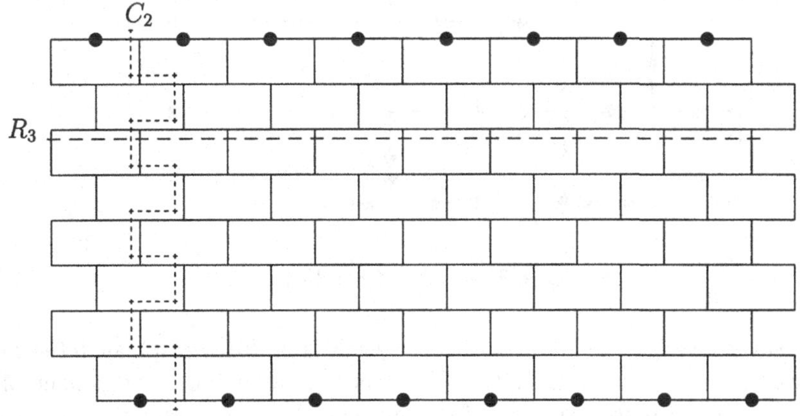

Figure 2: An elementary wall of height 8

An elementary wall of height 8 is depicted in Figure 2. An *elementary wall of height h* is similar. It is a piece of the hexagonal lattice consisting of h levels each containing h bricks. More precisely, an elementary wall of height h contains $h+1$ vertex disjoint paths, $R_1, \ldots, R_{h+1}$ which we call *rows*, and $h+1$ vertex disjoint paths, $C_1, \ldots, C_{h+1}$ which we call *columns*. The reader should be able to complete the definition by considering Figure 2, in which R_1 is the top row. (For fussy formalists: the first and last row, i.e. R_1 and R_{h+1}, both contain $2h+1$ vertices. All the other rows contain $2h+2$ vertices. All the columns contain $2h$ vertices. Column i joins the $(2i-1)$st vertex of R_1 with the $(2i-1)$st vertex of R_{h+1}; it contains the $(2i-1)$st and $2i$th vertex of every other row, as well as the edge between them. For $j \leq h$

and odd, each C_i contains an edge between the $(2i-1)$st vertex of R_j and the $(2i-1)$st vertex of R_{j+1}. For $j \leq h$ and even, each C_i contains an edge between the $2i$th vertex of R_j and the $2i$th vertex of R_{j+1}. These are all the edges of the wall.)

The *perimeter* of an elementary wall of height h is the cycle formed by $R_1 \cup C_1 \cup R_{h+1} \cup C_{h+1}$. Its *corners* are: $R_1 \cap C_1$, $R_1 \cap C_{h+1}$, $R_{h+1} \cap C_1$, and $R_{h+1} \cap C_{h+1}$.

A *wall of height h* is obtained from the elementary wall by replacing the edge set by a corresponding set of internally vertex disjoint paths whose interiors are vertex disjoint from the original elementary wall, see Figure 3. The rows, columns, corners, and perimeter of the wall correspond to the same objects in the original elementary wall. The *nails* of the wall are the vertices of degree three within it as well as its corners.

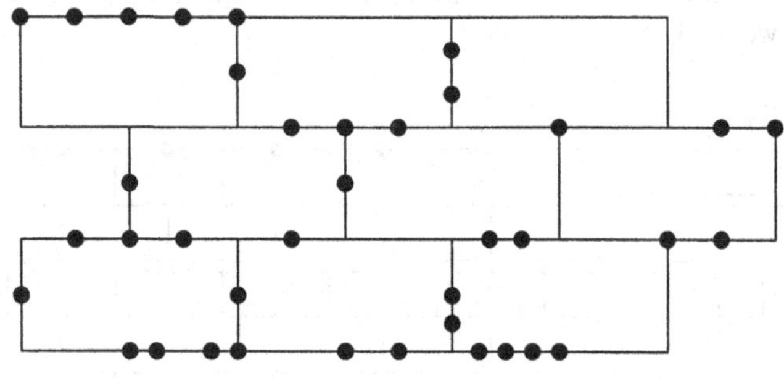

Figure 3: A wall of height 3

We say that a set S of vertices in a graph G is *k-linked* (k an integer) if for any set X of fewer than k vertices there is a (unique) component of $G-X$ containing more than half of the vertices of S. Obviously, the vertices of any clique C are $\lfloor |V(C)|/2 \rfloor$-linked. We note that the nails of any wall of height h can be shown to be $\lfloor h/2 \rfloor$-linked (since any set of at most $h/2$ vertices misses half the rows and the nails in these rows will clearly all be in the same component), even though this set may be independent. One measure of the connectivity of G which interests us is the largest k for which G contains a k-linked set. We shall denote this by linkedness(G), the *linkedness* of G.

We now introduce a measure of connectedness which is more abstract. To motivate it, let us consider a k-linked set S in a graph G. For each subset X of $V(G)$ with $|X| < k$, the *big component* of $G-X$ is that which contains more than half the vertices of S. Let β be the set of big components. Obviously any two elements of β intersect. Just as obviously, no set X of fewer than k vertices intersects every element of β.

By a *bramble*, we mean a set of connected subgraphs any two of which *touch*, that is intersect or are joined by an edge. We note that the vertices of any clique form a bramble. Since every row of a wall intersects every column, a wall of height h contains the bramble

$$\text{Crosses}_h = \{R \cup C \mid R \text{ is a row, } C \text{ is a column}\}.$$

The *order* of a bramble β is the minimum cardinality of a *hitting set* for β (i.e. a set X of vertices which intersects every element of β). Obviously the order of the bramble given by the vertices of a clique is just the cardinality of the clique. The order of the bramble Crosses_h is $h+1$. To see that it is at most $h+1$, we note that taking one vertex from each row yields a hitting set. To see that it is at least $h+1$, we note that if X is a set of less than $h+1$ vertices then X misses some row R and some column C and hence the corresponding element of Crosses_h. The *bramble number* of G, denoted $\text{BN}(G)$ is the maximum of the orders of its brambles. Our remarks in the last paragraph show that every k-linked set generates a bramble of order k and hence for every graph G: $\text{BN}(G) \geq \text{linkedness}(G)$. Before going any further, we prove that a partial converse holds.

Theorem 1.1 *For every graph G,* $\text{linkedness}(G) \leq \text{BN}(G) \leq 2\,\text{linkedness}(G)$.

Proof We know $\text{BN}(G) \geq \text{linkedness}(G)$. Therefore we need only show that $\text{linkedness}(G) \geq \text{BN}(G)/2$. Clearly, it suffices to show that if X is a minimum order hitting set for a bramble β then it is $\lceil |X|/2 \rceil$-linked. Suppose not, and let Y be a set of fewer than $|X|/2$ vertices such that no component of $G-Y$ contains more than half the vertices of X. Since $|Y| < |X|$, Y is not a hitting set for β, so there is a component U of $G-Y$ which contains an element of β. Since every two elements of β touch, it follows that every element of β intersects $U \cup Y$ (and hence U is unique). Since X is a hitting set for β, $X \cap U$ is a hitting set for those elements of β completely contained in U. Thus $(X \cap U) \cup Y$ is a hitting set for β, contradicting the minimality of $|X|$. ∎

Now, Theorem 1.1 shows that our two connectivity parameters are essentially measuring the same property of graphs. A much more difficult theorem shows that the property they measure is tied to the size of the largest wall in the graph. To wit:

Theorem 1.2 ([46](see also [38])) *Let h be the maximum of the heights of the walls in G. Then* $h+1 \leq \text{BN}(G) \leq 25^{34h^5}$.

We have already proven that $h+1 \leq \text{BN}(G)$. We need only prove that the second inequality holds. However, before proving this theorem characterizing graphs whose bramble number is high, we analyze the structure of graphs whose bramble number is low. We sketch a proof of Theorem 1.2 in Section 4 and give the details in Section 7.

1.3 Tree decompositions and tree width

A *tree decomposition* of a graph G consists of a tree T and for each node t of T, a subset W_t of $V(G)$ such that:

(i) For each vertex v of G, the set $S_v = \{t \mid t \text{ is a node of } T, v \text{ is in } W_t\}$ induces a non-empty subtree of T, and

(ii) For each edge of G with endpoints x and y, S_x intersects S_y.

We let $\mathcal{W} = (W_t \mid t \text{ is a node of } T)$, and speak of the tree decomposition $[T, \mathcal{W}]$. We use $S_v([T, \mathcal{W}])$ instead of S_v if this precision is necessary.

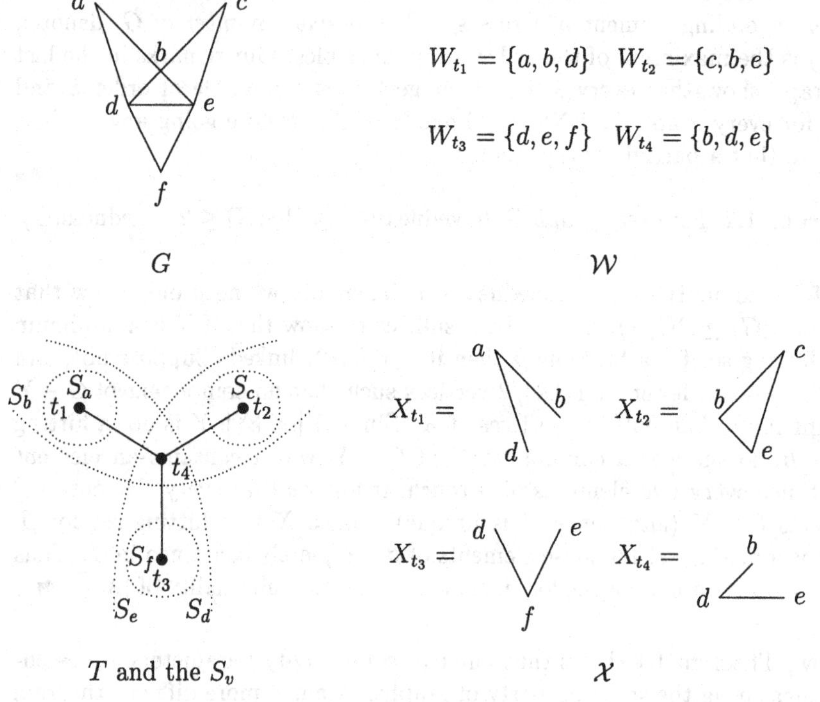

Figure 4: An example of a tree decomposition

As depicted in Figures 4 and 5, given a tree decomposition $[T, \mathcal{W}]$ of G, we can choose for each node t in T, a subgraph X_t of G with node set W_t such that each edge of G is in precisely one of these subgraphs. To do so, we place each edge e with endpoints x and y in X_t for some arbitrary element t of $S_x \cap S_y$. For any such set of subgraphs, we set $\mathcal{X} = (X_t \mid t \text{ is a node of } t)$ and speak of the tree decomposition $[T, \mathcal{X}]$. We justify this abuse of notation by remarking that we are really always considering a partition of the edge set and hence a tree decomposition of the second type. However, sometimes we simply

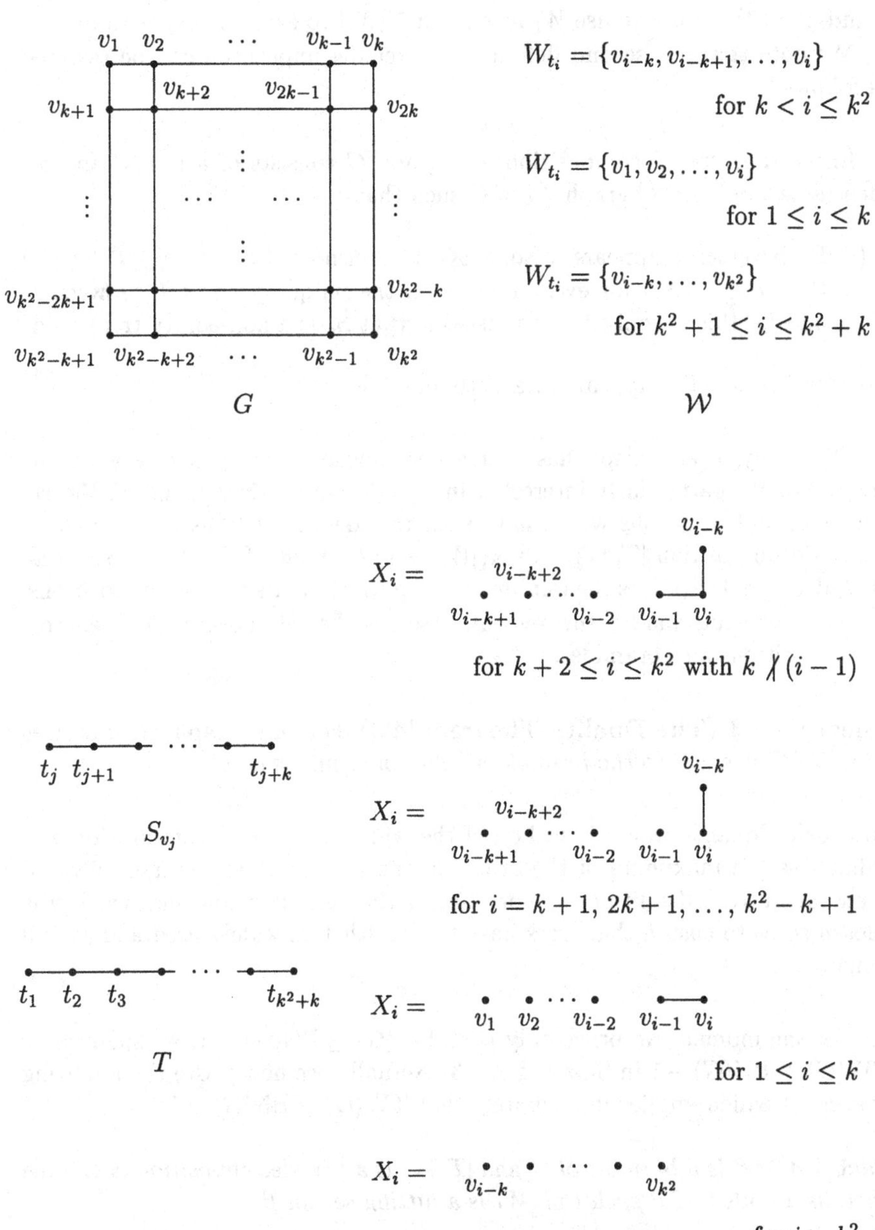

Figure 5: Another example of a tree decomposition

don't care which element of $S_x \cap S_y$ contains a particular edge with endpoints x and y. In this case we use W_t instead of $V(X_t)$ to ease our exposition.

We note that our second definition of tree decomposition can be restated as follows.

Definition A tree decomposition of a graph G consists of a tree T and for each node t of T, a subgraph X_t of G such that:

(i) Each vertex v appears in some X_t, furthermore if $v \in V(X_{t_1}) \cap V(X_{t_2})$ then $v \in V(X_s)$ for every node s on the unique path of T between t_1 and t_2 (this is equivalent to insisting that S_v is a non-empty tree), and

(ii) each edge of G appears in exactly one X_t.

Obviously, every graph has a tree decomposition using a tree with one node. We are particularly interested in tree decompositions in which the W_t are small and in graphs which have such tree decompositions. The *width* of a tree decomposition $[T, \mathcal{W}]$ is $\max\{|W_t| - 1,\ t \text{ a node of } T\}$. The *tree width* of G, denoted $\mathrm{TW}(G)$, is the minimum of the widths of its tree decompositions.

The following duality theorem underscores the relationship between tree decompositions and brambles.

Theorem 1.3 (The Duality Theorem [48]) *For any graph G, the tree width of G is equal to the bramble number of G minus one.*

Remark Equality would hold here if the width of a tree decomposition was defined as the maximum of the orders of the W_t, which seems more natural than the current definition. Unfortunately the current definition, which was chosen so as to ensure that trees have tree width 1, is widely used and so will remain with us.

For the moment we prove only that $\mathrm{TW}(G) \geq \mathrm{BN}(G) - 1$, we shall prove $\mathrm{TW}(G) \leq \mathrm{BN}(G) - 1$ in Subsection 2.3. Actually, we now prove the following statement which implies immediately that $\mathrm{TW}(G) \geq \mathrm{BN}(G) - 1$:

Fact 1.4 *If β is a bramble of G and $[T, \mathcal{W}]$ is a tree decomposition of G then there is a node t of T such that W_t is a hitting set for β.*

Proof We shall need the following two facts:

> **1.5 (The Helly Property For Trees)** *If $\mathcal{S}$ is a family of at least two subtrees of a tree T, every pair of which intersect, then $\bigcap\{S \mid S \in \mathcal{S}\}$ is non-empty.*

Proof We proceed by induction on $|V(T)|$. If T has only one node then there is nothing to prove. Otherwise, let t be a leaf of T with neighbour s. If any tree S in $\mathcal{S}$ is exactly t then $\bigcap\{S \mid S \in \mathcal{S}\}$ is t. Otherwise, $T - t$ is a tree and for each S in $\mathcal{S}$, $S - t$ is a subtree of $T - t$. Furthermore, every pair of trees in $\mathcal{S}' = \{S - t \mid S \in \mathcal{S}\}$ intersect because if t is in $S_1 \cap S_2$ for some S_1, $S_2 \in \mathcal{S}$ then so is s. So by induction the trees of $\mathcal{S}'$ have a non-empty common intersection and so do those of $\mathcal{S}$. ∎

Fact 1.6 *If C is a connected subgraph of a graph G and $[T, \mathcal{W}]$ is a tree decomposition then $S_C = \{t \mid W_t$ intersects $C\}$ induces a subtree of T.*

Proof This is true by definition if C has only one vertex. We proceed by induction on $|C|$. So, let C be a connected subgraph of G with at least two vertices and let v be a leaf of some spanning tree for C. Then, $C - v$ is connected and by induction, S_{C-v} is a subtree of T. S_v is also a subtree of T and since there is an edge from v to $C - v$, we know these two subtrees intersect. Thus the union of their vertex sets, which is precisely S_C, induces a subtree of T, as required. ∎

Now, let β be a bramble in a graph G and let $[T, \mathcal{W}]$ be a tree decomposition of G. For each B in β, we let S_B be the set $\{t \mid W_t$ intersects $B\}$. We know each S_B induces a subtree of T. Now, for every B and C in β, B touches C and so $S_B \cap S_C$ is non-empty. Thus, by the Helly property for trees, there is a node t in $\bigcap\{S_B \mid B \in \beta\}$. Now, W_t is a hitting set for β, as required. ∎

In Section 2, in addition to completing the proof of this duality theorem, we will discuss methods for solving difficult optimization problems on graphs of bounded tree width in polynomial, often linear, time. The idea is to generalize standard dynamic programming techniques from trees to bounded-width tree decompositions. These techniques work on trees because removing any edge disconnects the graph. We can generalize the techniques to graphs of bounded tree width because of the following analogous result which shows that the edges of a tree decomposition of G of width k correspond to cutsets of size at most k.

Definition Let $[T, \mathcal{W}]$ be a tree decomposition of a graph G. For any subtree S of T, by V_S we mean $\bigcup\{W_t \mid t$ is a node of $S\}$.

Lemma 1.7 *Let $[T, \mathcal{W}]$ be a tree decomposition of a graph G. Let rs be an arc of T and let R and S be the components of $T - rs$ containing r and s respectively. Then, $(V_R - W_s, V_S - W_r)$ is a partition of $V - (W_r \cap W_s)$ and furthermore, there is no edge of G between $V_R - W_s$ and $V_S - W_r$.*

Proof First, we note that for each vertex v of G, exactly one of the following holds: $S_v \subseteq R$ and hence $v \in V_R - W_s$, $S_v \subseteq S$ and hence $v \in V_S - W_r$, or S_v contains the arc rs and hence $v \in W_s \cap W_r$. Thus, $(V_R - W_s, V_S - W_r)$ is a partition of $V - (W_r \cap W_s)$. Now, if u is in $V_R - W_s$ and v is in $V_S - W_r$ then $S_u \subseteq R$ while $S_v \subseteq S$. Thus, $S_u \cap S_v = \emptyset$ and so $uv \notin E(G)$. ∎

Corollary 1.8 *Let $[T, \mathcal{W}]$ be a tree decomposition of a graph G and let t be a node of T with l neighbours $s_1, \ldots, s_l$. Let S_i be the component of $T - t$ containing s_i. Then $(V_{S_1} - W_t, \ldots, V_{S_l} - W_t)$ is a partition of $V - W_t$. Furthermore, for $1 \leq i < j \leq l$, there is no edge between $V_{S_i} - W_t$ and $V_{S_j} - W_t$.*

To complete this subsection, we present a number of technical results concerning tree decompositions.

Given a tree decomposition $[T, \mathcal{W}]$ of a graph G, if there are two adjacent nodes s and t of T such that $W_s \subseteq W_t$ then we can contract s and t into one new node $s * t$ to form a smaller tree T' (i.e. $V(T') = V(T) - s - t + s * t$ and $E(T') = E(T - s - t) \cup \{x(s * t) \mid xs \text{ or } xt \in E(T)\}$). It is easy to see that if we let $W'_r = W_r$ for each node r of T' except $s * t$, set $W'_{s*t} = W_t$, and set $\mathcal{W}' = (W'_r \mid r \text{ is in } T')$ then $[T', \mathcal{W}']$ is also a tree decomposition of G with the same width as $[T, \mathcal{W}]$. So, if we choose a tree decomposition $[T, \mathcal{W}]$ of G of minimum width which has no more nodes than any other tree decomposition of G of the same width then for any adjacent nodes s and t of T, we know that W_s neither is contained in nor contains W_t. We call a tree decomposition with the latter property *nice*.

We can prove by induction on the number of vertices in G that

1.9 *any nice tree decomposition $[T, \mathcal{W}]$ of G has no more nodes than G has vertices.*

Proof This is clearly true if T has only one node. If T has at least two nodes then we simply find a leaf l of T with neighbour t and delete the set Y of vertices of G which appear in W_l but not in W_t. $[T - l, \mathcal{W} - W_l]$ is a tree decomposition of $G - Y$, since $W_l - Y \subseteq W_t$. It is clearly nice. Furthermore, Y is non-empty since $[T, \mathcal{W}]$ is nice. The result follows. ∎

We can also use nice tree decompositions to prove

1.10 *a simple graph G of tree width at most k has minimum degree at most k. Hence if G has n vertices then it has at most kn edges.*

Proof We prove this by induction on the number of vertices in G. It is obviously true if G has a tree decomposition of width at most k using a one node tree. So, we assume that G has tree width at most k and let $[T, \mathcal{W}]$ be a nice tree decomposition of G of width at most k with at least two nodes. Again we consider a leaf l of T and the necessarily non-empty set Y of vertices which appear in W_l and in no other W_s. Clearly, each vertex y in Y has degree at most k in G as its neighbourhood is contained in W_l. The rest of (1.10) follows by induction. ∎

1.4 Tangles, separations, and canonical tree decompositions

The duality theorem discussed in the last section shows that the notions of brambles and tree-decompositions are intimately linked. We want to investigate this relationship further. To do so we will need to consider some special brambles called tangles. We will also need to look at tree decompositions from a new perspective, for which we will need to introduce the notion of separations. Having developed this machinery, we will be able to state our theorem concerning the existence, for every graph, of a tree decomposition whose nodes correspond to the maximally connected pieces of the graph.

A *separation* of G consists of an ordered pair (A, B) of subgraphs of G which have disjoint edge sets and whose union is G. The *order* of a separation (A, B), denoted $\text{ord}((A, B))$, is $|V(A) \cap V(B)|$. A and B are its *sides.*

Two separations (A, B) and (C, D) are *laminar* if either $A \subseteq C$ (and hence $D \subseteq B$), or $A \subseteq D$ (and hence $C \subseteq B$), or $B \subseteq C$ (and hence $D \subseteq A$), or $B \subseteq D$ (and hence $C \subseteq A$). A set of separations is laminar if every pair of separations within it is.

Note that any arc rs in a tree decomposition $[T, \mathcal{X}]$ corresponds to two separations (A, B) and (B, A) where, if T_1 and T_2 are the two components of $T - rs$, we have:

$$A = \bigcup\{X_t \mid t \text{ is in } T_1\}, \quad B = \bigcup\{X_t \mid t \text{ is in } T_2\}, \quad \text{and}$$
$$V(A \cap B) = V(X_r) \cap V(X_s).$$

We shall call these two separations, the separations *made by rs*. It is easy to verify that the set of separations made by the arcs of a tree decomposition is laminar. It is also straightforward to verify that any laminar set $\mathcal{S}$ of distinct separations in a graph G corresponds to a unique tree decomposition $[T_{\mathcal{S}}, X_{\mathcal{S}}]$ of G such that each separation S in $\mathcal{S}$ is made by precisely one arc of $T_{\mathcal{S}}$ and for each arc a of T_s one of the separations made by a is in $\mathcal{S}$ (this was first noted in [41]).

A bramble $\mathcal{T}$ is a *tangle* if for any triple $\{T_1, T_2, T_3\}$ of elements of $\mathcal{T}$ either

(i) $T_1 \cap T_2 \cap T_3$ is non-empty, or

(ii) there is an edge e such that all of T_1, T_2, and T_3 contain an endpoint of e.

Remark Note that (i) implies (ii) unless $T_1 \cap T_2 \cap T_3$ is a singleton component.

The *tangle number* of G, denoted $\text{TN}(G)$ is the maximum order of a tangle in G (the order of a tangle is its order as a bramble).

In [41], Robertson and Seymour define branch width and use a duality theorem relating the tangle number of a graph to its branch width to show that the bramble number of G is at most 3/2 of its tangle number. This bound is best possible, as is shown by the graphs K_k, $k \geq 1$ (see [41]). We shall show here:

1.11 *The bramble number of G is at most 3 times its tangle number.*

To do so, we need some definitions.

A *preference of order k* in a graph G is a function f mapping each set X of $k-1$ or fewer vertices of G to a component of $G-X$ such that if X_1 and X_2 are both sets of at most $k-1$ vertices of G then $f(X_1)$ and $f(X_2)$ touch.

A *strong preference of order k* in a graph G is a preference f such that if X_1, X_2, and X_3 are all sets of at most $k-1$ vertices of G then either $f(X_1) \cap f(X_2) \cap f(X_3)$ is non-empty or there is an edge of G which has an endpoint in each of these three sets.

Now, if β is a bramble of order k, then for each $X \subseteq V(G)$ with $|X| < k$ there is **exactly one** component of $G-X$ containing an element of β. We let $f_\beta(X)$ be this component. Clearly, f_β is a preference of order k.

Conversely, if f is a preference of order k then

$$\beta_f = \{f(X) \mid X \subseteq V(G),\ |X| < k\}$$

is a bramble. Furthermore, β_f has order at least k because for any set X of fewer than k vertices, X is not a hitting set for β_f as it does not intersect $f(X)$.

Similarly, if $\mathcal{T}$ is a tangle of order k then $f_{\mathcal{T}}$ is a strong preference of order k while if f is a strong preference of order k then β_f is a tangle of order at least k.

Proof of 1.11 Now, consider a bramble β of order k in a graph G, and the associated preference f_β. We consider the restriction f'_β of f_β to sets of size at most $\lceil k/3 \rceil - 1$ or less. We claim f'_β is a strong preference of order $\lceil k/3 \rceil$ in G and hence that $\beta_{f'_\beta}$ is a tangle of order $\lceil k/3 \rceil$ or more. This implies that the bramble number of a graph is indeed at most 3 times as great as its tangle number.

To see that f'_β is indeed a strong preference consider three sets X_1, X_2, X_3 in $V(G)$ each with at most $\lceil k/3 \rceil - 1$ elements. As $|X_1 \cup X_2 \cup X_3| < k$, there is an element B of β which fails to intersect $X_1 \cup X_2 \cup X_3$. But now, $B \subseteq f_\beta(X_1) = f'_\beta(X_1)$. Similarly, $B \subseteq f'_\beta(X_2)$ and, $B \subseteq f'_\beta(X_3)$. So, $f'_\beta(X_1) \cap f'_\beta(X_2) \cap f'_\beta(X_3)$ is non-empty and f'_β is a strong preference as claimed. So we have proven (1.11). ∎

In fact, we have shown that every bramble of order at least $3k$ generates a tangle of order k. Thus, we do not lose too much by restricting our attention to tangles. We will therefore do so, because as we are about to see, there is a natural definition of tangles in terms of separations whilst the same is not true of brambles.

A *bias of order k* in a graph G is a function f mapping each separation (A, B) of order less than k either to A or to B such that

(i) $V(f((A, B))) - V(A \cap B) \neq \emptyset$ (or equivalently: G contains at least $k+1$ vertices and if $V(B) = V(G)$ then $f((A, B)) = B$) and

(ii) if Sep_1 and Sep_2 are both separations of order at most k in G then $E(f(\text{Sep}_1)) \cap E(f(\text{Sep}_2))$ is non-empty.

A *strong bias of order* k in a graph G is a bias f such that if Sep_1, Sep_2, and Sep_3 are three separations of order at most k in G then $E(f(\text{Sep}_1)) \cap E(f(\text{Sep}_2)) \cap E(f(\text{Sep}_3))$ is non-empty.

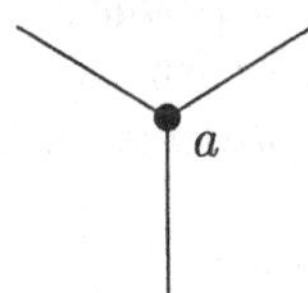

Figure 6: Graph showing that *bias* is not equivalent to *preference*

Now, a preference g of order ≥ 2 defines a bias f in a natural way: $f((A,B))$ will contain $g(V(A) \cap V(B))$ (one of the two sides will contain the big component by the definition of separation). Similarily, a strong preference defines a strong bias. If every bias arose from a preference in this manner then the two notions would be equivalent. Unfortunately, this is not the case. As an example, consider the graph in Figure 6. We define a bias of order two by setting $f((A,B))$ to be the side of (A,B) with two or more edges. By considering the separations (A,B) with $V(A \cap B) = a$, we can show that this bias does not arise from a preference in the manner discussed above. What makes tangles so useful is that every strong bias does arise from a strong preference, as we now show.

Lemma 1.12 *For every strong bias f of order k there is a strong preference g of order k such that $f((A,B))$ contains $g(V(A) \cap V(B))$.*

Proof Let f be a strong bias of order k in a graph G. We claim that for every set X of less than k vertices there is a component $g(X)$ of $G - X$ such that letting A be the subgraph consisting of $g(X)$ and the edges of G between $g(X)$ and X, we have: $f((A, G-A)) = A$. Using the fact that f is a bias, we see that, for every separation (R,S) of G with $V(R) \cap V(S) = X$, $f((R,S))$ contains $g(X)$. Since f is a strong bias, g is a strong preference of order k. Thus, f arises from the strong preference g.

To prove our claim, we assume the contrary and choose a set X for which the claim is false. We also choose a separation (A,B) with $V(A \cap B) = X$, $f((A,B)) = A$, and A minimal subject to these conditions.

Assume first that some edge e of A has both endpoints in X. Then, by the minimality of A, $f((A-e, B+e)) = B+e$. By the first condition on a bias, $f(e, G-e) = G-e$. But $E(A) \cap E(B+e) \cap E(G-e) = \emptyset$ contradicting the fact that f is a strong bias. So this case cannot occur.

Thus, there must be two components C_1 and C_2 of $G - X$ contained in A. We let $A_1 = A - V(C_1)$ and $A_2 = A - V(C_2)$. By the minimality of A,

$f((A_1, G-A_1)) = G-A_1$ and $f(A_2, G-A_2) = G-A_2$. Once again, we contradict the fact that f is a strong bias, since $E(G-A_1) \cap E(G-A_2) \cap E(A) = \emptyset$. This completes the proof of the claim and the lemma. ∎

We say that two brambles β_1 and β_2 are *distinguishable* if there is some $X \subseteq V(G)$ with $|X| < \text{ord}(\beta_1)$, $|X| < \text{ord}(\beta_2)$ such that $f_{\beta_1}(X) \neq f_{\beta_2}(X)$. Otherwise, the two brambles are *indistinguishable.* If β_1 and β_2 are distinguishable then any set $X \subseteq V(G)$ with $|X| < \text{ord}(\beta_1)$, $|X| < \text{ord}(\beta_2)$ such that $f_{\beta_1}(X) \neq f_{\beta_2}(X)$ is a (β_1, β_2)-*distinguisher.*

We call a bramble *maximal* if there is no bramble indistinguishable from it with higher order. We call a tangle *maximal* if there is no tangle indistinguishable from it with higher order.

We have seen that for every tree decomposition $[T, \mathcal{X}]$ of a graph G and every tangle $\mathcal{T}$ in G, there is a node t of T such that $V(X_t)$ is a hitting set for $\mathcal{T}$ (in fact this holds for all brambles, not just tangles).

We can now state the following companion result:

Theorem 1.13 (The Canonical Tree Decomposition Theorem [41]) *For any graph G, we can construct a tree decomposition T which has the following properties.*

(1) For each maximal tangle $\mathcal{T}$ of G, there is exactly one node $t(\mathcal{T})$ of T whose vertices form a hitting set for $\mathcal{T}$.

(2) If $\mathcal{T}_1$ and $\mathcal{T}_2$ are indistinguishable maximal tangles (and hence have the same order) then $t(\mathcal{T}_1) = t(\mathcal{T}_2)$.

(3) If $\mathcal{T}_1$ and $\mathcal{T}_2$ are distinguishable maximal tangles then $t(\mathcal{T}_1) \neq t(\mathcal{T}_2)$ and there is an arc st on the unique $t(\mathcal{T}_1)$ to $t(\mathcal{T}_2)$ path of T such that $V(X_s) \cap V(X_t)$ is a $(\mathcal{T}_1, \mathcal{T}_2)$-distinguisher of minimum order.

(4) For every node t of the tree decomposition, there is a maximal tangle $\mathcal{T}$ such that $t(\mathcal{T}) = t$.

Intuitively, this tree decomposition splits G up into its highly connected pieces using cutsets which are as small as possible.

We note that because each W_t is a hitting set for some tangle for which no other W_s is a hitting set, there is no $s \neq t$ such that W_t is contained in W_s. It follows that this tree decomposition is nice, and hence any set of distinguishable tangles contains at most n elements. This contrasts with the situation for brambles as a graph may contain exponentially many distinguishable maximal brambles. We delve more deeply into the properties which separate tangles from brambles in Section 3.

1.5 Graph minors

Readers who have been looking up the references will have noticed that the results discusssed in Subsection 1.3 come mostly from a paper entitled *Graph Minors X: Obstructions to tree decompositions,* and that a precursor of Theorem 1.2 appears in a paper entitled *Graph Minors V: Excluding a planar graph.* These are two of a long series of papers written by Robertson and Seymour (presently more than 25 papers in the series have been written or are in preparation) which contains a number of seminal results concerning graph minors. We discuss the role that tree decompositions and brambles played in the development of this theory in Section 5. For the moment, we will content ourselves with making a few definitions and stating the most important results obtained by Robertson and Seymour.

We *contract* an edge e in a graph G to obtain a new graph G_{xy} with $V(G_{xy}) = V(G) - x - y + (x * y)$ and

$$E(G_{xy}) = E(G - x - y) \cup \{z(x * y) \mid zx \text{ or } zy \in E(G)\}.$$

H is a *minor* of G if a graph isomorphic to H can be obtained from a subgraph of G by a sequence of edge contractions. This is equivalent to requiring that for each edge e of H there is some edge im(e) of G and for each vertex v of H there is some connected subgraph im(v) of G such that for an edge $e = uv$ of H, im(e) has one endpoint in im(u) and the other in im(v). We shall refer to such a structure as a *model* of H in G.

The results which we will discuss are:

1.14 (Wagner's Conjecture) *If G_1, G_2, G_3, ... is an infinite sequence of graphs then there exist $i \neq j$ such that G_i is a minor of G_j.*

1.15 *For any fixed graph H, there is a polynomial-time algorithm which solves the following decision problem:*

Problem: *H-Minor Containment*
Instance: *A graph G.*
Question: *Does G contain an H-minor?*

1.16 *For any fixed positive integer k, there is a polynomial-time algorithm which solves the following decision problem.*

Problem: *k Disjoint Rooted Paths*
Instance: *A graph G and two sets of vertices of G: $X = \{x_1, \ldots, x_k\}$, $Y = \{y_1, \ldots, y_k\}$.*
Question: *Are there k vertex disjoint paths P_1, ..., P_k in G such that P_i has endpoints x_i and y_i?*

These results have a variety of important corollaries. One of the most interesting is the following:

1.17 *If C is a class of graphs closed under the taking of minors then there is a polynomial-time recognition algorithm for C.*

Remark Minor Containment is clearly NP-complete if H is part of the input as determining if a graph has a Hamilton Cycle is a special case. Karp [22] proved that Disjoint Rooted Paths is NP-complete if k is part of the input. Lynch [28] showed this remains true even if we restrict our attention to planar graphs.

Remark The proof of (1.14) is found in [44]. The proofs of (1.15) and (1.16) are found in [42]. As we shall see, (1.16) implies (1.15).

Remark The algorithms of Robertson and Seymour for (1.15), (1.16), and (1.17) run in $\mathcal{O}(|V(G)|^3)$ time. Reed [34] developed similar algorithms for the same problems which run in $\mathcal{O}(|V(G)|^2)$ time.

We show now that (1.14) and (1.15) together imply (1.17). Let C be a class of graphs closed under the taking of minors. Then, there is a minimal set S of graphs such that $G \in C$ if and only if for each $H \in S$, G has no H-minor. This set is called the *obstruction set* for S. (1.14) implies that S is finite. (1.15) implies that for any fixed $H \in S$, we can check for H-minor containment in polynomial time. (1.17) follows.

The results we have just stated have hordes of implications, we mention just a few of them here. To do so, we need to define some minor closed classes of graphs. For any surface Σ, the class C_Σ of graphs embeddable in Σ is clearly closed under the taking of minors (in fact, minors were defined by Tutte [51] who used them extensively whilst studying planar graphs). Now, Kuratowski [24] showed that the obstruction set for planar graphs consists of the two graphs $K_{3,3}$ and K_5 depicted in Figure 7. Archdeacon and Huneke [3] proved that for any non-orientable surface Σ the obstruction set for C_Σ is finite. Clearly, (1.14) implies that for any surface Σ, the obstruction set for C_Σ is finite (this result actually appears in [40]). Furthermore, (1.17) implies that one can test for any fixed surface Σ whether G is embeddable in Σ in polynomial time and, by our remarks, in $\mathcal{O}(|V(G)|^2)$ time. Polynomial time algorithms for this problem had already been developed [18] but the exponent of their running times depended on Σ.

A graph is called *linklessly embeddable* if it can be embedded in three space so that no pair of disjoint cycles form a link (as do for example two consecutive links in a chain). Clearly, if a graph is linklessly embeddable so are all its minors. Thus, (1.17) implies that there is an algorithm to determine if G is linklessly embeddable which runs in polynomial time. Previously, we had not even known if any such algorithm existed. More recently, Robertson, Seymour, and Thomas [45] constructed the obstruction set for the linklessly embeddable graphs.

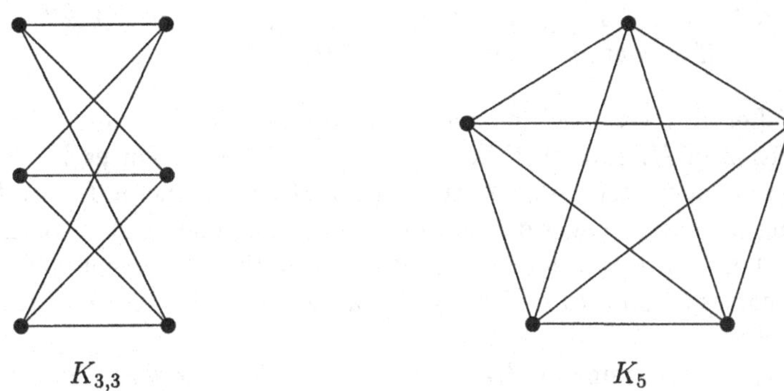

Figure 7: Kuratowski's forbidden minors

Other problems to which (1.17) can be applied include: gate matrix layout, topological bandwidth, disk dimension, and vertex integrity, see [16] and [17] for details.

We remark that (1.17) only implies the existence of an algorithm to determine if a graph is in a minor closed class $\mathcal{C}$. It does not tell us how to construct such an algorithm, because we do not know how to find the obstructions. If we had some absolute bound on the size of such obstructions, we could find the obstruction set and construct an algorithm. This fact has generated interest in bounding the size of the graphs in the obstruction set for particular classes of graphs. Tree decompositions play an important role here, see e.g. [47, 25].

In Section 5, we discuss the role tree decompositions and tangles play in the proofs of (1.14) and (1.16). We close this section by sketching a proof that (1.16) implies (1.15). We begin with a remark and some definitions:

Remark An algorithm for k Disjoint Rooted Paths implies an algorithm for the more general problem where we allow the the paths to share endpoints but insist that they are otherwise disjoint (thus X and Y may be multi-sets and may intersect). We simply make multiple copies of any vertex which appears more than once in $X \cup Y$.

Definitions To *subdivide* an edge e in a graph H, we replace it by a path of length two through a new vertex. A *subdivision* of a graph H, consists of a graph obtained from H by repeatedly sudividing edges. That is a graph in which each edge of H has been replaced by a path with the same endpoints such that these paths can share endpoints but are otherwise disjoint. We say G *contains a subdivision* of H if there is a subgraph of G isomorphic to a subdivision of H. We refer to such a subgraph as a *smodel* of H. We refer to the vertices of this smodel which correspond to the vertices of H as the *centres* of the smodel.

We note the following:

Lemma 1.18 *If H is a graph of maximum degree 3 then G has H as a minor if and only if G contains a subdivision of H.*

Proof Let H be a graph of maximum degree three. Assume G contains a subdivision of H, and let F be an smodel of H in G. For each edge e of H arbitrarily choose some edge of the path of H corresponding to e to be $\text{im}(e)$. Deleting the chosen edges decomposes F into components each containing one centre. For each vertex v of H we let $\text{im}(v)$ be the component containing the centre corresponding to v. This yields a model of H in G and hence H is a minor of G.

Now, if G contains an H-minor, then we shall consider the contractions performed to obtain H from a subgraph G' of G chosen so that $E(G')$ is minimal. It follows that we never contract an edge with an endpoint of degree one. We consider the contractions in reverse order and as a sequence of "decontractions". We claim that each graph we meet is a subdivision of H. In any decontraction, we take some vertex v of degree at most three and replace it by an edge uw so that u and w together are incident to at most three edges other than uw. It follows that one of them, say w, is incident to only one other edge, say xw. But now, the decontraction simply consists of subdividing the edge vx. Thus, G contains a subdivision of H. ■

Lemma 1.19 *For any graph H, there is a finite set Z_H of graphs such that G has H as a minor if and only if G contains a subdivision of some element of Z_H.*

Proof We again use the idea of decontracting edges. So we consider a minimal subgraph G' of G for which there exists a sequence of decontractions from H to a graph isomorphic to G'. We see that every time we decontract in F to obtain F' either we are simply subdividing an edge or $\sum_{v\in V(F')} d_{F'}(v) - 2 < \sum_{v\in V(F)} d_F(v) - 2$ (where $d_J(v)$ is the degree of v in J). It follows that we do only $\sum_{v\in V(H)} \max(0, d_H(v) - 2)$ decontractions which are not subdivisions. Since we can do the subdivisions after all the other decontractions, the result follows. ■

By (1.19), if we can test if G has a subdivision of any fixed H in polynomial time then we can test if G has any fixed F as a minor in polynomial time. To test if G has H as a subdivision we need only test, for each of the $\mathcal{O}(n^{|V(H)|})$ injections of $V(H)$ into $V(G)$, whether G has an smodel of H where the given injection specifies the centre corresponding to each vertex. To do this, however, we need only solve the extension of k Disjoint Rooted Paths which we remarked earlier is no more difficult than k Disjoint Rooted Paths. Thus, (1.16) does indeed imply (1.15).

We remark that in fact Robertson and Seymour actually proved (1.15) directly using the same technique they use to prove (1.16). They thereby obtain an $\mathcal{O}(n^3)$ algorithm for minor containment.

2 Graphs of bounded tree width

2.1 Using tree decompositions of bounded width

In this subsection, we describe how dynamic programming can be used to efficiently solve optimization problems on graphs of bounded tree width. Many of the algorithms we describe run in linear time. They all require as input a bounded width tree decomposition of the graph. Finding such a decomposition quickly is the subject of the next subsection.

To begin, we recall how to use dynamic programming to solve optimization problems on trees.

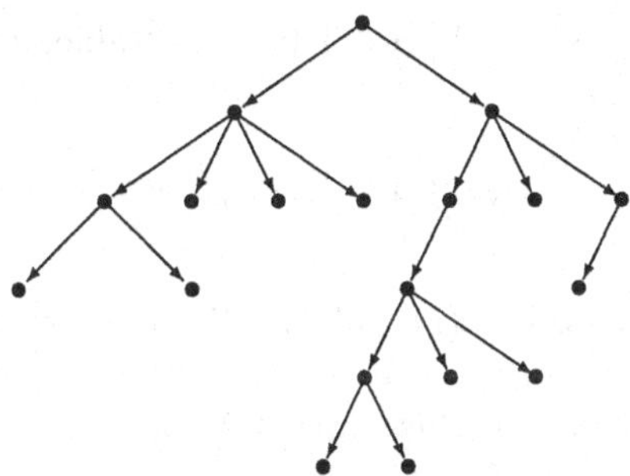

Figure 8: A rooted tree

Definitions We *root* a tree T at a node r by replacing each edge e of T by an arc with the same endpoints directed towards the component of $T - e$ not containing r. Thus, for each node t there is now a directed path from r to t. This gives us a *rooted tree* (T, r) with *root* r.

Definitions Let s and t be nodes of a rooted tree. If there is an arc from s to t then s is the *parent* of t and t is a *child* of s. If there is a directed path from s to t containing at least one arc then t is a *descendant* of s and s is an *ancestor* of t.

Definition Let s be a node of a rooted tree (T, r). We define T_s to be the rooted tree with root s consisting of s and all its descendants.

The following is a well known simple result:

Fact 2.1 *There is a linear-time algorithm which given a rooted tree produces an ordering of its nodes in which each node appears before all of its descendants.*

Reversing this order yields:

Fact 2.2 *There is a linear-time algorithm which given a rooted tree produces an ordering of its nodes in which each node appears after all of its descendants.*

It is this last fact which allows us to solve many optimization problems quickly on trees.

To illustrate, we consider finding a maximum weight stable set. More precisely, we assume we are given a tree T, and an integer weight w_t for each node t in T. We want to find MWS $= \max\{\sum_{v \in S} w(v) \mid S \subset V(T),\ S \text{ stable}\}$. To do so, we first root T at some arbitrary node r. Then, for each node s of T, we compute two parameters:

$$w_1(s) = \max\left\{\sum_{v\in S} w(v) \mid S \subset V(T_s),\ S \text{ stable},\ s \in S\right\}, \quad \text{and}$$

$$w_2(s) = \max\left\{\sum_{v\in S} w(v) \mid S \subset V(T_s), S \text{ stable},\ s \notin S\right\}.$$

We note that MWS $= \max(w_1(r), w_2(r))$. To compute w_1 and w_2, we first order the nodes of the tree so that every node appears after its descendants. We will consider the nodes in this order. We note that if s is a leaf, then $w_1(s) = w(s)$ and $w_2(s) = 0$. If s has descendants $t_1, \ldots, t_k$ then clearly $w_1(s) = w(s) + \sum_{i=1}^{k} w_2(t_i)$ and $w_2(s) = \sum_{i=1}^{k} \max(w_1(t_i), w_2(t_i))$. Thus, we can compute $w_1(s)$ and $w_2(s)$ in $\mathcal{O}(k)$ time. Since the sum of the number of descendants of the nodes in T is simply $|V(T)| - 1$, it follows that MWS can be computed in linear time.

Our technique for solving maximum weight stable set on a graph G given a tree decomposition of G is similar. The first step is to root the tree decomposition. A *rooted tree decomposition* of a graph G consists of a tree decomposition $[T, \mathcal{W}]$ of G and a rooted tree obtained by rooting T at some node r. For brevity's sake, we often use $[(T,r), \mathcal{W}]$ to denote this tree decomposition. It has the same width as $[T, \mathcal{W}]$. Recall that T_t is the rooted subtree consisting of t and all its descendants. We use G_t to denote the subgraph of G induced by $\{v \mid v \in W_s \text{ for some } s \in T_t\}$. We use $w(X)$ to denote $\sum_{x \in X} w(x)$. The key to our dynamic programming algorithm is the following corollary of Fact 1.8.

Fact 2.3 *For each node s in T, there are no edges between $G_s - W_s$ and $G - G_s$. Furthermore, for any two children t and t' of s, there are no edges between $G_t - W_s$ and $G_{t'} - W_s$.*

Now, in computing MWS for G given a rooted tree decomposition $[(T,r), \mathcal{W}]$ of G, we actually compute for each node t of G and each stable set S in W_t, the value $W(S,t) = \max\{w(S') \mid S' \subseteq V(G_t),\ S' \text{ stable},\ S = S' \cap W_t\}$.

Then, the solution to this instance of maximum weight stable set is simply $\max\{W(S,r) \mid S \subseteq W_r,\ S \text{ stable}\}$.

Now, if s is a leaf of T then for each stable set S in W_s, $W(S,s) = w(S)$. Fact 2.3 implies that if s has descendants $t_1, \ldots, t_k$ then for each stable set S in W_s, we can choose a maximum weight stable set of G_s whose intersection with W_s is S by choosing a maximum weight stable set in each G_{t_i} whose intersection with W_s is $S \cap W_{t_i}$. Thus,

$$W(S,s) = w(S) + \sum_{i=1}^{k} \max\{W(S_i,t_i) - w(S \cap S_i) \mid S_i \subseteq W_{t_i},\ S_i \text{ stable},\ S \cap W_{t_i} = S_i \cap W_s\}.$$

If $[T,\mathcal{W}]$ has width w then for each node s there are at most 2^{w+1} subsets S of W_s for which we may need to compute $W(S,s)$. Furthermore using the equation above, we see that if s has k descendants then we can compute $W(S,s)$ in $\mathcal{O}(k2^{w+1})$ time. We assume that $[T,\mathcal{W}]$ is a nice tree decomposition and hence has at most $|V(G)|$ nodes. It follows that our dynamic programming algorithm can be implemented in $\mathcal{O}(2^{w+1}n)$ time which is linear for any fixed w.

We remark that at the cost of some extra bookkeeping but no increase in the computational complexity of the algorithm, we can easily find a maximum weight stable set instead of just the weight of such an object.

Many other problems can be efficiently solved in graphs of bounded tree width by constructing a set of partial solutions corresponding to each node t of a bounded width tree decomposition. To do so, we need an efficient procedure for constructing the set of partial solutions corresponding to a node, given the sets corresponding to its descendants. Consider for example l Disjoint Rooted Paths for some fixed l. Thus, we have a graph G, a rooted tree decomposition $[(T,r),\mathcal{W}]$ of G, and subsets $X = \{x_1,\ldots,x_l\}$, $Y = \{y_1,\ldots,y_l\}$ of $V(G)$.

Now, consider a set $\mathcal{P} = \{P_1,\ldots,P_l\}$ of paths in G where P_i links x_i and y_i. The restriction of $\mathcal{P}$ to G_s for some s in T is a set $\mathcal{Q}$ of paths of G_s each element of which has its endpoints in $(X \cap G_s) \cup (Y \cap G_s) \cup W_s$. Thus $\mathcal{Q}$ has at most $l + |W_s| \leq l + w + 1$ elements. By a *path scheme* for s we mean a partition of $(X \cap G_s) \cup (Y \cap G_s) \cup W_s$ into $l + w + 1$ or fewer ordered sets. We say a set $\{R_1,\ldots,R_j\}$ of disjoint paths in G_s is a *realization* of the path scheme $\mathcal{O} = \{O_1,\ldots,O_j\}$ if for each i: $O_i \subseteq R_i$ and the elements of O_i appear along R_i in the given order. In this case we say that $\mathcal{O}$ is *realizable.* We note that the desired paths $P_1,\ldots,P_l$ exist in G if and only if there is a realizable path scheme $\{Q_1,\ldots,Q_j\}$ for r with $j \geq l$ and $\{x_i,y_i\} \subseteq Q_i$ for each i between 1 and l.

To solve the given instance of l Disjoint Rooted Paths, we shall compute the realizable path schemes for each node s of T. Since $|X \cup Y \cup W_s| \leq 2l+w+1$, there are at most $(2l+w+1)^{2l+w+1}$ path schemes for s. If s is a leaf we can determine which path schemes are realizable simply by considering the constant sized subgraph induced by W_s. If s is not a leaf, then any realization of a path scheme for s corresponds to a set of edges within W_s and a set of realizations of path schemes for the descendants of s. This fact allows us to compute all the realizable path schemes using dynamic programming in linear time if l and w are both fixed. In fact, with slightly more care we can solve l Disjoint Rooted Paths on graphs of bounded tree width in linear time even if l is part of the input.

Other well-known problems which can be solved in linear time given a bounded width tree decomposition of the input graph include: Clique, Hamilton Cycle, Chromatic Number, Domination Number, H-Minor Containment, and Bandwidth. In fact it has been shown that any problem which can be formulated as a certain kind of logical formula [13, 6] can be solved in linear time on graphs of bounded tree width. All of the problems mentioned so far fall into this class. However, there are problems which can be solved in polynomial time on graphs of bounded tree width which do not fit this paradigm. We close this section with a discussion of two such problems (see [8, 7] for some more examples of problems which can be efficiently resolved on graphs of bounded tree width).

Graph Isomorphism can also be solved in polynomial time on graphs of tree width at most w using dynamic programming, see [9]. However, the fastest algorithm known for this problem runs in $\mathcal{O}(n^{w+2})$ time. The algorithm considers two graphs G_1 and G_2 and a tree decomposition of G_1 of width w. For each node s of T, the algorithm computes which subgraphs of G_2 are isomorphic to G_s. The fact that W_s contains at most $w+1$ nodes allows us to consider at most $\mathcal{O}(n^{w+1})$ candidate subgraphs.

We now show that the chromatic index of a graph G can be computed in linear time given a rooted tree decompsition $[(T,r),\mathcal{W}]$ of G. The argument we present is due to McDiarmid and Reed, it is almost ten years old although it appears for the first time here. Recall that the *chromatic index* of a graph G, denoted $\chi'(G)$, is the minimum number of colours required to colour its edges so that any pair of edges which share an endpoint receive different colours. It is easy to see that the chromatic index of any graph G is at least its maximum degree which we denote $\Delta(G)$. Vizing [52] proved that in fact $\chi'(G) \leq \Delta(G)+1$. Thus to determine $\chi'(G)$, we need only determine if the edges of G permit a $\Delta(G)$ colouring.

A natural candidate for a partial solution corresponding to a node s of T, would be a Δ $(= \Delta(G))$ colouring of those edges of G_s incident to W_s which extends to a Δ edge colouring of G_s. However since Δ may be $n-1$, there can be exponentially many such colourings. Bodlaender [9] noted that it is sufficient to record the set of vertices of W_s incident to each colour in a Δ colouring of $E(G_s)$. He also noted that we do not need to record the names of the colours, it is sufficient to record for each subset of W_s, how many of the colour classes are incident precisely to this subset of W_s. He thereby restricted his attention to $\Delta^{2^{w+1}} = \mathcal{O}(n^{2^{w+1}})$ partial solutions at each node and derived a polynomial time algorithm.

McDiarmid and Reed took a different approach. Reproducing work of Vizing, they showed that we can restrict our attention to a subgraph of G of maximum degree less than $2w$. In such a graph, there are at most $2w(w+1)^{2w}$ $2w$-edge colourings of the edges incident to any W_s and a linear-time algorithm follows immediately. Crucial to this approach is the following lemma due to Vizing.

Lemma 2.4 (The Adjacency Lemma) *Let uv be an edge of a graph F of maximum degree at most k such that $d(u) + |\{w \mid w \in N(v),\ d(w) = k\}| \leq k$. Then $\chi'(F) \leq k$ if and only if $\chi'(F - uv) \leq k$. (This lemma is vacuously true unless $k = \Delta(F)$ in which case it is equivalent to: either $\Delta(F-uv) = \Delta(F)-1$ in which case $\chi'(F) = \Delta(F)$, or $\chi'(F) = \chi'(F - uv)$.)*

By a standard reduction of F we mean a graph obtained by repeatedly removing edges satisfying the conditions of the adjacency lemma for $k = \Delta(F)$. It is not difficult to see that there is in fact a unique standard reduction of any graph F, and that this graph either has the same maximum degree as F or has no edges.

Now, the colouring number of F, denoted $\delta^*(F)$, is the maximum over all the subgraphs of F of the minimum degree. Fact 1.10 implies that $\delta^*(F)$ is at most the tree width of F. Applying the adjacency lemma, we obtain:

Lemma 2.5 *If $\Delta(F) \geq 2\delta^*(F)$ then the standard reduction of F is empty and hence $\chi'(F) = \Delta(F)$.*

Proof Consider the standard reduction H of F. If H has no edges then $\chi'(F) = \Delta(F)$ and we are done. So, to prove the lemma, we need only show that if H has edges then $\Delta(F) \leq 2\delta^*(F)$. To this end, assume that H contains edges and hence $\Delta(H) = \Delta(F)$. Consider the set D of vertices of degree at least $\Delta(F)/2$ in H. Let v be any vertex of D. If v has a neighbour in $V(H) - D$ then by the adjacency lemma and our choice of H, v must have $\Delta(F)/2$ neighbours of degree $\Delta(F)$ in H. Thus, every such v has at least $\Delta/2$ neighbours in D. So the vertices of degree at least $\Delta(F)/2$ in H span a subgraph of minimum degree at least $\Delta(F)/2$. The result follows. ∎

McDiarmid and Reed noted that it is easy to compute the standard reduction G' of our graph G of tree width at most w using dynamic programming in linear time. If G' has no edges then the chromatic index of G is Δ, and our algorithm has no work to do. Otherwise, by Lemma 2.5 and the fact that $\delta^*(G)$ is at most w, we obtain that Δ is at most $2w$, and as remarked above dynamic programming can be used to compute $\chi'(G)$ in linear time.

To actually find an optimal colouring, we need to repeatedly rip out matchings which decrease the chromatic index. Such a matching can be found in linear time using dynamic programming. This yields a quadratic algorithm for edge colouring graphs of bounded tree width. Taking more care, we can obtain an algorithm which runs in $\mathcal{O}(n \log n)$ time. Can you find a linear time algorithm for this problem or a more efficient algorithm for Isomorphism Testing on graphs of bounded tree width?

2.2 Finding tree decompositions of bounded width

In the last section, we showed how various optimization problems could be solved efficiently in a graph, given a tree decomposition of the graph with bounded width. In this section, we describe, for each fixed k, an $\mathcal{O}(n^2)$ algorithm which given a graph G, either determines that G has tree width greater than k, or finds a tree decomposition of G of width at most $4k+1$.

To describe the algorithm we need to introduce the notion of separators. For our purposes, a *separator* for a graph G is a subset X of $V(G)$ such that no component of $G-X$ contains more than $\frac{2}{3}|V(G)-X|$ vertices. An *S-separator* for some $S \subseteq V(G)$ is a set X of vertices of G such that no component of $G-X$ contains more than $\frac{2}{3}|S-X|$ vertices of S.

The following fact links the notions of separators and tree decompositions:

Fact 2.6 *If G has tree width at most k then for all $S \subseteq V(G)$, G has an S-separator of order at most $k+1$.*

Proof Obviously every set S of at most $k+1$ vertices is itself an S-separator with at most $k+1$ vertices. So we consider only subsets of the vertices with order at least $k+2$. To begin, we define for each set X in $V(G)$, the set ADJ(X) which is $\{v \mid v \notin X$ and v is adjacent to some vertex in $X\}$. Now, for any set S of at least $k+2$ vertices of G, we define a set β_S of connected subgraphs of G as follows. A connected subgraph B of G is in β_S if and only if both $|\text{ADJ}(B)| \le k+1$ and $|B \cap S| > \frac{2}{3}|S - \text{ADJ}(B)|$. We will show that either G has an S-separator of order at most $k+1$ or β_S is a bramble of order at least $k+2$ and hence the tree width of G is at least $k+1$.

We show first that β_S is a bramble. To this end consider elements B_1 and B_2 of β_S. If B_1 and B_2 do not touch then $B_1 \subseteq G - \text{ADJ}(B_1) - B_2 - \text{ADJ}(B_2)$ and $B_2 \subseteq G - \text{ADJ}(B_2) - B_1 - \text{ADJ}(B_1)$. Thus, $|B_1 \cap S| + |B_2 \cap S| \le |S - \text{ADJ}(B_1) - \text{ADJ}(B_2)|$. However, $|B_1 \cap S| > \frac{2}{3}|S - \text{ADJ}(B_1)|$ and $|B_2 \cap S| > \frac{2}{3}|S - \text{ADJ}(B_2)|$. So $|B_1 \cap S| + |B_2 \cap S| > \frac{4}{3}|S - \text{ADJ}(B_1) - \text{ADJ}(B_2)|$. This contradiction shows that B_1 and B_2 touch and hence β_S is a bramble.

If H is a hitting set for β_S of order at most $k+1$ then no component of $G-H$ contains more than $\frac{2}{3}|S-H|$ vertices of S, for such a component would be in β_S, a contradiction. Thus, H is an S-separator. Otherwise, β_S has order at least $k+2$. This completes the proof of Fact 2.6. ∎

The algorithm we describe actually provides an algorithmic proof of:

Fact 2.7 *If G has an S-separator of order at most $k+1$ for all $S \subseteq V(G)$ then G has tree width at most $4k+1$.*

Combining these two facts, we see that we can determine the tree width of a graph to within essentially a factor of four simply by finding a smallest S-separator for each $S \subseteq V(G)$. Thus the notion of tree width is indeed intimately linked to that of separators.

Remark Planar separators have received much attention in the literature. A fundamental result, due to Lipton and Tarjan [26] is that every planar graph has a separator containing at most $2\sqrt{2|V(G)|}$ vertices. Alon, Seymour, and Thomas [2] improved the bound to $\frac{3}{2}\sqrt{2|V(G)|}$ vertices by studying an appropriate tangle. They also bounded the size of a minimum separator in a graph with no K_l minor for a fixed l, see [1].

Remark We note that if instead of the value $\frac{2}{3}|S - X|$ in the definition of separator we used $\frac{1}{2}|S|$ then a set S would have no separator of order less than k if and only if it were k-linked. The fraction $\frac{2}{3}$ is used for laminarity reasons, just as we often use tangles instead of brambles. We could prove similar results using the fraction $\frac{1}{2}$. This relationship between separators, a well studied notion [12], and both linkedness and bramble number, deserves further study.

Furthermore, we recall that, as we saw in the proof of Fact 2.6, a set S without a separator of order k defines a bramble of order k. So, Fact 2.7 also implies that the tree width of G exceeds its bramble number by at most a factor of four. We sharpen this result in the next section.

Our algorithm for finding tree decompositions recursively uses separators to decompose the graph into subgraphs and then pastes together tree decompositions of these subgraphs. In order to do the pasting we need the following, which the reader may find more intelligible if he is looking at Figure 9.

Fact 2.8 *Let X_1 and X_2 be two sets of vertices in a graph G with $X_1 \subseteq X_2$. Let $C_1, \ldots, C_l$ be the components of $G - X_1$. For i between 1 and l, let $Y_i = (X_2 \cap C_i) \cup X_1$, let $[T^i, \mathcal{W}^i]$ be a tree decomposition of $C_i \cup X_1$. Suppose that for each T^i there is a node t_i of T^i with $Y_i \subseteq W^i_{t_i}$. Let T be a tree obtained from $\bigcup_{i=1}^{l} T^i$ by adding a node t adjacent to each of $t_1, \ldots, t_l$. For $s \in T^i$, let $W_s = W^i_s$. Let $W_t = X_2$, and let $\mathcal{W} = (W_s \mid s \in T)$. Then $[T, \mathcal{W}]$ is a tree decomposition of G.*

Proof Each edge xy of G appears in some $X_1 \cup C_i$ so $\{x, y\} \subseteq W^i_s = W_s$ for some s in T^i. For each v in $C_i - X_2$, $\{s \mid v \text{ is in } W_s\} = \{s \mid v \text{ is in } W^i_s\}$ is a subtree of T because it is a subtree of T^i. For each v in $C_i \cap (X_2 - X_1)$, we have $\{s \mid v \text{ is in } W_s\} = \{s \mid v \text{ is in } W^i_s\} + t$ is a subtree of T because t_i is in the tree $\{s \mid v \in W^i_s\}$ and t is adjacent to t_i. For each v in X_1, $\{s \mid v \in W_s\}$ is the subtree of T consisting of t and for each i the subtree $\{s \mid v \text{ is in } W^i_s\}$ of T^i (which contains t_i). ∎

Remark We actually need only construct $[T^i, \mathcal{W}^i]$ for i with $C_i \not\subseteq X_2$ as for any $C_i \subseteq X_2$, $C_i \cup X_1$ is contained in W_t.

Without further ado, we present the algorithm.

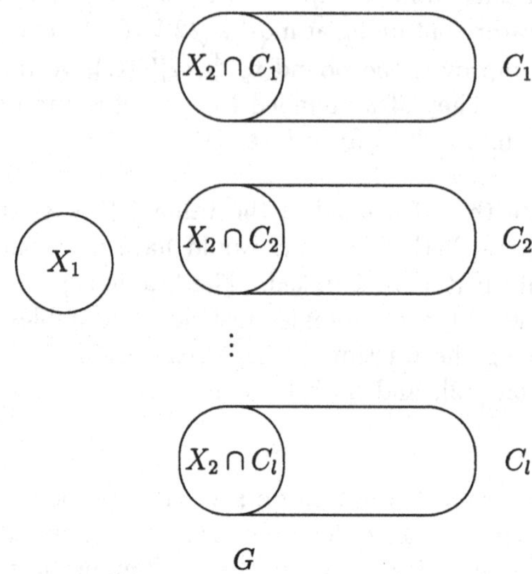
$X_2 \cap C_1$
C_1
X_1
$X_2 \cap C_2$
C_2
$X_2 \cap C_l$
C_l
G

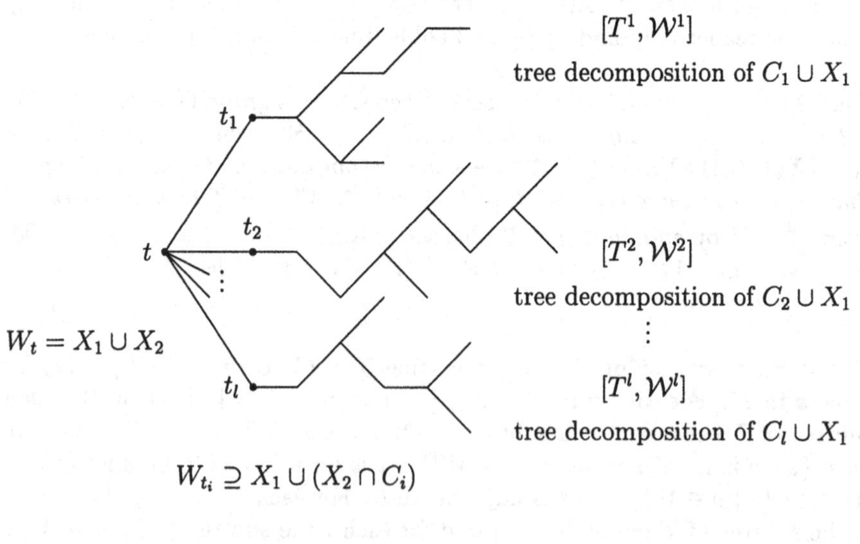
$[T^1, \mathcal{W}^1]$
tree decomposition of $C_1 \cup X_1$
t_1
t_2
t
$[T^2, \mathcal{W}^2]$
tree decomposition of $C_2 \cup X_1$
$W_t = X_1 \cup X_2$
$[T^l, \mathcal{W}^l]$
t_l
tree decomposition of $C_l \cup X_1$
$W_{t_i} \supseteq X_1 \cup (X_2 \cap C_i)$
$[T, \mathcal{W}]$

Figure 9:

Algorithm: k-Tree Finder

Input: A graph $G = (V, E)$ and a subset W of V with $|W| \leq 3k+1$,

Output: Either,

(i) a tree decomposition $[T, \mathcal{W}]$ of G of width at most $4k+1$ such that $W \subseteq W_t$ for some $t \in T$, or

(ii) a subset S of G such that G has no S-separator of order $k+1$ or less.

Description of Algorithm If G has $4k+2$ or fewer vertices then return a tree decomposition for G which uses a one node tree. Otherwise, arbitrarily add vertices to W until $|W| = 3k+1$. Next, attempt to find a W-separator X of order at most $k+1$ (using a procedure described below). If no such separator exists, return with output (ii) W. Otherwise, let $U_1, \ldots, U_l$ be the components of $G - X$ which are not contained in W. Let $G_i = X \cup U_i$ and let $W_i = (U_i \cap W) \cup X$. Note that for distinct i and j, $G_i \cap G_j \subseteq X$. Since X is a W-separator, $|W| \leq 3k+1$, and $|X| \leq k+1$, it follows that for each i, $|W_i| \leq 3k+1$. Thus, we can apply k-Tree Finder to (G_i, W_i) for each i. If it turns out that for some i there is an S_i such that G_i has no S_i-separator of order at most $k+1$ then clearly G also has no S_i-separator of order at most $k+1$ so we return (ii) S_i and stop. Otherwise, we find for each i, a tree decomposition $[T^i, \mathcal{W}^i]$ of G_i (we let W^i_t be the element of $\mathcal{W}^i$ corresponding to $t \in T^i$) and a distinguished node t_i of T^i such that $W_i \subseteq W^i_{t_i}$. In this case, we obtain a tree decomposition $[T, \mathcal{W}]$ of G by setting:

(a) $V(T) = \left(\bigcup_{i=1}^{l} V(T^i)\right) \cup \{t\}$,

(b) $E(T) = \left(\bigcup_{i=1}^{l} E(T^i)\right) \cup \left(\bigcup_{i=1}^{l} \{\overline{tt_i}\}\right)$ and

(c) $\mathcal{W} = \left(\bigcup_{i=1}^{l} \mathcal{W}^i\right) \cup (X \cup W)$ (that is, $W_t = X \cup W$ and $W_s = W^i_s$ for $s \in T^i$).

It follows from Fact 2.8 that $[T, \mathcal{W}]$ is indeed a tree decomposition. It is easy to verify that it has width at most $4k+1$ so we return (i) $([T, \mathcal{W}], t)$.

Now, the key step in this recursive procedure is finding a W-separator of order at most $k+1$. We then separate our problem into l smaller practically disjoint subproblems. As we shall see, this implies that we consider at most $\mathcal{O}(n)$ subproblems. In fact, we shall prove:

Claim 2.9 *Whilst applying k-Tree Finder to a graph G, we apply the algorithm to at most* $\max(1,\ 2|V(G)| - 6k - 3)$ *subproblems (including the original one). Hence, if we return a tree decomposition then the associated tree has at most* $\max(1,\ 2|V(G)| - 6k - 3)$ *nodes.*

We now consider the time necessary, during a particular application of the algorithm, to split the problem into subproblems and then combine the solutions to these subproblems. We shall show that we can do this in $\mathcal{O}(|E(G)|)$ time. Clearly, it then follows from Claim 2.9 that the algorithm runs in $\mathcal{O}(|V(G)||E(G)|)$ time. Note that the difficult part of splitting the problem into subproblems is determining if G has a small W-separator and finding one if it does. If this can be done in $\mathcal{O}(|E(G)|)$ time then so can the whole splitting process. Furthermore, building $[T, \mathcal{W}]$ from $\{[T^1, \mathcal{W}^1], \ldots, [T^l, \mathcal{W}^l]\}$ takes $\mathcal{O}(|T|)$ time which by Claim 2.9 is $\mathcal{O}(n)$. So, the fact that the algorithm runs in $\mathcal{O}(|E(G)||V(G)|)$ time follows from:

Claim 2.10 *Given a set S of at most $3k+1$ vertices in a graph G, we can determine if G has an S-separator with $k+1$ or fewer vertices in* $\mathcal{O}(k * 3^{3k+1} * |E(G)|)$ *time.*

To reduce the time complexity to $\mathcal{O}(n^2)$ we add a preprocessing step which counts the number of edges of G. If G has more than kn edges then by (1.10) its tree width is at least $k+1$. We could just return with this fact and stop. If we want to return a set S with no $(k+1)$-separator then we can run k-Tree Finder on a subgraph G' of G consisting of some $kn+1$ of its edges.

To complete our analysis of the algorithm it remains only to prove the two claims.

Proof of Claim 2.9 We prove the claim by induction on $|V(G)|$. If $|V(G)| \leq 4k+2$ then we make no recursive calls to the algorithm and the bound trivially holds. So, we assume that $|V(G)|$ is at least $4k+3$ and the bound holds for all graphs with fewer vertices than G. Now, if the algorithm fails to find a W-separator then again we make no recursive calls and the bound trivially holds. Otherwise, we find a W-separator X with $k+1$ vertices and create proper subgraphs $G_1, \ldots, G_l$ of G for some $l \geq 1$, and apply k-Tree Finder to each of these subgraphs in turn. Clearly, if n_i is the number of subproblems considered when we apply k-Tree Finder to (G_i, W_i) then we consider $1 + \sum_{i=1}^{l} n_i$ subproblems throughout our application of k-Tree Finder to (G, W). By the induction hypothesis, $n_i \leq \max(1,\ 2|V(G_i)| - 6k - 3)$. If $l = 1$ then since X is a W-separator and $|W| = 3k+1$, we have that $|V(G_1)| < |V(G)|$. The claim follows immediately by induction. So, we can assume that $l \geq 2$. Note that for any distinct i and j, $V(G_i) - W_i$ and $V(G_j) - W_j$ are disjoint non-empty subsets of $V(G) - W$. Furthermore, $n_i \leq \max(1,\ 2|V(G_i)| - 6k - 3) \leq 2|V(G_i) - W_i| - 1$. It follows that $1 + \sum_{i=1}^{l} n_i \leq 1 + 2|V(G) - W| - l \leq 2|V(G)| - 6k - 3$, as required. ∎

Proof of Claim 2.10 Consider a set S of at most $3k+1$ vertices of a graph G. First, we note that G has an S-separator with $k+1$ or fewer vertices if and only if there is a partition of $V(G)$ into A, B, and X such that the vertices of each component of $G-X$ are contained either in A or in B, $|X| \leq k+1$, $|A \cap S| \leq \frac{2}{3}|S-X|$, and $|B \cap S| \leq \frac{2}{3}|S-X|$. (Obviously, if such a partition exists then X is an S-separator. Conversely, given an S-separator X we can order the components of $G-X$ as $U_1, U_2, \ldots, U_l$ so that $|U_i \cap S| \geq |U_{i+1} \cap S|$ and then obtain our partition by setting $A = \bigcup_{i=1}^{j} U_j$ where j is the minimal integer for which $|\bigcup_{i=1}^{j+1} S \cap U_j| \geq \frac{2}{3}|S-X|$.) So, we will attempt to find such a partition rather than looking for an S-separator directly.

We do this by determining separately for each of the 3^{3k+1} choices of three disjoint sets S_A, S_B, S_C whose union is S whether or not there is such a partition of G with $S_A = A \cap S$, $S_B = B \cap S$, and $S_C = X \cap S$. If there is no choice of S_A, S_B, and S_C for which the desired corresponding partition of G exists then obviously G has no S-separator of order at most $k+1$. If for some choice of S_A, S_B, and S_C there is a corresponding partition of G then we will find an S-separator X with $X \cap S = S_C$ when considering this partition.

So, we turn our attention to a particular choice of S_A, S_B, and S_C. We first ensure that $|S_A| \leq 2|S_B|$, $|S_B| \leq 2|S_A|$ and $|S_C| \leq k+1$. We then check in $\mathcal{O}(k|E(G)|)$ time, using standard alternating path techniques, whether there are $k+2-|S_C|$ internally vertex disjoint paths from S_A to S_B in $G-S_C$. If there is no such set of paths then we find a set X of $k+1$ vertices with $X \cap S = S_C$ such that there are no S_A to S_B paths in $G-X$. Letting A be the union of those components of $G-X$ which contain an element of S_A and letting $B = V(G) - A - X$ yields a partition of G which shows that X is an S-separator. Conversely, if there is a partition of G corresponding to this choice of S_A, S_B and S_C then there cannot be $k+2-|S_C|$ internally vertex disjoint S_A to S_B paths in $G-S_C$ so we will find some S-separator. ∎

This completes our proof that k-Tree Finder runs in $\mathcal{O}(n^2)$ time. Now, Robertson and Seymour originally introduced k-Tree Finder in [36]. The implementation given in that paper has running time bounded by a polynomial whose exponent is a fast growing function of k. In [42] they presented the current version. Reed [32] by introducing further technical complications, speeded up the algorithm. His version runs in $\mathcal{O}(n\log(n))$ time. We remark that although these algorithms give bounded width tree decompositions of graphs of tree width at most k, they do not determine the tree width of the graph exactly.

In [36], Robertson and Seymour pointed out that since the class of graphs with tree width at most k is minor-closed (this is left as an exercise for the reader), they could test if a graph has tree width exactly k by testing for H-minor containment for each graph H in the obstruction set for this class. They noted that given the tree decomposition of width at most $4k+1$ returned by k-Tree Finder they could solve each such problem in linear time. They proved, by applying a special case of (1.14) that they had already resolved, that this obstruction set was finite. It follows that they can determine, for a fixed k, if a graph has tree width at most k by first applying k-Tree Finder and then applying a linear-time algorithm. Note since the obstruction set for graphs of tree width at most k is not known, they proved only the existence of an algorithm. They gave no way to construct one. In 1987, Arnborg, Corneil, and Proskurowski [4] independently constructed an $\mathcal{O}(n^{k+2})$ algorithm for determining, for fixed k, if G has tree width at most k and constructing a width k decomposition if one exists. They also showed that determining if G has tree width k is NP-complete if k is part of the input. In 1993, Arnborg et al. [5] gave a linear time algorithm to determine if the tree width of a graph is exactly k, for k fixed. However, this algorithm required much more than linear space (reading in unwritten memory is permitted) and if we actually wanted to find the tree decomposition $\mathcal{O}(|V(G)|^2)$ time would be required. Bodlaender and Kloks [11] developed a straightforward method for testing the tree width of a graph given a bounded width tree decomposition of it. This method runs in linear time, and actually constructs an optimal tree decomposition. Bodlaender [10], by combining this algorithm with some novel techniques developed a linear time algorithm to determine, for fixed k, if G has tree width k and to construct a tree decomposition of width at most k if one exists.

2.3 A duality theorem

Recall that a bramble is a set of connected subgraphs each two of which *touch*, that is intersect or contain distinct endpoints of some edge. A set H of vertices in G is a *hitting set* for a set of subgraphs of G, if each of the subgraphs intersects H. The *order* of a bramble $\mathcal{B}$ is the minimum of the orders of the hitting sets for $\mathcal{B}$. The *bramble number* of G, denoted $\mathrm{BN}(G)$, is the maximum of the orders of its brambles. In this section, we prove:

Theorem 2.11 ([48]) *The tree width of G is exactly one less than its bramble number.*

Now, as we saw in the introductory section on tree decompositions (see Fact 1.4), for any tree decomposition $[T, \mathcal{W}]$ of G and any bramble $\mathcal{B}$ in G, there is a $t \in T$ such that W_t is a hitting set for $\mathcal{B}$. Thus, $\mathrm{TW}(G) \geq \mathrm{BN}(G) - 1$; consider a maximum order bramble and a minimum width tree decomposition. It remains only to prove the converse. We actually prove a stronger statement which characterizes when a bramble can be extended to a bramble with order at least a given θ. To wit,

Lemma 2.12 *For any bramble $\mathcal{B}$ in G, either*

(i) there is a bramble $\mathcal{B}'$ of order θ with $\mathcal{B} \subseteq \mathcal{B}'$, or

(ii) *there is a tree decomposition $[T, \mathcal{W}]$ for G such that if t is a node of T with $|W_t| \geq \theta$ then t is a leaf and W_t is not a hitting set for $\mathcal{B}$.*

We remark that this lemma implies that $\text{TW}(G) \leq \text{BN}(G) - 1$; consider a bramble with no elements. We also observe that (1.4) implies that at most one of the two possibilities discussed in Lemma 2.12 can occur.

Proof of Lemma 2.12 Consider a graph G. We shall show that one of (i) or (ii) holds for each bramble $\mathcal{B}$ in G by assuming the contrary and deriving a contradiction. So, we choose a bramble $\mathcal{B}$ which satisfies neither (i) nor (ii) and which has fewer hitting sets with at most $\theta - 1$ vertices than any other such counterexample to the lemma. If there are no such small hitting sets for $\mathcal{B}$ then it is a bramble of order θ and thus (i) holds for $\mathcal{B}$, a contradiction. So, we let H be a minimal order hitting set for $\mathcal{B}$ and note that $|H| \leq \theta - 1$.

The idea of the proof is quite simple. Note first that $H \neq V(G)$ as otherwise there is a tree decomposition of G satisfying (ii) which uses a one node tree. So, we let $C_1, \ldots, C_l$ be the components of $G - H$. For each C_i, we shall find a tree decomposition $[T^i, \mathcal{W}^i]$ of $H \cup C_i$ satisfying:

(a) if t is a node of T^i with $|W^i_t| \geq \theta$ then t is a leaf and W^i_t is not a hitting set for $\mathcal{B}$, and (*)

(b) there is a leaf t_i of T^i with $W^i_{t_i} = H$.

We then form a tree decomposition $[T, \mathcal{W}]$ of G by taking a copy of each of these tree decompositions, adding a node t adjacent to $\{t_1, \ldots, t_l\}$, and setting $W_t = H$, as discussed in Fact 2.8. Now, each $[T^i, \mathcal{W}^i]$ satisfies (*)(a) so $[T, \mathcal{W}]$ shows that condition (ii) in the lemma holds for $\mathcal{B}$, a contradiction.

Thus to prove Lemma 2.12 and the theorem we need only show that for each component C_i of H, there is a tree decomposition $[T^i, \mathcal{W}^i]$ of $H \cup C_i$ satisfying (*). In doing so, we consider two possibilities. The first is that C_i fails to touch some element B of $\mathcal{B}$. In this case, let T^i be a tree with one edge st, let $W_s = C_i \cup H - B$, let $W_t = H$ and let $\mathcal{W}^i = \{W_s, W_t\}$. Since C_i does not touch B, $[T^i, \mathcal{W}^i]$ is a tree decomposition of $H \cup C_i$. Clearly, it satisfies (*).

The second possibility is that C_i touches every element of $\mathcal{B}$ and thus $\mathcal{B}+C_i$ is a bramble. In this case we remark that, since H is a hitting set for $\mathcal{B}$ but not $\mathcal{B}+C_i$, by our choice of $\mathcal{B}$, one of (i) or (ii) holds for $\mathcal{B}+C_i$. If (i) holds for $\mathcal{B}+C_i$ then it also holds for $\mathcal{B}$, a contradiction. Thus, (ii) holds for $\mathcal{B}+C_i$, so we consider a tree decomposition $[T', \mathcal{W}']$ of G such that for every node t of T' with $|W'_t| \geq \theta$, t is a leaf and W'_t is not a hitting set for $\mathcal{B}+C_i$. Now, we can assume that for some leaf t in T', W'_t fails to intersect C_i but is a hitting set for $\mathcal{B}$, as otherwise $[T', \mathcal{W}']$ would show that $\mathcal{B}$ satisfies (ii), a contradiction. We will show that we can transform $[T', \mathcal{W}']$ into a tree decomposition of $H \cup C_i$ satisfying (*). To do so, we need the following lemmas which we prove in a moment.

Lemma 2.13 *Let $\mathcal{L}$ be a bramble of order k in a graph F. Let H_1 and H_2 be hitting sets for $\mathcal{L}$. Then there are k vertex disjoint paths between H_1 and H_2 in F.*

Lemma 2.14 *Let $[T, \mathcal{W}']$ be a tree decomposition of a graph G. Let X be a cutset of G and let C be a component of $G - X$. Suppose there is some r in T such that W'_r is contained in $G - C$ and there are $|X|$ vertex disjoint paths of G between X and W'_r. Then there is a tree decomposition $[T, \mathcal{W}]$ of $X + C$ such that:*

(i) for each s in T, $|W_s| \leq |W'_s|$, and

(ii) for each leaf s of T except r, $W_s \subseteq W'_s$.

Now, $\mathcal{B}$ clearly has order $|H|$ and both H and W'_t are hitting sets for $\mathcal{B}$. Thus, Lemma 2.13 implies that there are $|H|$ vertex disjoint paths between H and W'_t. So, we can apply Lemma 2.14 to the tree decomposition $[T', \mathcal{W}']$ with $X = H$, $C = C_i$, and $r = t$. This yields the desired tree decomposition $[T, \mathcal{W}]$ of $H \cup C_i$ satisfying $(*)$ (to see this note that for any leaf s of T' with $|W'_s| \geq \theta$, either W'_s is not a hitting set for $\mathcal{B}$, or W'_s does not intersect C_i in which case W_s is contained in H and hence has at most $\theta - 1$ elements). Thus, to complete the proof of Lemma 2.12 and our theorem we need only prove Lemmas 2.13 and 2.14.

Proof of Lemma 2.13: Let F, $\mathcal{L}$, k, H_1, and H_2 be as in the statement of the lemma. Note that every element of $\mathcal{L}$ intersects both H_1 and H_2 and therefore contains a path from H_1 to H_2. Thus, if X is a set of vertices of G such that there is no path between $H_1 - X$ and $H_2 - X$ in $G - X$ then X is a hitting set for $\mathcal{L}$. So, any such X contains at least k vertices and the lemma follows by Menger's theorem. ∎

Proof of Lemma 2.14 Let G, $[T, \mathcal{W}']$, X, C, t, and W'_r be as in the statement of the lemma. Let $k = |X|$ and let $P_1, \ldots, P_k$ be k vertex disjoint paths from X to W'_r. Let x_i be the element of X contained in P_i. Note that since X is a cutset, C is a component of $G - X$, and none of W'_r is in C we know that no P_i intersects C. Now, we know that $S_{P_i} = \bigcup\{S_v \mid v \in V(P_i)\}$ is a subtree of T containing S_{x_i} and r. Thus there is a path Q_i of S_{P_i} with one endpoint r and the other in S_{x_i}. Since Q_i and S_{x_i} are intersecting subtrees of T, $S_{x_i} \cup Q_i$ is a subtree of T. Now, we obtain our tree decomposition $[T, \mathcal{W}]$ of $X + C$ by setting, for each s in T, $W_s = (W'_s \cap (X + C)) \cup \{x_i \mid s \text{ is on } Q_i\}$. To see that we do indeed obtain a tree decomposition note first that for each vertex v of C, $\{s \mid v \text{ is in } W_s\} = S_v([T, \mathcal{W}'])$ and so is a subtree of T. Furthermore, for v in X we have that $\{s \mid v \text{ is in } W_s\} = S_v([T, \mathcal{W}']) \cup Q_i$ and hence is a subtree of T. Finally, for any edge $e = xy$ of $X + C$, there is some s such that W'_s contains $\{x, y\}$ and therefore so does W_s.

We note that for each leaf s of T apart from r if Q_i contains s then s is the endpoint of Q_i in $S_{x_i}([T,\mathcal{W}'])$ and hence W'_s contains x_i. Thus, $W_s \subseteq W'_s$. Furthermore, if s is a node of T then

$$W_s - X \subseteq W'_s \cap C \subseteq W'_s - P_1 - P_2 - \cdots - P_l$$

and

$$\begin{aligned} |\{x_i \mid x_i \in W_s \cap X\}| &= |\{i \mid P_i \text{ intersects } W'_s\}| \\ &\leq |\{y \mid y \text{ is on some } P_i,\, y \in W'_s\}|. \end{aligned}$$

Thus $|W_s| \leq |W'_s|$ as claimed. This shows that $[T,\mathcal{W}]$ is a tree decomposition with the properties we desire. ■

That completes the proof of Lemma 2.12. ■

3 Untangling tangles

We have two objectives in this section. One is to gain a deeper understanding of tangles, in particular which of their properties make them better behaved than other brambles. First however we will prove (1.13) which we restate below for the reader's convenience.

The Canonical Tree Decomposition Theorem *Given a graph G, we can construct a tree decomposition T which has the following properties.*

(1) For each maximal tangle $\mathcal{T}$ of G, there is exactly one node $t(\mathcal{T})$ of T whose vertices form a hitting set for $\mathcal{T}$.

(2) If $\mathcal{T}_1$ and $\mathcal{T}_2$ are indistinguishable maximal tangles (and hence have the same order) then $t(\mathcal{T}_1) = t(\mathcal{T}_2)$.

(3) If $\mathcal{T}_1$ and $\mathcal{T}_2$ are distinguishable maximal tangles then $t(\mathcal{T}_1) \neq t(\mathcal{T}_2)$ and there is an arc st on the unique $t(\mathcal{T}_1)$ to $t(\mathcal{T}_2)$ path of T such that $V(X_s) \cap V(X_t)$ is a $(\mathcal{T}_1,\mathcal{T}_2)$-distinguisher of minimum order.

(4) For every node t of the tree decomposition, there is a maximal tangle $\mathcal{T}$ such that $t(\mathcal{T})= t$.

Proof Our first step in constructing this tree decomposition is to define an equivalence relation $\sim$ on the set of maximal tangles of G such that $\mathcal{T}_1 \sim \mathcal{T}_2$ precisely if $\mathcal{T}_1$ and $\mathcal{T}_2$ are indistinguishable (and hence by the definition of maximal, $\text{ord}(\mathcal{T}_1) = \text{ord}(\mathcal{T}_2)$). To simplify matters, we want to choose a representative tangle from each equivalence class of $\sim$. To this end, note that for any tangle $\mathcal{T}$, $\beta_{f(\mathcal{T})}$ is a tangle of the same order as $\mathcal{T}$ which is indistinguishable from $\mathcal{T}$; in fact, $\beta_{f(\mathcal{T})} = \{Y \mid Y \text{ is a component of } G-X \text{ for some } X \text{ with fewer than } \text{ord}(\mathcal{T}) \text{ vertices and } Y \text{ contains an element of } \mathcal{T}\}$. So, we define a tangle $\mathcal{T}$ to be *canonical* if each element Y of $\mathcal{T}$ is a component of $G - X$ for some X with $|X| < \text{ord}(\mathcal{T})$. Obviously, if $\mathcal{T}$ is canonical then $\beta_{f(\mathcal{T})} = \mathcal{T}$. Furthermore, if $\mathcal{T}_1 \sim \mathcal{T}_2$ then $\beta_{f(\mathcal{T}_1)} = \beta_{f(\mathcal{T}_2)}$. Thus, we see that for any tangle $\mathcal{T}$, $\beta_{f(\mathcal{T})}$ is the unique canonical tangle in the equivalence class of $\sim$ containing $\mathcal{T}$. We let $\mathcal{T}_G$ be the set of canonical maximal tangles. By the above remarks, $\mathcal{T}_G$ contains exactly one representative from each equivalence class of $\sim$.

We are going to choose for each pair of tangles $\mathcal{T}_1$ and $\mathcal{T}_2$ in $\mathcal{T}_G$ a minimal $(\mathcal{T}_1, \mathcal{T}_2)$-distinguisher and then use these distinguishers to construct the desired tree decomposition. To do so, we recall that $V(G) = \{v_1, \ldots, v_n\}$ and $E(G) = \{e_1, \ldots, e_m\}$ are indexed sets.

If $\mathcal{T}_1$ and $\mathcal{T}_2$ are two tangles of G and the ordered pair (A, B) is a separation of G whose order is less than $\min(\text{ord}(\mathcal{T}_1), \text{ord}(\mathcal{T}_2))$ and one of the two graphs $A - X$ or $B - X$ completely contains an element of $\mathcal{T}_1$ whilst the other completely contains an element of $\mathcal{T}_2$ then we say that (A, B) is a $(\mathcal{T}_1, \mathcal{T}_2)$-*separation.* Note that if (A, B) is a $(\mathcal{T}_1, \mathcal{T}_2)$-separation then $V(A) \cap V(B)$ is a $(\mathcal{T}_1, \mathcal{T}_2)$-distinguisher. Conversely, it is easy to see that if X is a $(\mathcal{T}_1, \mathcal{T}_2)$-distinguisher then there is at least one $(\mathcal{T}_1, \mathcal{T}_2)$-separation (A, B) with $X = V(A) \cap V(B)$, we simply need to assign each component of $G - X$ to one of A or B so that the two components which completely contain an element of one of the two tangles are assigned to different sides of the separation.

The *canonical* $(\mathcal{T}_1, \mathcal{T}_2)$*-separation* for two distinguishable $\mathcal{T}_1$ and $\mathcal{T}_2$ is defined to be the unique $(\mathcal{T}_1, \mathcal{T}_2)$-separation (A, B) which:

(i) has minimum order,

(ii) subject to (i), lexicographically minimizes $V(A) \cap V(B)$ (where S is lexicographically less than R if the lowest indexed element of $S \cup R$ which is not in $S \cap R$ is in S),

(iii) subject to (ii), lexicographically minimizes $V(A)$,

(iv) subject to (iii), maximizes $|E(A)|$ (this is equivalent to requiring that all of the edges of G with both endpoints in $V(A) \cap V(B)$ are in A.)

We let $\mathcal{S}_G$ be the set of precisely those separations which are canonical $(\mathcal{T}_1, \mathcal{T}_2)$-separations for some pair $(\mathcal{T}_1, \mathcal{T}_2)$ of tangles in $\mathcal{T}_G$. We shall show that $\mathcal{S}_G$ is a set of laminar separations.

As noted in the introduction, this implies that there is a unique tree decompostion $[T_{\mathcal{S}_G}, X_{\mathcal{S}_G}]$ of G such that each separation S in $\mathcal{S}_G$ is made by precisely one arc of $T_{\mathcal{S}_G}$ and for each arc a of $T_{\mathcal{S}_G}$ one of the separations made by a is in $\mathcal{S}_G$. To complete the construction we have been describing, we simply need to show that the tree decomposition $[T_{\mathcal{S}_G}, X_{\mathcal{S}_G}]$ satisfies properties (1)–(4). Forthwith the details.

Lemma 3.1 *For every graph G, the set $\mathcal{S}_G$ is laminar.*

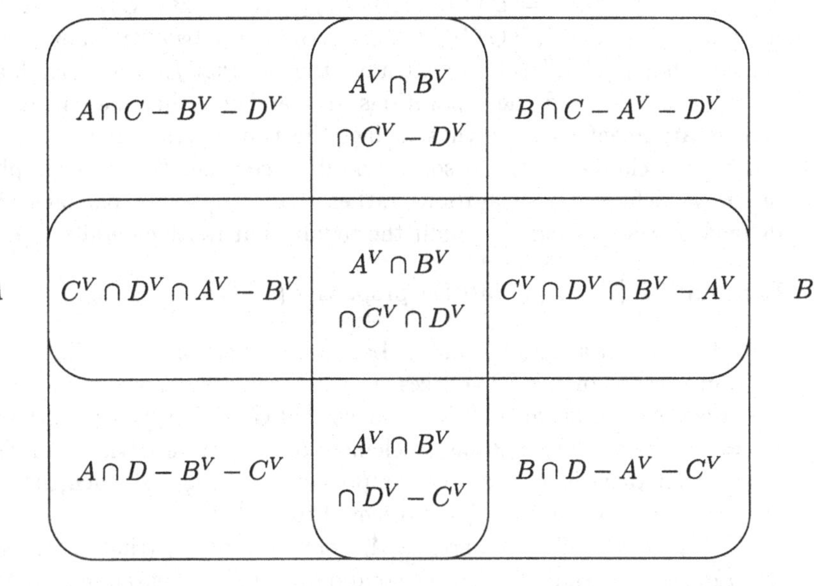

Figure 10: Here F^V denotes $V(F)$

Proof Let (A, B) be a (T_i, T_j)-separation of minimum order and let (C, D) be a (T_k, T_l)-separation of minimum order (see Figure 10). Choose the labels so that (where (i) the letters in the next equation and in the right hand sides of the following three equations correspond to vertex sets and (ii) $V(F \cap H) = V(F) \cap V(H)$):

$$|(C \cap D) \cap (A - B)| \leq \qquad (*)$$
$$\min\{|(C \cap D) \cap (B - A)|,\ |(A \cap B) \cap (C - D)|,\ |(A \cap B) \cap (D - C)|\}$$

Now,

$$\text{ord}((A, B)) = |(A \cap B) \cap (D - C)| + |(A \cap B) \cap (C - D)| + |A \cap B \cap C \cap D|,$$

$$\text{ord}((A \cap C, B \cup D)) = |(C \cap D) \cap (A - B)| + |(A \cap B) \cap (C - D)| + |A \cap B \cap C \cap D|,$$

$$\text{ord}((A \cap D, B \cup C)) = |(A \cap B) \cap (D - C)| + |(C \cap D) \cap (A - B)| + |A \cap B \cap C \cap D|.$$

Thus, if $(*)$ holds with strict inequality then both $(A \cap C, B \cup D)$ and $(A \cap D, B \cup C)$ have order smaller than (A, B) and so cannot be (T_i, T_j)-separations. By swapping the tangle indices if necessary, we can ensure that $f_{\mathcal{T}_j}((A,B)) = B$. Thus, the definition of a bias tells us that $f_{\mathcal{T}_j}((A \cap C, B \cup D)) = B \cup D$. Hence, $f_{\mathcal{T}_i}((A \cap C, B \cup D)) = B \cup D$. Similarly, $f_{\mathcal{T}_i}((A \cap D, B \cup C)) = B \cup C$. But these two facts along with the fact that $f_{\mathcal{T}_i}((A,B)) = A$ contradict the fact that $f_{\mathcal{T}_i}$ is a strong bias.

If $(*)$ is tight and the separations are canonical then we can use a similar argument to show that in fact the two separations are indeed laminar as claimed. To do so, we need to consider the lexicographic order we defined on separations rather than simply the partial order defined by their orders. We omit the tedious but routine details. ■

Lemma 3.2 *$[T_{\mathcal{S}_G}, X_{\mathcal{S}_G}]$ satisfies properties (1)–(4).*

Proof Consider a tangle $\mathcal{T}_i$ in $\mathcal{T}_G$. Let A_i be the set of arcs of $T_{\mathcal{S}_G}$ which contain fewer than $\text{ord}(\mathcal{T}_i)$ vertices. For each arc a in A_i, we direct $a = st$ towards the component S of $T_{\mathcal{S}_G} - a$ such that $G_S - V(X_s) \cap V(X_t)$ (recall $G_S = \bigcup\{X_t \mid t \in S\}$) contains an element of $\mathcal{T}_i$. Now, a trivial induction proves that there is at least one component U_i of $T_{\mathcal{S}_G} - A_i$ such that all the arcs of A_i incident to U_i are oriented towards U_i.

We claim that all of the arcs of A_i point towards U_i (that is for each arc a in A_i, the head of a is in the component of $T - a$ containing U_i) and hence U_i is the only component of $T_{\mathcal{S}_G} - A_i$ with this property. Otherwise, there must exist two arcs $a_1 = s_1t_1$ and $a_2 = s_2t_2$ of A_i such that the subtree S_1 containing the head of a_1 and the subtree S_2 containing the head of a_2 are disjoint. Now, both $G_{S_1} - (V(X_{s_1}) \cap V(X_{t_1}))$ and $G_{S_2} - (V(X_{s_2}) \cap V(X_{t_2}))$ contain elements of $\mathcal{T}_i$, but these two subgraphs do not touch, a contradiction which proves the claim.

Next, we claim that U_i contains exactly one node. Otherwise, there is an arc a of U_i. By definition, one of the separations made by a is the canonical $(\mathcal{T}_j, \mathcal{T}_k)$-separator for some two tangles in $\mathcal{S}_G$. Furthermore, this separation has order at least the order of $\mathcal{T}_i$ as a is not in A_i. So, we have:

none of the separations made by the arcs of A_i are $(\mathcal{T}_j, \mathcal{T}_k)$-separations, but $A_i \subseteq A_j \cap A_k$. (!)

Let a' be the arc of A_i which makes the canonical $(\mathcal{T}_i, \mathcal{T}_j)$-separation. By (!), if we consider our orientations of A_i, A_j, and A_k, we see a' has the same orientation as an arc of A_j and as an arc of A_k. However, the orientation of a differs in these two orientations. So, for one of these two sets we obtain a contradiction of the fact that for every l, all the arcs of A_l point to U_l.

So, $|U_i| = 1$ and we let t_i be the node of U_i. Now, our orientation of the edges incident to t_i shows that no other node of $T_{\mathcal{S}_G}$ can contain a hitting set for $\mathcal{T}_i$, so t_i must. Thus for any $T_i \in \mathcal{T}_G$, (1) holds with $t(\mathcal{T}_i) = t_i$. Furthermore, the orientation of the edges incident to t_i also shows that for any $T' \sim T_i$, t_i is the unique node of T such that $V(X_{t_i})$ is a hitting set for T'. Thus (1) holds, as does (2).

That $[T_{\mathcal{S}_G}, X_{\mathcal{S}_G}]$ satisfies (3) follows immediately. To see this note first that, by our construction, for any two tangles $\mathcal{T}_i$ and $\mathcal{T}_j$ in $\mathcal{S}_G$ there is some arc st of $T_{\mathcal{S}_G}$ which corresponds to the canonical $(\mathcal{T}_i, \mathcal{T}_j)$-separation. Thus, $V(X_s) \cap V(X_t)$ is a minimal $(\mathcal{T}_i, \mathcal{T}_j)$-distinguisher. Furthermore, since the subtree of $T_{\mathcal{S}_G} - a$ "containing" an element of $\mathcal{T}_i$ is distinct from the subtree of $T_{\mathcal{S}_G} - a$ "containing" an element of $\mathcal{T}_j$, it follows from considering our orientations of A_1 and A_2 that st must be on the $t(\mathcal{T}_i)$-$t(\mathcal{T}_j)$ path of G.

Proving that $[T_{\mathcal{S}_G}, X_{\mathcal{S}_G}]$ satisfies (4) is slightly more difficult. We remark first that by our orientation of the A_i every arc on the $t(\mathcal{T}_i)$-$t(\mathcal{T}_j)$ path of $T_{\mathcal{S}_G}$ which contains fewer than $\min(\text{ord}(\mathcal{T}_i), \text{ord}(\mathcal{T}_j))$ vertices corresponds to a $(\mathcal{T}_i, \mathcal{T}_j)$-separation. Furthermore, by the definition of canonical separation, it is easy to see that for each arc a of $T_{\mathcal{S}_G}$ precisely one of the separations made by a is a canonical separation.

Now, choose a node t of $T_{\mathcal{S}_G}$, and then choose, from amongst all the arcs incident to t, an arc st and canonical separation (A, B) made by st which:

(i) has maximum order,

(ii) subject to (i), lexicographically maximizes $V(A) \cap V(B)$

(iii) subject to (ii), lexicographically maximizes $V(A)$.

Now, (A, B) is a canonical $(\mathcal{T}_i, \mathcal{T}_j)$-separation for some pair of tangles $(\mathcal{T}_i, \mathcal{T}_j)$ in $\mathcal{T}_G$. If t is neither $t(\mathcal{T}_i)$ nor $t(\mathcal{T}_j)$ then there is another arc a of the $t(\mathcal{T}_i)$-$t(\mathcal{T}_j)$ path of $T_{\mathcal{S}_G}$ incident to t. By our remark, and condition (i) in the choice of st, a also corresponds to a $(\mathcal{T}_i, \mathcal{T}_j)$-separation. But now, our choice of st and the existence of such an a ensure that st does not correspond to the canonical $(\mathcal{T}_i, \mathcal{T}_j)$-separation, a contradiction. Thus, either $t = t(\mathcal{T}_i)$ or $t = t(\mathcal{T}_j)$, proving (4). This completes the proof of our lemma and hence the theorem. ∎

This completes the proof of the Canonical Tree Decomposition Theorem. ∎

Note that $[T_{\mathcal{S}_G}, X_{\mathcal{S}_G}]$ is a nice tree decomposition because each of its nodes t is a hitting set for a tangle for which none of the neighbours of t are a hitting set. It follows that a graph G has at most $|V(G)|$ indistinguishable tangles.

This is in stark contrast with the situation for arbitrary brambles. For, as we show in a moment, there are graphs which have exponentially many indistinguishable maximal brambles. In fact, as we show now, even in simple graphs like cycles the number of indistinguishable maximal brambles can be a cubic function of the number of vertices.

To see this, consider a cycle C. Let $\{P_1, P_2, P_3\}$ be a set of three vertex disjoint paths which partition the vertex set of C. Obviously, $\{P_1, P_2, P_3\}$ is a bramble, and since every cycle has tree width two, it is a maximal bramble. Furthermore, we claim that if $\{Q_1, Q_2, Q_3\}$ is another such set of paths then $\{P_1, P_2, P_3\}$ and $\{Q_1, Q_2, Q_3\}$ are distinguishable. To prove this claim, we note first that we can relabel $\{P_1, P_2, P_3\}$ and $\{Q_1, Q_2, Q_3\}$ so that P_1 does not contain any of Q_1, Q_2, or Q_3. Next, we let x_2 be the endpoint of P_2 adjacent to a vertex of P_1, and x_3 be the endpoint of P_3 adjacent to a vertex of P_1. Now, it is easy to see that $\{x_2, x_3\}$ is a distinguisher for these two brambles. This shows that a cycle on n vertices has $\binom{n}{3}$ indistinguishable maximal brambles.

Consider now, a $(10n+2)\times n$ cylinder. That is, a planar graph G consisting of n cycles, $C_1, \ldots, C_n$, each with $10n+2$ vertices and $10n+2$ vertex disjoint paths $P_1, \ldots, P_{10n+2}$ each of which has n vertices, one on each cycle (see Figure 11). We shall define a number of preferences of order $2n+1$ in this graph. To begin we define one, f. To this end let Y be any set of at most $2n$ vertices. There is at least one component of $G-Y$ which completely contains a P_i. If there is only one such component then we let $f(Y)$ be this component. In this case we say Y is *tractable*. It is easy to verify that if Y_1 and Y_2 are both tractable then $f(Y_1)$ and $f(Y_2)$ touch.

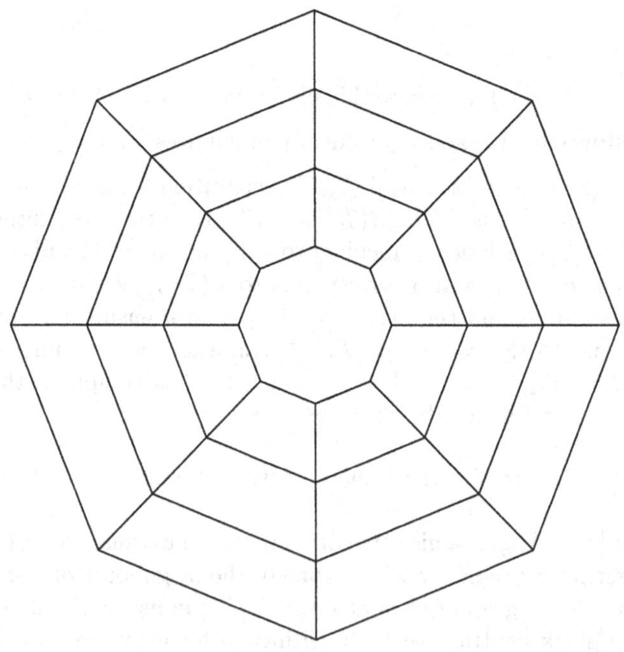

Figure 11: An 8×4 cylinder

If there are two components of $G-Y$ containing a P_i then Y must have order $2n$ and intersect each cycle in precisely two vertices. In this case, if there is a cycle C_i such that one component of $C_i - Y$ contains more than $5n+1$ vertices of C_i then we choose the minimal i for which this holds and set $f(Y)$ to be the component of $G-Y$ containing most of C_i. Note that for any $j \leq i$, both components of $C_j - Y$ must contain $5n$ vertices of C_j. Thus, any two of the preferred components selected in this part of the procedure touch. Also if Y is not tractable and we have chosen $f(Y)$ then $f(Y)$ clearly touches $f(Z)$ for every tractable Z.

Let $\mathcal{Y}$ be the family of those sets Y for which we have not yet specified $f(Y)$. Clearly, if Y is in $\mathcal{Y}$ then, for every j, $C_j - Y$ consist of two components each of size $5n$. This fact allows us to show that both components of $G-Y$ meet all the $f(Y)$ selected so far as well as both components of $G-Y'$ for any other Y' in $\mathcal{Y}$. It follows that we can choose our preference f by arbitrarily picking one of the two components of $G-Y$ to be $f(Y)$ for each $Y \in \mathcal{Y}$. Furthermore, any two such preferences yield distinguishable brambles; the set Y on which we made a different choice is a distinguisher. It is not hard to verify that there are more than 3^n elements in $\mathcal{Y}$. Hence we have created 2^{3^n} distinguishable brambles (which are, incidentally, maximal).

The reason that we cannot mimic the proof of (1.13) to prove that there are at most $|V(G)|$ indistinguishable brambles in a graph G is that the analogue of (3.1) does not hold for the distinguishers of brambles. That is, the advantage that tangles have over brambles is the laminarity of the cutsets which separate them. We now present a lemma which proves the laminarity of the cutsets separating certain sets of vertices from a tangle. To begin, we need some definitions.

Definition Let $\mathcal{T}$ be a bramble in a graph G of order k. A set X of vertices of G is $\mathcal{T}$-linked if $|X| \leq k-1$ and for every hitting set H for $\mathcal{T}$ there are $|X|$ vertex disjoint paths from X to H.

Remark If $\mathcal{T}$ is a bramble and Y is a set of fewer than $\text{ord}(\mathcal{T})$ vertices then $Y \cup f_{\mathcal{T}}(Y)$ is a hitting set for $\mathcal{T}$. So if X is a $\mathcal{T}$-linked set then:

for every set Y of vertices with $|Y| < |X|$, we have $f_{\mathcal{T}}(Y) \cap X \neq \emptyset$. ($)

Actually, a set X of at most $\text{ord}(\mathcal{T}) - 1$ vertices is $\mathcal{T}$-linked precisely if ($) holds for X. To see this note that if Y separates X from some hitting set H for $\mathcal{T}$ and $|Y| < |X| < \text{ord}(\mathcal{T})$ then Y must also separate X from $f_{\mathcal{T}}(Y)$.

Definition We say that Y *separates* X from $\mathcal{T}$ if $|Y| < \text{ord}(\mathcal{T})$ and X is disjoint from $f_{\mathcal{T}}(Y)$. In this case, Y is a *cutter* for X (with respect to $\mathcal{T}$).

Lemma 3.3 (The Closest Cutter Lemma) *Let $\mathcal{T}$ be a tangle of order at least $h+1$ in a graph G. Let Z be a $\mathcal{T}$-linked set in G with h vertices. Let $\mathcal{F}$ be the family of those subsets of h vertices of G separating Z from $\mathcal{T}$ (note* *that $\mathcal{F}$ is non-empty, as it contains Z). Then there is some element X^* of $\mathcal{F}$ such that for every other X in $\mathcal{F}$: $f_{\mathcal{T}}(X^*) \subseteq f_{\mathcal{T}}(X)$. Furthermore, X^* is a $\mathcal{T}$-linked set, and there are h vertex disjoint paths from Z to X^*.*

Proof Consider h, $\mathcal{T}$, G, Z, and $\mathcal{F}$ as in the statement of the lemma. Assume, for a contradiction that there exist distinct X_1 and X_2 in $\mathcal{F}$ such that there is no X in $\mathcal{F}$ with $f_{\mathcal{T}}(X) \subset f_{\mathcal{T}}(X_1)$, and no X in $\mathcal{F}$ with $f_{\mathcal{T}}(X) \subset f_{\mathcal{T}}(X_2)$. Then, let $\beta_1 = f_{\mathcal{T}}(X_1)$ and let $\beta_2 = f_{\mathcal{T}}(X_2)$. Let $X_3 = (X_1 \cap \beta_2) \cup (X_1 \cap X_2) \cup (X_2 \cap \beta_1)$ and let $X_4 = (X_1 - \beta_2) \cup (X_2 - \beta_1)$. Obviously, $|X_3| + |X_4| = |X_1| + |X_2| = 2h$. Furthermore, X_4 separates $B_1 \cup B_2$ from Z and hence separates Z from $\mathcal{T}$. So, since Z is $\mathcal{T}$-linked, $|X_4| \geq h$. Hence, $|X_3| \leq h$ and $\beta_3 = f_{\mathcal{T}}(X_3)$ exists. Now, note that X_3 also separates Z from $\mathcal{T}$. For, we know that either $\beta_1 \cap \beta_2 \cap \beta_3$ is non-empty or there is an edge e with an endpoint in each of β_1, β_2, and β_3. However, in the latter case, by the definition of X_3, the endpoint of e in β_3 must be in $\beta_1 \cap \beta_2 \cap \beta_3$. So, $\beta_1 \cap \beta_2 \cap \beta_3$ is non-empty and, again by the definition of X_3, we obtain that $\beta_3 \subseteq \beta_1 \cap \beta_2$. So, indeed β_3 does not intersect Z.

But now, X_3 contradicts either the fact that Z is $\mathcal{T}$-linked or our choice of X_1 and X_2. This contradiction proves the existence of the desired X^*. Now, assume that X^* is not $\mathcal{T}$-linked. In this case, there is some T in $\mathcal{T}$ and a set Y of at most $h-1$ vertices separating X^* from T. Note that T is disjoint from X^* and thus must be contained in $f_{\mathcal{T}}(X^*)$. So there is no path in $G - X^*$ from Z to T and hence Y also separates Z from T. But this contradicts the fact that Z is $\mathcal{T}$-linked. So, we see that X^* is $\mathcal{T}$-linked. It follows that there are h vertex disjoint paths from Z to X^* as both these sets are $\mathcal{T}$-linked. ∎

We note that just as with the analogue of (3.1), the closest cutter lemma fails to hold for the brambles in a cycle. To see this consider a cycle of length four with vertices $\{v_1, v_2, v_3, v_4\}$, appearing in that cyclic order around the cycle. Now, $\beta = \{\{v_1\}, \{v_2\}, \{v_3, v_4\}\}$ is a bramble in this graph. Furthermore, $X = \{v_3, v_4\}$ is a β-linked set. However, both $\{v_4, v_2\}$ and $\{v_1, v_3\}$ are cutters for X. This shows that the closest cutter lemma does not hold for this bramble.

We now prove two lemmas about $\mathcal{T}$-linked sets. The first requires us to define a new connectivity invariant tied to the bramble number.

Definition We say a set X of vertices is *well-linked* if for every pair A and B of subsets of X with $|A| = |B|$ there are $|A|$ vertex disjoint paths between A and B. We define the well-linkedness of G, denoted $\mathrm{WL}(G)$, to be the size of the largest well-linked set in G.

Remark Note that we do not insist that A and B are disjoint thus if we required the paths to have no internal vertices in X we would have an equivalent definition.

Lemma 3.4 $\mathrm{BN}(G) \leq \mathrm{WL}(G) \leq 4\,\mathrm{BN}(G)$.

Proof To prove the first inequality, we show that any minimum hitting set H for a bramble β is a well-linked set. Otherwise, by Menger's Theorem, there are two equal sized subsets A and B of H and a set Y of fewer than $|A|$ vertices separating A and B. By symmetry, we can assume that $A \cap f_\beta(Y)$ is empty. Now, we know that $Y \cup (H \cap f_\beta(Y))$ is a hitting set for β contained in $(H - A) \cup Y$. However, this contradicts the minimality of H.

To prove the second inequality, we show that any well-linked set X is (k_X)-linked where $k_X = \lceil |X|/4 \rceil$. Since every k-linked set defines a bramble of order k, the result follows. So, assume for a contradiction that there is a set W of fewer than k_X vertices such that no component of $G - W$ contains more than half the vertices of X. This implies that we can partition $G - W$ into two sets C and D each of which contains more than a quarter of the vertices of X. (To see this, let U be the component of $G - W$ which has maximum intersection with X. Set $C = U$ and recursively add more components to C until C contains more than a quarter of the vertices of X.) Now, choose a subset A of k_X vertices of $X \cap C$ and a subset B of $D \cap X$ of the same size. The sets A, B, and W contradict our assumption that X is well-linked. ∎

Lemma 3.5 *Let β be a bramble. Every β-linked set is well-linked.*

Proof We mimic the first half of the proof of the last lemma to show that if X is not well-linked and has cardinality less than $\text{ord}(\beta)$ then there is a subset A of X and a set Y of smaller cardinality than A such that A is disjoint from $f_\beta(Y)$. Then $Y \cup (X \cap f_\beta(Y))$ separates X from the hitting set $Y \cup f_\beta(Y)$. Thus X is not β-linked. ∎

Lemma 3.6 *If $\mathcal{T}$ is a tangle then every $\mathcal{T}$-linked set extends to a $\mathcal{T}$-linked set of order* $\text{ord}(\mathcal{T}) - 1$.

Proof Let X be a $\mathcal{T}$-linked set of order at most $\text{ord}(\mathcal{T}) - 2$. Let X^* be the closest cutter for X. Let y be any element of $f_\mathcal{T}(X^*)$. We claim that $X+y$ is a $\mathcal{T}$-linked set. Otherwise, there would be a set Z of fewer than $|X+y|$ vertices separating $X+y$ from $\mathcal{T}$. Since X is $\mathcal{T}$-linked, we know $|Z| = |X|$. Since X^* is the closest cutter for X, we know $f_\mathcal{T}(X^*) \subseteq f_\mathcal{T}(Z)$. Thus $y \in f_\mathcal{T}(Z)$ contradicting the fact that Z separates $X+y$ from $\mathcal{T}$. ∎

Finally, to close this section, we present two more definitions.

Definition Let $\mathcal{T}_1$ be a tangle in a graph G, and let $\mathcal{T}_2$ be a tangle in a subgraph of G. Obviously $\mathcal{T}_2$ is also a tangle in G. We say $\mathcal{T}_2$ is *conformal* with $\mathcal{T}_1$ if the order of $\mathcal{T}_1$ is at least that of $\mathcal{T}_2$, and they are indistinguishable in G.

Definition Let $\mathcal{T}$ be a tangle in a graph G. Let Z be a set of $k < \text{ord}(\mathcal{T})$ vertices of G. Define $\mathcal{T}/Z$ to be $\{T \in \mathcal{T} \mid T \cap Z = \emptyset\}$. Then $\mathcal{T}/Z$ is a tangle of order at least $\text{ord}(\mathcal{T}) - k$ in $G - Z$. Furthermore, it is conformal with $\mathcal{T}$.

4 Excluding walls

In this section, we give a brief sketch of the proof of Theorem 1.2, which states that if a graph G has no wall of height h then it has no bramble of order greater than 25^{34h^5}. The proof of this theorem is long and fairly technical; we delay discussing its details until Section 7. We will however discuss it briefly in order to present a strengthening which we will need in the next two sections. To begin, we recall that if a graph has a bramble of order $3k$ then it has a tangle of order k.

The main idea of the proof is to start with a large $\mathcal{T}$-linked set for some huge order tangle $\mathcal{T}$ and then grow a set $\mathcal{P}$ of paths from this set "towards" the tangle. We use a subset of these paths to form the rows of our wall. We shall show that we can also find a set $\mathcal{C}$ of disjoint paths between the appropriate elements of $\mathcal{P}$ which we use to form the columns. In order to construct the paths of $\mathcal{C}$ we will need to repeatedly use a lemma which shows that given two vertices s and t in a $\mathcal{T}$-linked set X we can find a path P between s and t in $G-(X-s-t)$ and a tangle T' in $G-P$ whose order is not much smaller than that of $\mathcal{T}$ and such that X is a $\mathcal{T}'$-linked set. The precise lemma we need is (see also Figure 12):

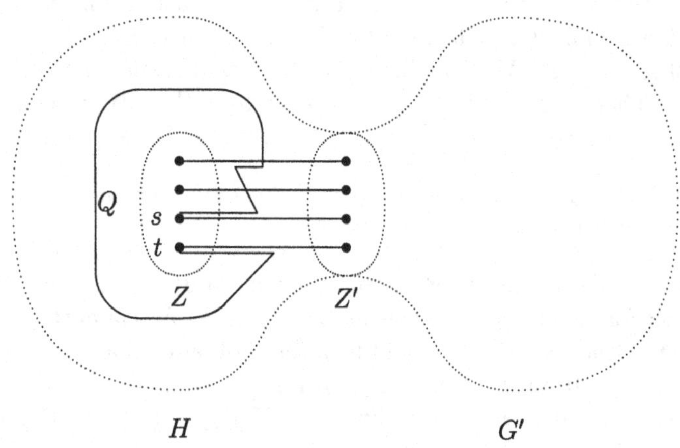

Figure 12:

Lemma 4.1 **([46])** *Let h and θ be integers with $\theta \geq h \geq 2$. Let $\mathcal{T}$ be a tangle in a graph G of order at least $24(h+\theta)+7$, and let Z be a $\mathcal{T}$-linked set of h vertices of G. Let s and t be two vertices of Z. Then, there are subgraphs H and G' of G such that setting $Z' = V(H) \cap V(G')$ we have:*

(i) $|Z'| = h$ and there are h vertex-disjoint paths $P_1, \dots, P_h$ in H from Z to Z' and an s to t path Q in $H - Z'$ such that for each i, $Q \cap P_i$ is a (possibly empty) path, and

(ii) there is a tangle $\mathcal{T}'$ in G' of order θ and conformal with $\mathcal{T}$ such that Z' is a $\mathcal{T}'$-linked set.

Repeated applications of this lemma and some heavy slogging allow us to build up a column C_1 from various paths like Q. By iterating the column building procedure, we can knit together a wall W of height h, adding the columns one by one. The reader may find that Figure 13 aids his intuition. Note that the set of rows is obtained by concatenating a family of sets of subpaths which are obtained via Lemma 4.1. We remark that both the set of the initial endpoints and the set of final endpoints of each set in this family is a $\mathcal{T}$-linked set. Now, for each $i > 1$, we let H_i be the connected subgraph consisting of C_i and for each row R_j the portion of R_j strictly between $R_j \cap C_{i-1}$ and $R_j \cap C_i$. Our remark implies that each H_i contains a $\mathcal{T}$-linked set Z_i containing h vertices one on each row. This fact can be used to obtain the strengthening of Theorem 1.2 stated as Theorem 4.2 below.

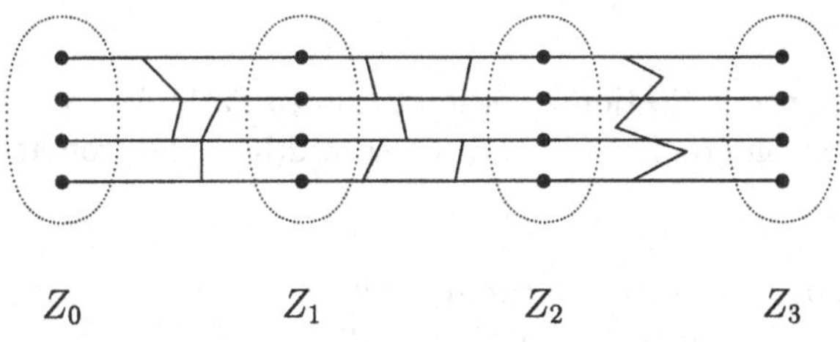

Figure 13:

Theorem 1.2 states that given a large tangle $\mathcal{T}$ in a graph G, we can use $\mathcal{T}$ to find a high wall W. We actually show that we can find such a wall which is not separated from $\mathcal{T}$ by any small order cutset. To be more precise, we need a definition. We say a wall W of height h is *attached* to a tangle $\mathcal{T}$ if $\mathcal{T}$ has order at least h and for every set X of fewer than h vertices $f_{\mathcal{T}}(X)$ is the unique component of $G - X$ containing a row (and hence a column) of W. We can extend Theorem 1.2 to prove:

Theorem 4.2 *Let h be an integer. Let $\mathcal{T}$ be a tangle of order at least 25^{34h^5-1} in a graph G. Then there is a wall W of height h attached to $\mathcal{T}$.*

Proof Idea Construct the wall W as discussed above. Let X be any set of at most h vertices of G. Clearly, X fails to intersect at least one of the H_i defined above, and hence there is some i such that Z_i is in a component U of $G - X$. Now, as Z_i is a $\mathcal{T}$-linked set, U must be $f_{\mathcal{T}}(X)$. X also fails to intersect some row R_j. Thus, since R_j intersects Z_i, R_j is contained in $f_{\mathcal{T}}(X)$. So, $f_{\mathcal{T}}(X)$ is indeed the component of $G - X$ completely containing a row of W. ∎

5 Graph minors revisited

We remind the reader that two of the most important results concerning graph minors proved by Robertson and Seymour were:

(1.14) Wagner's Conjecture *In any infinite sequence $G_1, G_2, \ldots$ of graphs, there exist $i \neq j$ such that G_i is a minor of G_j.*

(1.16) *There is a polynomial-time algorithm to resolve k Disjoint Rooted Paths for any fixed k (Robertson and Seymour's algorithm runs in $\mathcal{O}(|V(G)|^3)$ time, Reed improved this to $\mathcal{O}(|V(G)|^2)$.)*

We now investigate the role that tree decompositions played in the proof of these results. We consider (1.16) first and we begin with a definition.

Definition Let (G, X, Y) be an instance of k Disjoint Rooted Paths. A vertex v is *irrelevant* (with respect to (G, X, Y)) if the desired paths exist in G if and only if they exist in $G - v$.

Now, as mentioned in Section 2, Disjoint Rooted Paths for fixed k is easy to solve in linear time on graphs of bounded tree width. Robertson and Seymour [42] proved:

Theorem 5.1 *For every k there is an h_k such that if (G, X, Y) is an instance of k Disjoint Rooted Paths and W is a wall of height h_k in G then there is an irrelevant vertex v in W. Furthermore, such a wall and corresponding irrelevant vertex can be found in polynomial time. (Robertson and Seymour's algorithm runs in $\mathcal{O}(|V(G)|^2)$ time, Reed improved this to $\mathcal{O}(|E(G)|)$).*

Now, obviously having found an irrelevant vertex v for (G, X, Y) we can restrict our attention to $(G - v, X, Y)$. Robertson and Seymour repeatedly apply Theorem 5.1 and delete the irrelevant vertex it returns until the graph they are considering contains no high wall. Theorems 1.1, 1.2, and 1.3 imply that such a graph has tree width at most $25^{34h_k^5}$ and hence we can solve the k Disjoint Rooted Paths problem using dynamic programming.

We now briefly sketch the methods Robertson and Seymour use to prove Theorem 5.1. This is not a digression as it leads into our discussion of the proof of Wagner's Conjecture. They use two different approaches depending on whether or not they can find a large clique minor in G.

To simplify our discussion of the first technique, we consider an instance (G, X, Y) of k Disjoint Rooted Paths such that G contains not just a large clique minor but, in fact, a large clique. So, let C be a clique in G with $2k+1$ vertices. We will show that some vertex of C is irrelevant.

By Menger's theorem we can find (in $\mathcal{O}(kn)$ time, using standard techniques) either a set $\mathcal{Q}$ of $2k$ vertex disjoint paths between C and $X \cup Y$ or a set Z of fewer than $2k$ vertices such that Z separates $X \cup Y$ from C. In the first case, we can insist that the paths in $\mathcal{Q}$ are internally disjoint from C, we simply shorten any path which contains a vertex of C in its interior. Now, the paths in $\mathcal{Q}$ along with an appropriately chosen set of edges from C yield the desired k paths between X and Y, see e.g. Figure 14. Thus the (unique) vertex of C which is not the endpoint of any path of $\mathcal{Q}$ is irrelevant. (In fact, given this situation, we can simply exit the algorithm and return the desired paths.)

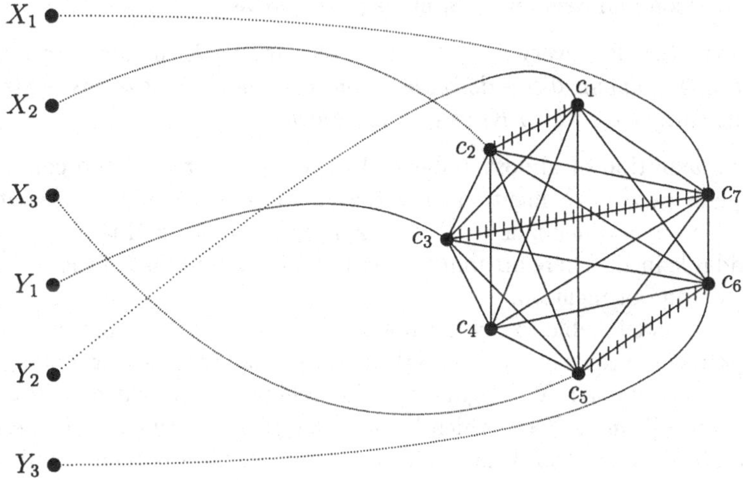

Figure 14: c_4 is irrelevant

Suppose then that we find a set Z of at most $2k-1$ vertices which separates $X \cup Y$ from the component U of $G - Z$ which contains $C - Z$. In fact, we can choose an algorithm which returns such a set Z with minimal cardinality as well as a set $\mathcal{R}$ of $|Z|$ vertex disjoint paths between Z and C. Again, we can take these paths to be internally disjoint from C. We claim that any vertex of C which is not an endpoint of one of these paths is irrelevant. To see this note that if the desired paths exist in G then their restriction to the graph induced by $Z \cup V(U)$ is a set of paths with endpoints in Z. But, as in the previous case, any such set of paths will still exist if we delete all of $V(C)$ except those vertices on the paths in $\mathcal{R}$. This implies the claim.

A similar approach yields:

Lemma 5.2 ([42]) *Let (G, X, Y) be an instance of k Disjoint Rooted Paths and let C be a model of K_{6k+3} in G. Then there is a vertex of C which is irrelevant. Furthermore, given C, we can find such a vertex in $\mathcal{O}(k|E(G)|)$ time.*

Showing that there is an irrelevant vertex if G has no clique minor is more complicated. It relies on a characterization of graphs with no large clique minor. To give some of the flavour of this characterization, we discuss graphs without K_l as a minor for $l \in \{5, 6\}$. We recall that no planar graph contains K_5 as a minor. Similarly, for every l and every surface Σ in which K_l is not embeddable, no graph embedded in Σ contains a clique minor of order l. This, however, is not the whole story. Consider, for example, a graph G obtained from a wall W by adding a vertex adjacent to all of W. This graph contains no K_6 minor because W contains no K_5 minor. However, for every surface Σ, if W is high enough then G will not be embeddable on Σ.

Robertson and Seymour [43] managed to prove:

Theorem 5.3 *For every l, if G is a graph with no K_l minor then for every node t of the canonical tree decomposition $[T, \mathcal{X}]$ of G, X_t is almost embeddable in some surface on which K_l cannot be drawn.*

The definition of almost embedded depends on l and is too complicated to explain here. We mention only that if for some bounded size set Z of points (the bound depends on l), $G - Z$ is embeddable in Σ then G is almost embeddable in Σ. There are many other allowable extensions of the embedding which we do not mention.

We will not be able to say much about the proof of (5.3) since we have not even stated it properly. We remark however that the starting point for the proof is Theorem 4.2. We begin by finding a high wall W which is attached to the canonical tangle $\mathcal{T}$ for which W_t $(= V(X_t))$ is a hitting set. If the rest of the graph is attached to W in a sufficiently non-planar way then G will contain a K_l minor. Otherwise, we can extend the embedding of W in the plane to obtain an almost embedding of X_t in a surface in which K_l cannot be drawn.

Now, (5.3) is important because it allows Robertson and Seymour to apply theorems about graphs actually embedded on surfaces to graphs without a large clique minor. For example, Robertson and Seymour [39] proved (using techniques we will not discuss):

Theorem 5.4 *For every surface Σ and integer k there is an integer $h(\Sigma, k)$ such that the following holds. Let (G, X, Y) be an instance of k Disjoint Rooted Paths such that G is embedded in Σ. Let W be a wall of height $h(\Sigma, k)$ in G. Then, there is a vertex v in W which is irrelevant, and, given W, we can find such a vertex in $\mathcal{O}(k|E(G)|)$ time.*

Using (5.3), Robertson and Seymour [42] extended this to obtain:

Theorem 5.5 *For every integer k there is an $h(k)$ such that the following holds. Let (G, X, Y) be an instance of k Disjoint Rooted Paths such that G contains no clique minor of order $6k + 3$. Let W be a wall of height $h(k)$ in G. Then, there is a vertex v in W which is irrelevant, and, given W, we can find such a vertex in $\mathcal{O}(k|E(G)|)$ time. More strongly: there is an $\mathcal{O}(k|E(G)|)$-time algorithm which given an instance (H, A, B) of k Disjoint Rooted Paths such that H contains a wall of height $h(k)$, finds such a wall as well as either a model of K_{6k+3} in H or an irrelevant vertex in W.*

Combining (5.5) and (5.2) yields (5.1). This completes our discussion of Robertson and Seymour's algorithm for Disjoint Rooted Paths, we turn now to their proof of Wagner's conjecture.

We note first that Kruskal [23] proved

5.6 *Wagner's conjecture holds for trees.*

Robertson and Seymour, using techniques of Nash-Williams [31] extended this to prove:

5.7 *Wagner's conjecture holds for graphs of bounded tree width.*

Now, it is intuitively obvious and not difficult to show:

5.8 *For any planar graph H, there is an elementary wall W containing H as a minor.*

Proof Idea The most intuitive approach is to draw H in the plane and then to approximate this drawing by paths in a sufficiently fine hexagonal mesh. ■

From (5.7) and (5.8), we obtain immediately:

Theorem 5.9 ([37]) *Wagner's conjecture holds for planar graphs.*

Proof Consider an infinite sequence G_1, G_2, ... of planar graphs. We want to show that there exists an $i \neq j$ such that G_i is a minor of G_j. So, we can assume that for $j \geq 2$, G_j does not have G_1 as a minor. By (5.8), there exists an h such that every wall of height h contains G_1 as a minor. So, we know that for $j \geq 2$, G_j does not contain a wall of height h and hence has tree width at most 25^{34h^5}. Now, by (5.7), there are $i, j \geq 2$, $i \neq j$ such that G_i is a minor of G_j. ■

Robertson and Seymour [40], by bootstrapping with (5.9) and then applying induction, were able to prove:

Theorem 5.10 *For every surface Σ, Wagner's Conjecture holds for graphs embedded on Σ.*

With (5.10) in hand, we turn to the proof of Wagner's conjecture for general graphs. So, consider an infinite sequence G_1, G_2, ... of graphs. We want to show that there exists an $i \neq j$ such that G_i is a minor of G_j. Let $l = |V(G_1)|$. For $j \geq 2$, let $[T^j, \mathcal{X}^j]$ be the canonical tree decomposition of G_j. We can assume that for $j \geq 2$, G_j does not have G_1 as a minor, and hence does not have K_l as a minor. Applying (5.3), we deduce that, for $j \geq 2$ and every node t of T^j, X_t^j is almost embeddable in a surface in which K_l cannot be embedded. This result permitted Robertson and Seymour to extend (5.10) to show that the set of graphs $\{X_t^j \mid j \geq 2,\ t \in V(T^j)\}$ satisfies Wagner's conjecture. Thus, the graphs G_j, $j \geq 2$ can be thought of as trees where the vertices are labelled from a set of graphs satisfying Wagner's Conjecture. Robertson and Seymour, extending Kruskal's theorem once again, managed to show that Wagner's Conjecture must hold for such a set of graphs, and hence there is an $i \neq j$ such that G_i is a minor of G_j.

In the brief account we have given so far, we have glossed over the techniques Robertson and Seymour used to extend Kruskal's theorem (5.6) to obtain first (5.7) and then Wagner's conjecture in general. We close this section by presenting a porous precis of their procedure. One motivation for doing so is simply to deepen the reader's understanding of the approach taken by Robertson and Seymour. Another is that we will need to present some results on tree decompositions which are of interest in their own right.

As we saw in Section 2, one property of trees which makes them easy to work with is that they can be rooted. It is this fact which permitted Kruskal to prove (5.6). In particular, he considered *rooted minors.* We say a rooted tree (T_1, r_1) is a rooted minor of a rooted tree (T_2, r_2) if there is a model of T_1 in T_2 such that $r_2 \in \text{im}(r_1)$. Kruskal actually proved that in any infinite sequence (T_1, r_1), (T_2, r_2), ... of rooted trees there exists $i \neq j$ such that (T_i, r_i) is a rooted minor of (T_j, r_j). Robertson and Seymour extended ideas found in Nash-Williams's proof of this theorem. Crucial to his proof is:

5.11 *If (T, r) is a rooted tree, and s is a descendant of t in T, then any rooted tree which is a rooted minor of (T_s, s) is also a rooted minor of (T_t, t).*

Proof Given a model of a rooted tree (S, p) in (T_s, s), with $s \in \text{im}(p)$, we add the path of T between s and t to $\text{im}(P)$ to obtain a model of (S, p) in (T_t, t) with $t \in \text{im}(p)$. ∎

In order to extend the ideas in Nash-Williams's proof (the details of which we omit) to prove Wagner's conjecture for graphs of bounded tree width, we need to develop a notion of rooted minors for such graphs. We will define this concept in a moment but make some remarks about our definition now. Consider two graphs G_1, G_2 and corresponding rooted tree decompositions $[(T^1, r^1), \mathcal{X}^1]$ and $[(T^2, r^2), \mathcal{X}^2]$. In order for $[(T^1, r^1), \mathcal{X}^1]$ to be a rooted minor of $[(T^2, r^2), \mathcal{X}^2]$ we shall insist that $|V(X^1_{r^1})| = |V(X^2_{r^2})|$ and that there is a model of G_1 in G_2 such that for each vertex v of $X^1_{r^1}$ there is a vertex of $X^2_{r^2}$ in $\text{im}(v)$. Now, in order to mimic Nash-Williams's proof, we need an analogue of (5.11). In order to develop such an analogue, we need to impose more conditions on the tree decomposition and the rooted minors. Specifically:

Definition An *ordered rooted tree decomposition* $[(T, r), \mathcal{X}]$ for G consists of a rooted tree, along with a subgraph X_t of G and corresponding *ordered* set W_t (which is an ordering of $V(X_t)$) for each node t of the rooted tree such that these subgraphs satisfy the standard axioms for tree decompositions.

Definition Consider two graphs G_1, G_2 and corresponding ordered rooted tree decompositions $[(T^1, r^1), \mathcal{X}^1]$ and $[(T^2, r^2), \mathcal{X}^2]$ such that $|W^1_{r1}| = |W^2_{r2}| = k$. Let $\{x_1, \ldots, x_k\}$ be the ordering of W^1_{r1} and let $\{y_1, \ldots, y_k\}$ be the ordering of W^2_{r2}. We say that $[(T^1, r^1), \mathcal{X}^1]$ is an *ordered rooted minor* of $[(T^2, r^2), \mathcal{X}^2]$ if there is a model of G_1 in G_2 such that for each x_i in W_{r_1}, y_i is in $\text{im}(x_i)$.

Now, Thomas proved the following beautiful lemma, which as we show below, implies an analogue of (5.11) for ordered rooted minors of certain ordered rooted tree decompositions (Robertson and Seymour proved a weaker version of this lemma which they themselves called a clumsy substitute for it.)

Lemma 5.12 ([49]) *Let G be a graph of tree width w. Then there is a tree decomposition $[T, \mathcal{X}]$ of G of width w such that the following property holds for every two nodes t_1 and t_2 of T:*

> *Let P be the unique subpath of T between t_1 and t_2. If there is no t on P such that $|W_t| \leq \min(|W_{t_1}|, |W_{t_2}|)$ then for every pair of equal sized sets S_1 and S_2 with $S_i \subseteq W_{t_i}$ there are $|S_1|$ vertex disjoint paths between S_1 and S_2 in G.*

Corollary 5.13 *Let G be a graph of tree width w. Then there is an ordered rooted tree decomposition $[(T, r), \mathcal{X}]$ of G of width w such that if s is a descendant of t with $|W_s| = |W_t|$, and for every node f on the s-t path of T we have $|W_f| \geq |W_t|$ the following holds:*

> *Any ordered rooted tree decomposition which is a rooted ordered minor of $[(T_s, s), \{X_p \mid p \in V(T_s)\}]$ is also a rooted ordered minor of $[(T_t, t), \{X_p \mid p \in V(T_t)\}]$.*

Proof of Corollary 5.13 Let G be a graph of tree width w. Let $[T, \mathcal{X}]$ be the tree decomposition G satisfying the condition given in Lemma 5.12. Arbitrarily choose a root r of T to obtain a rooted tree decomposition $[(T, r), \mathcal{W}]$. Now, we order the vertices of each W_t as follows. We begin at the root, and consider each node before any of its descendants. For a given node a, with $|W_a| = k$ say, if there is no ancestor b of a with $|W_b| \leq k$ then we arbitrarily order the nodes of W_a. Similarily if the first ancestor b of a with $|W_b| \leq k$ encountered on the path from a up to r satisfies $|W_b| < k$ then we order W_a arbitrarily. Otherwise, we let b be the ancestor of a such that $|W_b| = k$ and such that no interior node c of the a to b path in T has $|W_c| = k$. We find the set $\mathcal{P}$ of k paths between W_a and W_b in G guaranteed to exist by Lemma 5.12. Now, for each path P in $\mathcal{P}$, if the endpoint of P in W_b is the ith vertex in the ordering of W_b then the endpoint of P in W_a will be the ith vertex in the ordering of W_a. The paths in $\mathcal{P}$ ensure that any ordered rooted minor of W_a is an ordered rooted minor of W_b. Corollary 5.13 follows by repeatedly applying this fact to work our way up from s to t. ∎

Remark The condition in Corollary 5.13 that there is no node f on the s-t path such that W_f has fewer than $|W_t|$ elements is obviously necessary, for any such W_f is a cutset in G which may well prevent us from extending our rooted minors. The same holds true for the corresponding condition in Lemma 5.12.

Corollary 5.13 allows us to extend Nash-Williams's proof technique to prove Wagner's Conjecture for graphs of bounded tree width. We turn now to the proof of Wagner's Conjecture for arbitrary graphs. As already noted, we can assume that for some fixed l we are considering graphs with no K_l minor. Once again, we want to consider rooted minors, this time using the canonical tree decompositions. Unfortunately, there are two significant complications. The first is that the W_t in such a tree decomposition may not be linked. The second is that the size of the W_t may be arbitrarily large so that, even if we consider one vertex trees, if we choose an infinite sequence of decompositions where every W_t is a different size we will not find a tree decomposition which is a rooted minor of another in the sequence.

We deal with these problems by considering rooted minors rooted not at the W_t but in the arcs of the tree, i.e. at $W_s \cap W_t$ for two adjacent nodes s and t of the tree. This deals quite adequately with the first problem. The fact that the arcs of our tree decompositions correspond to canonical separators allows us easily to prove an analogue of (5.13) for the cutsets corresponding to the arcs. However, these arc cutsets may still be arbitrarily large. To deal with this problem, we simply restrict our attention to the arcs in the canonical tree decompositions which correspond to cutsets with at most $f(k)$ vertices, for a suitably chosen $f(k)$. This yields a new tree decomposition which corresponds to a subset of our original set of laminar separations. We can extend (5.3) to prove that for each node t of this new tree decomposition we still have that X_t is almost embeddable in some surface in which K_l is not embeddable and we are therefore in a position to mimic Nash-Williams's proof.

As the above discussion demonstrates, the canonical tree decomposition theorem plays a key role in the proof of Wagner's Conjecture. We expect that it will have many other uses as it seems to be an extremely natural way of decomposing a graph into its highly connected pieces.

6 Packing and covering

Let $\mathcal{F}$ be a family of graphs. An *$\mathcal{F}$-packing* in a graph G is a set of vertex disjoint subgraphs of G, each of which is isomorphic to a member of $\mathcal{F}$. The $\mathcal{F}$-packing number of G, denoted $p_{\mathcal{F}}(G)$, is the maximum cardinality of an $\mathcal{F}$-packing in G. An *$\mathcal{F}$-cover* is a set X of vertices such that $G - X$ contains no subgraph isomorphic to a member of $\mathcal{F}$. The $\mathcal{F}$-covering number of G, denoted $c_{\mathcal{F}}(G)$ is the minimum cardinality of an $\mathcal{F}$-cover for G. Since a cover must contain a vertex from every subgraph of a packing, $p_{\mathcal{F}}(G) \leq c_{\mathcal{F}}(G)$.

Erdős and Pósa [15] proved:

6.1 *There exists a constant μ such that for the family $\mathcal{C}$ of all cycles, $c_{\mathcal{C}}(G) \le \mu p_{\mathcal{C}}(G) \log p_{\mathcal{C}}(G)$.*

A family $\mathcal{F}$ of graphs is said to have the *Erdős-Pósa property* if there is an integer-valued function f such that $c_{\mathcal{F}}(G) \le f(p_{\mathcal{F}}(G))$ for every graph G.

Brambles are an extremely effective tool for proving that a given family of graphs has the Erdős-Pósa property. For example, Thomassen [50] used them to show that for every integer m, the family of cycles whose length is divisible by m has the Erdős-Pósa property. Reed [33] used them to prove that the family of odd cycles has a similar though weaker property (as we shall see the family of odd cycles does not have the Erdős-Pósa property). He was thereby able to resolve a conjecture of Erdős [20] concerning odd cycle covers. Reed, Robertson, Seymour, and Thomas [35] used similar techniques to prove that the directed cycles in a directed graph satisfy the Erdős-Pósa property. This settled a 25 year old conjecture due to Younger [53].

It seems likely that the theory of brambles will have many other applications to packing and covering. The analogous directed theory will probably also be very useful, once it has been fully developed.

We turn now to a more detailed discussion of those applications of brambles to packing and covering mentioned above. In doing so, we will find it convenient to define for every family $\mathcal{F}$ of graphs the function $f_{\mathcal{F}}$ where for each non-negative integer k, $f_{\mathcal{F}}(k)$ is either $\max\{c_{\mathcal{F}}(G) \mid G \text{ satisfies } p_{\mathcal{F}}(G) = k\}$ or ∞ if no such maximum exists. Then, $\mathcal{F}$ has the Erdős-Pósa property if and only if $f_{\mathcal{F}}$ is integer valued. We note further that for any $\mathcal{F}$, $f_{\mathcal{F}}(0) = 0$.

6.1 Cycles and a useful lemma

In this subsection, we show how brambles (actually, their companions: well-linked sets) arise when considering the Erdős-Pósa property. We then show how they can be used to prove that the Erdős-Pósa property holds for the family of all cycles.

Suppose that some family $\mathcal{F}$ of connected graphs does not have the Erdős-Pósa property. Then consider the smallest integer k such that $f_{\mathcal{F}}(k)$ is infinite. We know $k \ge 1$. Since $f_{\mathcal{F}}(k) = \infty$, there is a graph G with $p_{\mathcal{F}}(G) = k$ and $c_{\mathcal{F}}(G) > 4f_{\mathcal{F}}(k-1)$. We can apply the following lemma to such a graph.

Lemma 6.2 (The Key Lemma) *Consider a family $\mathcal{F}$ of connected graphs, an integer $k \ge 1$ such that $f_{\mathcal{F}}(k-1)$ is finite, and any graph G with $p_{\mathcal{F}}(G) = k$ and $c_{\mathcal{F}}(G) > 4f_{\mathcal{F}}(k-1)$. Then every minimum $\mathcal{F}$-cover H in G is an $\lfloor |H|/4 \rfloor$ linked set.*

Proof We need to show that for every set X of at most $|H|/4$ vertices of G, some component of $G - X$ contains more than half the vertices of H. So, consider some such set X. Since X is too small to be a hitting set, there must be a component U of $G - X$ containing an element of F. Then, $p_{\mathcal{F}}(G - U)$ is at most $k-1$ so there is an $\mathcal{F}$-cover H' for $G-U$ of size at most $f_{\mathcal{F}}(k-1) < |H|/4$. Now, since H is an $\mathcal{F}$-cover, $H \cap U$ is an $\mathcal{F}$-cover for U. Furthermore, since each element of $\mathcal{F}$ is connected, X contains a vertex of every subgraph of G which (i) is isomorphic to an element of $\mathcal{F}$, and (ii) intersects both U and $G-U$. Thus, $X \cup H' \cup (H \cap U)$ is an $\mathcal{F}$-cover for G. By the minimality of H, it follows that $|H \cap U| \geq |H| - |X| - |H'| > |H|/2$. This completes the proof of the lemma. ∎

To demonstrate the power of this lemma, we apply it to prove:

Theorem 6.3 *If we let $\mathcal{C}$ be the family of all cycles, then $f_{\mathcal{C}}(k) \leq 4{*}25^{34(k+1)^5}$.*

Proof Assume the theorem is false and consider the smallest integer k for which the bound claimed for $f_{\mathcal{C}}(k)$ fails to hold. Obviously, $k \geq 1$. Then there is a graph G with $p_{\mathcal{C}}(G) = k$ and $c_{\mathcal{C}}(G) > 4 * 25^{34(k+1)^5} > 4f_{\mathcal{C}}(k-1)$. Thus, by the key lemma, G contains a $25^{34(k+1)^5}$-linked set. Now, Theorems 1.1, 1.2, and 1.3 imply that G contains a wall of height $k+1$. But this wall clearly contains $k+1$ vertex disjoint cycles, a contradiction. ∎

Remark Erdős and Pósa's original proof gave the much better bound on $f_{\mathcal{C}}(k)$ of $\mu k \log(k)$ for a constant μ. As we shall see, however, the technique used in our proof can be extended to situations which had previously proven intractable. For example, Robertson and Seymour noted that essentially the same proof shows that for any set S of planar graphs, the family $\mathcal{F}$ of graphs which contain some element of S as a minor has the Erdős-Pósa property (Erdős and Pósa's result is a special case where S consists of one graph, a loop).

6.2 Even cycles

As shown by Thomassen [50], the technique of the last section can be used to prove that the family of even cycles has the Erdős-Pósa property. Actually, Thomassen proved that for every integer m, the family of cycles with length $0 \bmod m$ satisfy the Erdős-Pósa property. In this section, we sketch Thomassen's proof. We will use the term *m-cycles* to denote the family of cycles of length $0 \bmod m$.

The result which allowed Thomassen to apply the key lemma to the m-cycles was the following.

6.4 *For any h and m (h even), there is a $g(h,m)$ such that if W is a wall of height $g(h,m)$ then W contains a wall W' of height h such that all the paths of W' between nails have length $0 \bmod m$ (or equivalently all the paths of W' between two nails and containing no nails in their interior have length $0 \bmod m$).*

Such a W' contains h vertex disjoint cycles. Each of these must be an m-cycle. Thus, we obtain:

6.5 *For any h and m (h even), there is a $g(h,m)$ such that if W is a wall of height $g(h,m)$ then W contains h vertex disjoint m-cycles.*

Now, we define $f(h,m)$ recursively insisting only that $f(0,m)=0$ and for $h \geq 2$: $f(h,m) \geq 4f(h-1,m)+4$ and $f(h,m) \geq 4 * 25^{34g(h+1,m)^5}$. Mimicking the proof of (6.3), we obtain:

6.6 *a graph either has a packing of h vertex disjoint m-cycles or a covering of the m-cycles of size at most $f(h,m)$.*

Thus the Erdős-Pósa property does indeed hold for the m-cycles for every integer m.

We remark that Thomassen, in his original proof, considered a tree decomposition of G rather than a k-linked set. We have presented the proof in terms of k-linked sets and brambles for three reasons. The first is to acquaint the reader with these less familiar notions. The second is that k-linked sets seem to generalize to directed graphs more easily than tree decompositions. The third is that, as we shall see in the next section, in some applications it is important that a minimum cover H is $\lfloor |H|/4 \rfloor$-linked. Just knowing that G contains a k-linked set for large k may not suffice.

6.3 Odd cycles

In this subsection, we discuss packings and coverings of odd cycles. To begin, we show that the odd cycles do not have the Erdős-Pósa property. Our examples are similar to, but slightly different from those given by Lovász and Schrijver [27]. We remark that Dejter and Neumann-Lara [14] have shown that for any k other than 0, the cycles of length $k \bmod m$ do not satisfy the Erdős-Pósa property, by generalizing these examples.

Consider the elementary wall W_h of height h. W_h is bipartite with a unique bipartition, (A,B) say. Both the top and bottom row of W_h contain $2h+1$ vertices. Consider the graph H_h obtained from W_h by first adding a matching between the vertices of degree two in the top and bottom rows apart from the corners so that the $2i$th vertex in the top row is joined to the $(2h+2-2i)$th vertex in the bottom row (see e.g. Figure 15) and then subdividing each of these auxiliary edges, thereby creating h auxiliary vertices. Note that each auxiliary vertex is joined to two vertices of W_h on different sides of the bipartition (A,B). So, the parity of any cycle in H_h depends on the number of auxiliary vertices it contains. One can use this to show that H_h does not contain two vertex disjoint odd cycles. (One way of doing this is to extend the planar embedding of W_h to an embedding of H_h in the projective plane with all faces even. It follows that any odd cycle of H_h is non-null homotopic. However, any two non-null homotopic cycles in the projective plane intersect.)

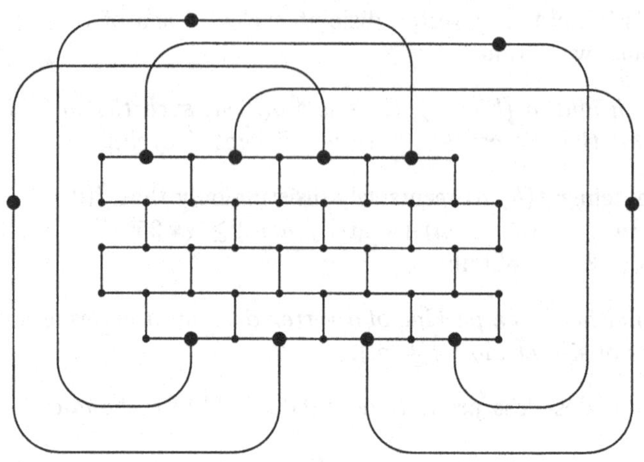

Figure 15: H_4

We claim that H_h also contains a set of $\lfloor h/2 \rfloor$ odd cycles, such that each vertex is in at most 2 of these odd cycles. It follows that H_h has no odd cycle cover with less than $\lfloor h/4 \rfloor$ vertices and hence the odd cycles do not have the Erdős-Pósa property.

To prove our claim, we let v_i be the auxiliary vertex joined to the $2i$th vertex in the top row, call this a_i and the $(2h+2-2i)$th vertex in the bottom row, call this b_i. For $2 \leq i \leq h$, we let P_i be the path between a_i and b_i obtained by concatenating (as shown in Figure 16):

(i) the subpath of R_1 between a_i and the last nail before it on R_1 which we note is in C_i,

(ii) the subpath of C_i between this nail and some element x_i of R_i,

(iii) the subpath of R_i between x_i and some element y_i of $R_i \cap C_{h+2-i}$,

(iv) the subpath of C_{h+2-i} between y_i and the first nail of R_{h+1} after b_i, and

(v) the subpath of R_{h+1} between this nail and b_i.

Now, $P_i + v_i$ induces an odd cycle C_i. Furthermore, a vertex is clearly in at most two of $P_2, \ldots, P_{\lfloor h/2 \rfloor + 1}$ as it is in at most one column and one row. This completes the proof of our claim.

In a similar spirit, we define a more general class of counterexamples, the Escher walls.

Definition Let W be a bipartite wall of height h with bipartition (X, Y). Let $A = \{a_1, \ldots, a_h\}$ be a subset of $R_1 \cap X$ labelled so that a_i appears before a_{i+1}

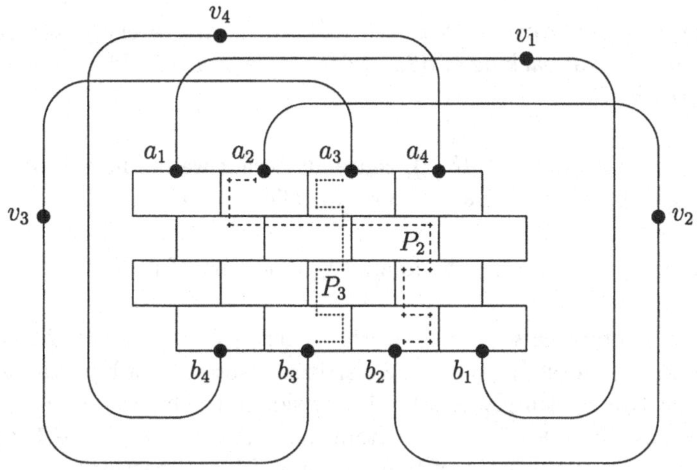

Figure 16:

on this row. Let $B = \{b_1, \ldots, b_h\}$ be a subset of $R_{2h+1} \cap X$ labelled so that b_i appears after b_{i+1} on this row. Let A be such that for $1 \leq i < h$ there is some internal vertex of the subpath of R_1 between a_i and a_{i+1} which is a nail. Similarly, let B be such that for $1 \leq i < h$ there is some internal vertex of the subpath of R_{2h+1} between b_i and b_{i+1} which is a nail. Let $P_1, \ldots, P_h$ be vertex disjoint paths such that P_i has endpoints a_i and b_i, is internally vertex disjoint from W, and has an odd number of edges. Then $W \cup P_1 \cup \cdots \cup P_h$ forms an *Escher wall* of height h.

The same proof techniques show that an Escher wall of height h contains no 2 vertex disjoint cycles but does contain $\lfloor h/2 \rfloor$ odd cycles using each vertex at most twice and hence no odd cycle cover with fewer than $\lfloor h/4 \rfloor$ vertices. Thus, the Escher walls provide a counterexample to the statement:

> the Erdős-Pósa property holds for the odd cycles.

The result we discuss in this section shows that, in a certain sense, they are the only counterexamples. To wit,

Theorem 6.7 *For every h and k there is an $f(h,k)$ such that every graph contains either k vertex disjoint odd cycles, an Escher wall of height h, or an odd cycle cover with at most $f(h,k)$ vertices.*

This theorem has two interesting consequences. First, it implies the following result relating half-integral packings and coverings of odd cycles.

Definition A *half-integral packing* of odd cycles in G consists of a set S of odd cycles such that each vertex of G is in at most two elements of S. The size of the packing is $|S|/2$.

Corollary 6.8 *For every k there is a $g(k)$ such that every graph contains either a half-integral packing of odd cycles of size k or an odd cycle cover with at most $g(k)$ vertices.*

Proof Simply set $g(k) = f(4k, k)$. Since an Escher wall of height $4k$ contains a half-integral packing of size k, the result follows. ■

Theorem 6.7 also implies the following conjecture of Erdős.

Definition A graph is *k-quasi bipartite* if every subgraph H of G contains a stable set with at least $|V(H)|/2 - k$ vertices. (Note that a bipartite graph is 0-quasi bipartite, we simply pick the larger side of the bipartition of H. Since an odd cycle with $2l + 1$ vertices contains no stable set of size $l + 1$ it follows that a graph G is 0-quasi bipartite if and only if it is bipartite.)

Conjecture 6.9 (Erdős [20]) *For all k there is an $l(k)$ such that every k-quasi bipartite graph contains an odd cycle cover of size at most $l(k)$.*

Remark Erdős's original conjecture concerned *k-near bipartite* graphs in which only the even H needed to contain large stable sets. Since every k-near bipartite graph is $(k + 1)$-quasi bipartite, the conjectures are equivalent.

Proof of Conjecture 6.9 We need the following result which is proved using brute force.

> **Observation 6.10** *For every k there is an $h(k)$ such that an Escher wall of height $h(k)$ is not k-quasi bipartite.*

We also need the trivial fact that a set of $2k + 1$ vertex disjoint odd cycles is not k-quasi bipartite. Now, Erdős's Conjecture follows from Theorem 6.7 with $l(k) = f(h(k), 2k + 1)$. ■

The proof of (6.7) uses the fact that under certain conditions, a minimal hitting set H for the odd cycles in a graph G is a $\lfloor |H|/4 \rfloor$-linked set. However, we actually consider a bipartite graph obtained from G and H as follows.

Definition Let $H = \{x^1, \ldots, x^{|H|}\}$ be an odd cycle cover in a graph G. Let (A, B) be a bipartition of $G - H$. For each vertex x of H, we define two new vertices x_A and x_B. We let $H' = \bigcup\{x_A, x_B \mid x \in H\}$. We define a new graph $G' = G'(H, (A, B))$ with vertex set $V - H \cup H'$ and edge set

$$E(G - H) \cup \{x_A b \mid b \in B,\ xb \in E(G)\}$$
$$\cup \{x_B a \mid a \in A,\ xa \in E(G)\} \cup \left\{x^i_A x^j_B \mid x_i x_j \in E(G),\ i < j\right\}.$$

Now, we prove (6.7) for each h in turn. Thus, we restrict our attention to the class of graphs with no Escher wall of height h and prove that the Erdős-Pósa property holds for this class of graphs. Hence, we can apply the Key Lemma to prove that a minimum hitting set H for an appropriate counter-example G is $\lfloor |H|/4 \rfloor$-linked. It turns out that similar techniques allow us to prove that H' is $\lfloor |H'|/4 \rfloor$-linked in G'. Now, (4.2) implies the existence of a large wall W in G' *attached to the tangle defined by this* $\lfloor |H'|/4 \rfloor$*-linked set.* Thus, for every small cutset X the component of $G - X$ completely containing a row of W also contains at least half the vertices of H'. It is this fact which allows us to find a set of paths between some elements of H' and the nails on the perimeter of some subwall W' of W such that these paths together with W' either yield an Escher wall of height h in G (a contradiction), or contain the desired k vertex disjoint odd cycles. The proof is too complicated to discuss further, we simply remark that the following easy observation links it to the Disjoint Rooted Paths problem and allows us to apply techniques Robertson and Seymour developed for that problem.

Observation 6.11 *Let $x^1, \ldots, x^l$ be l elements of H. let $P_1, \ldots, P_l$ be vertex disjoint paths of G' such that P_i has endpoints x^i_A, x^i_B and is internally disjoint from H'. Then, these l paths correspond to l vertex disjoint odd cycles in G.*

6.4 Directed cycles

In 1966, Gallai [19] conjectured that there is an integer n such that any directed graph D without two vertex disjoint directed cycles contains a set X of at most n vertices such that $D - X$ is acyclic. In 1973, Younger [53] conjectured that for every k there is an $f(k)$ such that every directed graph contains either k vertex disjoint directed cycles or a set X of at most $f(k)$ vertices such that $D - X$ is acyclic. In 1991, McCuaig [29] proved Gallai's conjecture, i.e. Younger's conjecture for $k = 2$, with $n = 3$. Until recently, Younger's conjecture has remained open for all other values of k except the trivial cases $k = 0$ and $k = 1$.

In 1995, Reed, Robertson, Seymour and Thomas [35] proved Younger's conjecture. To do so, they considered directed analogues of well-linked sets and grids.

Definition A set S of vertices of a digraph D is *well-linked* if for every pair X and Y of equal-sized disjoint subsets of S there exist $|X|$ vertex disjoint paths from X to Y in $D-(S-X-Y)$. (Note that by symmetry we also have paths from Y to X. Note also that the definition is the same if we do not insist that X and Y are disjoint.)

Definition A directed wall of height h (see Figure 17) consists of two sets of $h+1$ disjoint paths $R_1, \ldots, R_{h+1}$ and $C_1, \ldots, C_{h+1}$ such that each R_i and C_j intersect in a non-empty path and furthermore:

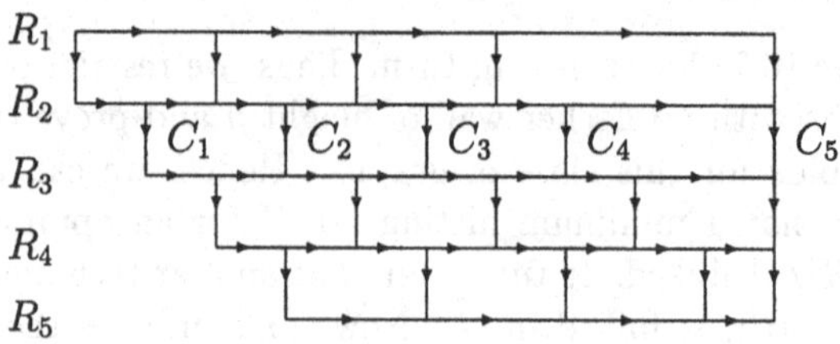

Figure 17: A directed wall of height 4

(i) Each R_i intersects the C_j in increasing index order,

(ii) Each C_j intersects the R_i in increasing index order.

For convenience, we also define $f_{dc}(k)$ to be the maximum cardinality of a minimum directed cycle cover over all graphs G without $k+1$ vertex disjoint cycles, or ∞ if this maximum is undefined.

Although one cannot prove that a minimum hitting set of the directed cycles in a directed graph is well-linked, the following fact is true and can be proved along the same lines as the Key Lemma:

Fact 6.12 *Let D be a digraph without $k+1$ vertex disjoint directed cycles. Let H be a minimum hitting set for the directed cycles of D. Let X and Y be any two disjoint equal-sized subsets of H, each containing at least $f_{dc}(k-1)$ vertices. Then there are $|X|$ vertex disjoint paths from X to Y in $D-(H-X-Y)$.*

Thus, we see that a minimum hitting set for the directed cycles in a directed graph is well-linked modulo its small subsets. Now, Reed, Robertson, Seymour and Thomas were able to show that for any h and l, if a digraph contains a sufficiently large set which is well-linked modulo its subsets of size at most l then it contains either

(i) h vertex disjoint directed cycles, or

(ii) a directed wall W of height h as well as h vertex disjoint paths $Q_1, \ldots, Q_h$ such that Q_i is a path from the last vertex of C_i to the first vertex of C_i (which is not neccesarily internally vertex disjoint from W).

They also proved, using straightforward induction arguments, that for every k there is an $f(k)$ such that if a graph contains a directed wall of height $f(k)$ and corresponding paths as in (ii) then it contains k vertex disjoint cycles.

Combining these three results yields a proof of Younger's conjecture.

The proof suggests that the directed analogue of tree width may have many applications. Unfortunately, at the moment even the correct definition of directed tree width seems hard to determine.

However, the following conjecture, a directed analogue of (1.2), seems to be of interest and its proof may well lead to a better understanding of global directed connectivity.

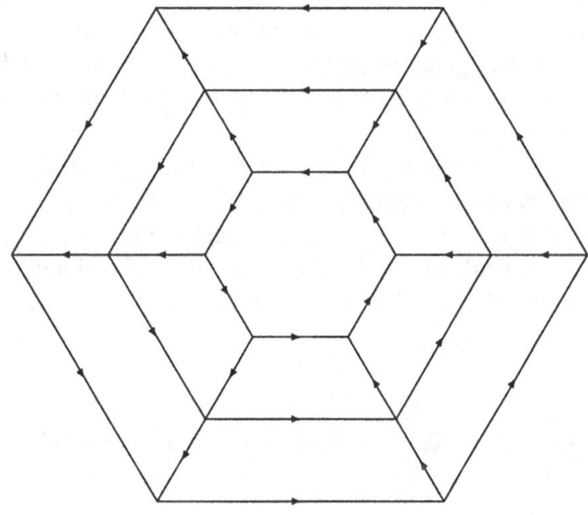

Figure 18: A directed cylinder of height 3

Definition A directed cylinder of height h (see Figure 18) consists of a set of $2h$ disjoint paths $R_1, \ldots, R_{2h}$ and a set of h disjoint directed cycles $C_1, \ldots, C_h$ such that each R_i and C_j intersect in a non-empty path and furthermore:

(i) If $i \leq h$ then R_i intersects the C_j in increasing index order,

(i) If $i > h$ then R_i intersects the C_j in decreasing index order,

(ii) For each C_j there is an arc e_j of C_j such that the path $P_j = C_j - e_j$ intersects the R_i in increasing index order.

Conjecture 6.13 *For every h there is an $f(h)$ such that if a directed graph contains a well-linked set of order $f(h)$ then it contains a directed cylinder of height h.*

We present this conjecture at the end of our discussion of brambles and tree decompositions. However, we hope that it is the beginning of a rich theory of directed brambles and directed tree decompositions.

7 Building a wall

In this section, we prove Theorem 1.2 which states that if a graph G has no wall of height g then it has no bramble of order greater than $25^{34g^5} - 1$. The proof we present is long and fairly technical. We foist the details on the reader for three reasons. The first is that they require an application of the Closest Cutter Lemma discussed in Section 3 and hence illustrate the importance of the laminarity of cutters. The second is that, as we saw in the last two sections, the theorem has many important applications. The third is that our proof although similar to that given in [46], avoids the use of a result from matroid theory, and uses tangles rather than preferences. We hope that this makes it more accessible to the reader.

As mentioned in Section 4, the main idea of the proof is to start with a large $\mathcal{T}$-linked set for some huge order tangle $\mathcal{T}$ and then grow a set $\mathcal{P}$ of paths from this set "towards" the tangle. These paths shall form the rows of our wall. We shall show that we can also finds a set $\mathcal{C}$ of disjoint paths between the elements of $\mathcal{P}$ which we will use to form the columns. We turn now to the details.

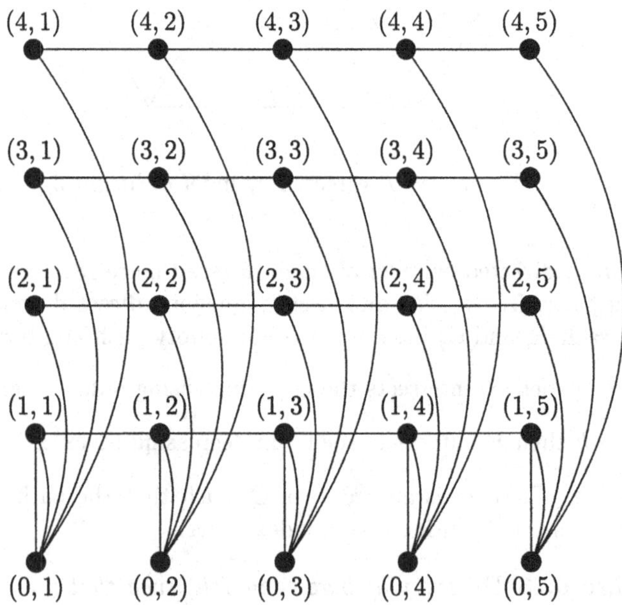

Figure 19: $L_{4,5}$

We begin with a definition. We denote by $L_{h,k}$ the graph with vertex set $\{(x,y) \mid 0 \leq x \leq h,\ 1 \leq y \leq k\}$ where $(0,y)$ is adjacent to (x,y) for $1 \leq y \leq k$, $1 \leq x \leq h$ and (x,y) is adjacent to $(x,y+1)$ for $1 \leq x \leq h$, $1 \leq y \leq k-1$ (see Figure 19).

We note that G has an $L_{h,k}$ minor if and only if there are vertex disjoint paths $P_1, \ldots, P_h$ in G and vertex disjoint trees $T_1, \ldots, T_k$ in $G - \bigcup_{i=1}^{h} V(P_i)$

such that for i between 1 and h, there are vertices $v_{i,1}, \ldots, v_{i,k}$ appearing in the given order along P_k with $v_{i,j}$ adjacent to a vertex of T_j.

We shall show that, for $g \geq 2$, if G is a graph with no wall of height g and G has a bramble of order at least $25^{(4gh^2+14h^2-19h)k}$ then G has an $L_{h,k}$ minor. We leave it as an exercise to the reader to prove that every graph containing L_{g+1,g^2} as a minor contains a wall of height g (the paths $P_1, \ldots, P_{g+1}$ correspond to the rows of the wall, we use the T_i to construct the columns). Thus, it follows that if G has a bramble of order at least 25^{34g^5} then G contains a wall of height g.

To describe how we find the desired $L_{h,k}$ minor, we need to introduce *expanders*, *concentrators*, and *h-steps*.

Definition Let r and s be integers with $1 \leq r \leq s$. An (r,s)-expander (H, X_1, X_2) consists of a graph H, two disjoint sets X_1 and X_2 of vertices of H with $|X_1| = r$, and $|X_2| = s$, such that for any subset Y of r vertices of X_2 there are r vertex disjoint paths from X_1 to Y in $H - (X_2 - Y)$.

Definition Let r and s be integers with $1 \leq r \leq s$. An (s,r)-concentrator (H, X_1, X_2, T) consists of a graph H, two sets X_1 and X_2 of vertices of H with $|X_1| = s$, and $|X_2| = r$, and a tree T in $H - X_2$ such that for some subset Z of r vertices of X_1 there are r vertex disjoint paths from Z to X_2 in $H - V(T) - (X_1 - Z)$ each of which contains some vertex adjacent to a vertex in T.

Definition An r-step (H, X_1, X_2, T) consists of a graph H, two sets X_1 and X_2 of vertices of H with $|X_1| = |X_2| = r$, and a tree T in $H - X_1 - X_2$ such that there are r vertex disjoint paths from X_1 to X_2 in $H - V(T)$ each of which contains some vertex adjacent to a vertex in T.

One way of constructing an r-step is to concatenate an (r,s)-expander and an (s,r)-concentrator.

We actually prove:

Lemma 7.1 *If G contains no wall of height g and has a bramble of order at least $25^{(4gh^2+14h^2-19h)k}$ then we can find, within G, a sequence of h-steps (H_1, X_0, X_1, T_1), (H_2, X_1, X_2, T_2), $\ldots$, (H_k, X_{k-1}, X_k, T_k) such that:*

(i) for $1 \leq i \leq k-1$, $H_i \cap H_{i+1} = X_i$, and

(ii) if $|i-j| > 1$ then $H_i \cap H_j = \emptyset$.

Now, in each $H_i - T_i$ there is a set $\mathcal{P}_i$ of h vertex disjoint paths from X_{i-1} to X_i each containing a vertex adjacent to some vertex in T_i. We let F be the subgraph of G with vertex set $\left(\bigcup_{j=1}^{k} V(T_j)\right) \cup (\bigcup V(P) \mid P \text{ is in some } \mathcal{P}_i)$. Then, clearly, F contains an $L_{h,k}$ minor.

In order to find the desired h-steps, we start with a tangle $\mathcal{T}$ of large order and a $\mathcal{T}$-linked set X_0. Each time we find an h-step: (H_i, X_{i-1}, X_i, T_i), we will find a tangle $\mathcal{T}_i$ in $G - H_1 - \cdots - H_i$ such that X_i is a $\mathcal{T}_i$-linked set. That is, we will prove the following two lemmas.

Lemma 7.2 *Let θ, h and n be integers with $\theta > n \geq h \geq 2$. Let $\mathcal{T}$ be a tangle of order $\theta 25^{nh}$ in a graph G, and let Z be a $\mathcal{T}$-linked set of order h in G. Then, there are subgraphs H and G' of G such that setting $Z' = V(H) \cap V(G')$ we have:*

(i) (H, Z, Z') is an (h, n)-expander, and

(ii) there is a tangle $\mathcal{T}'$ in G' of order θ and conformal with $\mathcal{T}$ such that Z' is a $\mathcal{T}'$-linked set.

Lemma 7.3 *Let g, h and θ be integers with $g \geq 2$, $h \geq 3$ and $\theta \geq 20gh$. Let $n = (2g+7)(2h-5)+2$ (note that $\theta > n$). Let $\mathcal{T}$ be a tangle of order $2\theta 25^{4g(h-1)+n}$ in a graph G containing no wall of height g. Let Z be a $\mathcal{T}$-linked set of order n in G. Then, there are subgraphs H and G' of G such that setting $Z' = V(H) \cap V(G')$ we have:*

(i) there is a tree T in $H - Z'$ such that (H, Z, Z', T) is an (n, h)-concentrator, and

(ii) there is a tangle $\mathcal{T}'$ in G' of order θ and conformal with $\mathcal{T}$ such that Z' is a $\mathcal{T}'$-linked set.

Now, as we have remarked, if (H, X_1, X_2) is an (h, n)-expander and if (F, X_2, X_3, T) is an (n, h)-concentrator such that $V(H) \cap V(F) = X_2$, then $(H \cup F, X_1, X_3, T)$ is an h-step. Thus, Lemmas 7.2 and 7.3 taken together imply

Lemma 7.4 *Let θ, g and h be integers with $g \geq 2$ and $h \geq 3$ and $\theta \geq 20gh$. Let $n = (2g+7)(2h-5)+2$ (note $\theta > n$). Let $\mathcal{T}$ be a tangle of order $2\theta 25^{nh+4g(h-1)+n}$ in a graph G containing no wall of height g. Let Z be a $\mathcal{T}$-linked set of order h in G. Then, there are subgraphs H and G' of G such that setting $Z' = V(H) \cap V(G')$ we have:*

(i) there is a tree T in $H - Z'$ such that (H, Z, Z', T) is an h-step, and

(ii) there is a tangle $\mathcal{T}'$ in G' of order θ and conformal with $\mathcal{T}$ such that Z' is a $\mathcal{T}'$-linked set.

Now, by (2.13) we know that every proper subset of a minimum order hitting set for $\mathcal{T}$ is a $\mathcal{T}$-linked set. Thus, if G contains a tangle $\mathcal{T}$ of order at least θ then it contains a $\mathcal{T}$-linked set of order at least θ. This fact, combined with k applications of Lemma 7.4 and our earlier remarks about finding an $L_{h,k}$ minor in a sequence of k h-steps yields

Theorem 7.5 *Let g, h, and k be integers with $g \geq 2$ and $h \geq 3$. Let $n = (2g+7)(2h-5)+2$. If $\mathcal{T}$ is a tangle of order at least $20gh25^{(nh+4g(h-1)+n+1)k}$ in a graph G which contains no wall of height g then G has an $L_{h,k}$ minor.*

Since L_{g+1,g^2} contains a wall of height h we obtain:

Corollary 7.6 *If G has a tangle of order at least 25^{34g^5-1} then it contains a wall of height h.*

Proof For $g \geq 2$ this follows from (7.5). If a graph has no wall of height 1, then it contains no cycle of length 6 or greater. This implies that each of its blocks is a subdivision of a multigraph with at most 5 vertices. It is easy to construct a tree decomposition of any such graph of width at most 5. The corollary follows. ∎

Since every bramble of order $3k$ defines a tangle of order k, this yields:

Corollary 7.7 *If G has a bramble of order at least 25^{34g^5} then it contains a wall of height h.*

It remains only to prove Lemmas 7.2 and 7.3. The following two technical results whose proofs appear at the end of this section, are the keys to these two lemmas.

Lemma 7.8 *Let $\theta \geq 2$ be an integer. Let $\mathcal{T}$ be a tangle of order at least $24\theta+7$ in a graph G. Let $\{s,t\}$ be a $\mathcal{T}$-linked set in G. Then there is a path P between s and t such that there is a tangle in $G - V(P)$ of order θ, conformal with $\mathcal{T}$.*

Lemma 7.9 *Let C be a connected subgraph of a graph G, let h be an integer, and let $\mathcal{T}$ be a tangle of order greater than $2h$ in $G - C$ which we call $\mathcal{T}_0$ when thinking of it as a tangle in G. Let Z be a $\mathcal{T}_0$-linked set of order h which intersects C. Then G contains two subgraphs G' and H such that setting $Z' = V(H) \cap V(G')$ we have:*

(i) $|Z'| = h$ and there are h vertex disjoint paths in H between Z and Z',

(ii) there is a tangle $\mathcal{T}'$ in G' of order $\mathrm{ord}(\mathcal{T}') - h$ and conformal with $\mathcal{T}$ such that Z' is a $\mathcal{T}$-linked set, and

(iii) $G' \cap C = \emptyset$.

To illustrate the importance of Lemmas 7.8 and 7.9 we use them to deduce the following two lemmas, from which Lemma 7.2 follows almost immediately.

Lemma 4.1 *Let h and θ be integers with $\theta > h \geq 2$. Let $\mathcal{T}$ be a tangle in a graph G of order at least $24(h+\theta)+7$, and let Z be a $\mathcal{T}$-linked set of order h in G. Let s and t be two vertices of Z. Then, there are subgraphs H and G' of G such that setting $Z' = V(H) \cap V(G')$ we have:*

(i) $|Z'| = h$ and there are h vertex-disjoint paths $P_1, \ldots, P_h$ in H from Z to Z' and an s to t path Q in $H - Z'$ such that for each i, $Q \cap P_i$ is a (possibly empty) path, and

(ii) there is a tangle $\mathcal{T}'$ in G' of order θ and conformal with $\mathcal{T}$ such that Z' is a $\mathcal{T}'$-linked set.

Proof By Lemma 7.8, there is a path C between s and t and a tangle $\mathcal{T}_1$ in $G - V(C)$ of order $h + \theta$ conformal with $\mathcal{T}$. So, we have that (in G) $\mathcal{T}_1$ is indistinguishable from $\mathcal{T}$. We claim that (in G) Z is a $\mathcal{T}_1$-linked set. Otherwise, there is a set X with fewer than $|Z|$ vertices separating Z from some hitting set H for $\mathcal{T}_1$. Since $\mathrm{ord}(\mathcal{T}_1) > |Z| > |X|$, there is some component U of $G - X$ containing an element T of $\mathcal{T}_1$. Since T intersects H, $U \cap Z = \emptyset$. Since $\mathcal{T}$ is indistinguishable from $\mathcal{T}_1$, we know there is an element of $\mathcal{T}$ contained in U. It follows that every element of $\mathcal{T}$ intersects $U + X$, and hence there is a hitting set H' for $\mathcal{T}$ in $U + X$. But then X is a cutset separating Z from H'. This contradicts the fact that Z is $\mathcal{T}$-linked, proving the claim.

Since $\mathrm{ord}(\mathcal{T}_1) > 2|Z|$, we may apply Lemma 7.9. We deduce that there are subgraphs H and G' of G such that setting $Z' = V(H) \cap V(G')$, we have:

(i) $|Z'| = h$ and there are h vertex disjoint paths $P_1, \ldots, P_h$ in H between Z and Z',

(ii) there is a tangle $\mathcal{T}'$ in G' of order $\mathrm{ord}(\mathcal{T}') - h = \theta$ and conformal with $\mathcal{T}$ such that Z' is a $\mathcal{T}$-linked set, and

(iii) $G' \cap C = \emptyset$.

Now, we let Q be a path in $H - Z'$ between s and t which minimizes $|E(Q) - \bigcup_1^k E(P_i)|$. Such a choice is possible because $Q = C$ is one possibility. Clearly, Q intersects each P_i in a path, as required. ∎

Lemma 7.10 *Let θ and n be integers with $\theta > n \geq 2$. Let $\mathcal{T}$ be a tangle of order at least $\theta 25^n$ in a graph G, and let Z be a $\mathcal{T}$-linked set in G of order n. Then, there are two subgraphs H and G^* of G such that setting $Z^* = V(H) \cap V(G^*)$, we have:*

(i) $|Z^*| = n$, and there are n vertex disjoint paths from Z to Z' in H,

(ii) there is a tangle $\mathcal{T}^*$ in G^* of order θ and conformal with $\mathcal{T}$ such that Z^* is a $\mathcal{T}^*$-linked set, and

(iii) Z is in one component of $H - Z^*$.

Proof We prove that for each r between 0 and $n-1$ there are subgraphs H_r and G'_r of G such that setting $Z_r = V(H_r) \cap V(G'_r)$ we have that: $H_{r-1} \subseteq H_r$, $|Z_r| = n$ and there are n vertex-disjoint paths from Z to Z_r, there is a tangle $\mathcal{T}_r$ in G'_r of order $\theta 25^{n-r}$ and conformal with $\mathcal{T}$ such that Z_r is a $\mathcal{T}_r$-linked set, and Z is in the union of $n-r$ components of $H_r - Z_r$. The lemma follows as we simply set $Z^* = Z_{n-1}$, $G^* = G'_{n-1}$, and $\mathcal{T}^* = \mathcal{T}_{n-1}$.

We begin by setting $Z_0 = Z$, $G'_0 = G$, $\mathcal{T}_0 = \mathcal{T}$, and H_0 to be the graph induced by Z. For $r \geq 1$, having obtained $Z_{r-1}, H_{r-1}, G'_{r-1}$ and $\mathcal{T}_{r-1}$, we obtain Z_r, H_r, G'_r and $\mathcal{T}_r$ by applying Lemma 4.1 to Z_{r-1}, H_{r-1}, G'_{r-1} and $\mathcal{T}_{r-1}$ and a pair of vertices s and t of Z_{r-1} which are in different components of H_{r-1} (unless all of Z_{r-1} is in the same component of H_{r-1} in which case s and t are chosen arbitrarily from H_{r-1}). This yields two subgraphs H and G' in G_r, and a tangle $\mathcal{T}'$ in G'. We set $H_r = H \cup H_{r-1}$, $G'_r = G'$, $Z_r = V(H) \cap V(G')$, and $\mathcal{T}_r = \mathcal{T}'$. ∎

We now deduce Lemma 7.2.

Proof of Lemma 7.2 Let h, θ, n, G, $\mathcal{T}$ and Z be as in the statement of the lemma. We know that Z is contained in a $\mathcal{T}$-linked set Z_0 of order n. We let $\mathcal{T}_0 = \mathcal{T}$ and $G_0 = G$. We shall find a sequence of sets $Z_1, \ldots, Z_h$, two sequences of graphs $H_1, \ldots, H_h$ and $G_1, \ldots, G_h$ and a sequence of tangles $\mathcal{T}_1, \ldots, \mathcal{T}_h$ such that for $i \in \{1, \ldots, h\}$:

(i) $V(H_i) \cap V(G_i) = Z_i$,

(ii) $H_i \subseteq G_{i-1}$,

(iii) $|Z_i| = n$ and there are n vertex disjoint paths from Z_0 to Z_i in $\bigcup_{j=1}^{i} H_j \subseteq G - G_i$,

(iv) Z_{i-1} is contained in one component of $H_i - Z_i$, and

(v) $\mathcal{T}_i$ is a tangle of order at least $\theta 25^{(h-i)n}$ in G_i conformal with $\mathcal{T}$ and Z_i is a $\mathcal{T}_i$-linked set.

We find the desired Z_i, H_i, G_i, $\mathcal{T}_i$ using h applications of Lemma 7.10.

We claim that setting $Z' = Z_h$, $G' = G_h$, $H = \bigcup_{i=1}^{h} H_i$, $\mathcal{T}' = \mathcal{T}_n$ proves the lemma. We need only show that for every set Y in Z' with $|Y| = h$ there are h vertex disjoint paths from Z to Y in H. If the paths did not exist then there would be a set X of at most $h-1$ vertices which separates Z from Y. Now, we know there are n vertex disjoint paths from Z_0 to Z' in H. Clearly, there is one such path P which is disjoint from X and has an endpoint in Z, and another P' which is disjoint from X and has an endpoint in Y. Furthermore, there is some H_l with $H_l \cap X = \emptyset$. However, both P and P' contain a vertex of Z_{l-1} and Z_{l-1} lies in one component of H_l, and hence of $H - X$. This contradicts our assumption that X separates Z from Y in H. ∎

The proof of Lemma 7.3 requires a bit more work. In particular, we will need the following lemma.

Lemma 7.11 *Let θ, g, and n be integers with $\theta > n \geq (2g+5)$. Let $\mathcal{T}$ be a tangle of order at least $\theta 25^{4g}$ in a graph G which contains no wall of height g. Let Z be a $\mathcal{T}$-linked set of order n in G and let $\{v_1, \ldots, v_{2g+5}\}$ be some ordered set of vertices of Z. Then there are graphs F and G^* in G such that setting $Z^* = V(F) \cap V(G^*)$, we have:*

(i) $|Z^*| = n$ and there is a set $\mathcal{P}$ of n vertex disjoint paths from Z to Z^* in F,

(ii) there is some $l \in \{3, \ldots, 2g+3\}$ such that letting P_1 be the element of $\mathcal{P}$ which has v_l as an endpoint, there is some other element P_2 of $\mathcal{P}$ which has neither v_{l-1} nor v_{l+1} as an endpoint and a path Q of $F - \bigcup\{V(P) \mid P \in \mathcal{P} - P_1 - P_2\}$ which has an endpoint on P_1 and an endpoint on P_2,

(iii) there is a tangle $\mathcal{T}^*$ in G^* of order at least θ and conformal with $\mathcal{T}$ for which Z^* is a $\mathcal{T}^*$-linked set.

Proof We define $Z_0 = \{z_1^0, \ldots, z_n^0\} = Z$, $\mathcal{T}_0 = \mathcal{T}$, and $G_0 = G$. For $m \in \{1, \ldots, 2g+5\}$, we define k_m so that $z_{k_m}^0 = v_m$. We apply Lemma 4.1, $4g-1$ times to prove the existence of a sequence of graphs $G_1 \supset \cdots \supset G_{4g-1}$, a sequence of tangles $\mathcal{T}_1$, ..., $\mathcal{T}_{4g-1}$, and a sequence of sets of vertices $Z_1, \ldots$, Z_{4g-1} with $Z_i = \{z_1^i, \ldots, z_n^i\}$, such that for $1 \leq i \leq 4g-1$

(i) $Z_{i-1} \subseteq G_{i-1} - G_i$ and there is a set $\mathcal{P}_i = \{P_{i,1}, \ldots, P_{i,n}\}$ of n vertex disjoint paths in $G_{i-1} - (G_i - Z_i)$ from Z_{i-1} to Z_i such that $P_{i,j}$ links z_j^{i-1} to z_j^i,

(ii) $\mathcal{T}_i$ is a tangle of order at least $\theta 25^{4g-i}$ in G_i conformal with $\mathcal{T}$ and Z_i is a $\mathcal{T}_i$-linked set of G_i,

(iii) there is a path Q_i of $G_{i-1} - G_i$ between $z_{k_{g+3}}^{i-1}$ and $z_{k_1}^{i-1}$ whose intersection with every element of $\mathcal{P}_i$ is a (possibly empty) path.

For each j in $1, \ldots, n$ we let R_j be the path from z_j^0 to z_j^{4g-1} obtained by concatenating $\{P_{i,j} \mid 1 \leq i \leq 4g-1\}$. Now, if for any Q_i there is a subpath Q of Q_i which has its endpoints on R_{k_l} and R_r with $l \in \{3, \ldots, 2g+3\}$ and $r \notin \{k_{l-1}, k_l, k_{l+1}\}$ then setting $Z^* = Z_{4g-1}$, $\mathcal{T}^* = \mathcal{T}_{4g-1}$, $G^* = G_{4g-1}$, $F = G - (G_{4g-1} - Z_{4g-1})$, and $\mathcal{P} = \{R_j \mid 1 \leq j \leq n\}$ shows that the lemma holds.

So, we can assume that each Q_i either intersects $R_{k_{g+3}}$, $R_{k_{g+4}}$, ..., $R_{k_{2g+3}}$ in the given order before intersecting any other R_j, or intersects $R_{k_{g+3}}$, $R_{k_{g+2}}$, ..., R_{k_3} in the given order before intersecting any other R_j. Now, either there are $2g$ values of i such that Q_i satisfies the first condition or there are $2g$ values of i such that Q_i satisfies the second condition. In the first case, the subgraph consisting of these Q_i along with $R_{k_{g+3}}, \ldots, R_{k_{2g+3}}$ clearly contains a wall of height g. Similarily, in the second case, there is a wall of height g in G. But G contains no wall of height g so the lemma holds. ∎

Proof of Lemma 7.3 Let g, h, n, θ, G, $\mathcal{T}$ and Z be as in the statement of Lemma 7.3. Set $Z_0 = Z$, $G_0 = G$, and $\mathcal{T}_0 = \mathcal{T}$. By Lemma 7.10, we can find an H_1, G_1, Z_1, $\mathcal{T}_1$, and T_1 such that $V(H_1) \cap V(G_1) = Z_1$, T_1 is a tree in $H_1 - Z_1$, $|Z_1| = n$, there are n vertex-disjoint paths $P_1, \ldots, P_n$ in H_1 from Z_0 to Z_1 such that T_1 intersects each P_i and $\mathcal{T}_1$ is a tangle in G_1 of order $2\theta 25^{4g(h-1)}$ conformal with $\mathcal{T}$ such that Z_1 is a $\mathcal{T}_1$ linked set. By choosing T_1 so as to minimize $E(T) - \bigcup_{i=1}^{n} E(P_i)$ we can ensure that T_1 intersects each P_i in a path. Furthermore we can also ensure that each leaf l of T_1 is $T \cap P_i$ for some i.

We prove by induction on r that for each r between 1 and $h-1$ we can find subgraphs H_r and G_r of G such that setting $Z_i = V(H_i) \cap V(G_i)$ we have:

(1) $H_{r-1} \subseteq H_r$,

(2) $|Z_r| = n$ and there is a set $\mathcal{P}_r$ of n vertex disjoint paths $P_{r,1}, \ldots, P_{r,n}$ of H_r between Z and Z_r,

(3) there is a tree T_r in $H_r - Z_r$ with at least $r+1$ leaves intersecting each $P_{r,i}$ in a non-empty path such that each leaf of T_r is the intersection of T_r with some $P_{r,j}$, and

(4) there is a tangle $\mathcal{T}_r$ of order $2\theta 25^{4g(h-r-1)}$ in G_r and conformal with $\mathcal{T}$ such that Z_r is a $\mathcal{T}_r$-linked set.

This implies Lemma 7.3 as can be seen by letting $v_1, \ldots, v_h$ be h leaves of T_{h-1}, letting z_i be the vertex of Z_{h-1} which is on the path in $\mathcal{P}_{h-1}$ which contains v_i, letting $T = T_{h-1} - \{v_1, \ldots, v_h\}$, letting $Z' = \{z_1, \ldots, z_h\}$, letting $G' = G_{h-1} - (Z_{r-1} - Z')$, letting $H = H_{h-1}$, and letting $\mathcal{T}' = \mathcal{T}_{h-1}/(Z_{h-1} - Z')$.

Now, we see that the result holds for $r = 1$. To prove the existence of H_r and G_r satisfying (1)–(4) given H_{r-1} and G_{r-1}, we will apply Lemma 7.11 to Z_{r-1} in G_{r-1}. To this end, consider the tree T_{r-1}. We can assume that T_{r-1} has no more than r leaves as otherwise, we simply set $T_r = T_{r-1}$, $G_r = G_{r-1}$, $H_r = H_{r-1}$, $\mathcal{T}_r = \mathcal{T}_{r-1}$, and $P_{r,j} = P_{r-1,j}$. So, T_{r-1} has exactly r leaves and can therefore be partitioned into at most $2r-3$ paths, all the internal vertices of which have degree 2 in T_{r-1}. Now, one of these paths R intersects at least $n/(2r-3) \geq 2g+7$ of the paths in $\mathcal{P}_{r-1}$, and hence there are at least $2g+5$ paths of $\mathcal{P}_{r-1}$ whose intersection with T_{r-1} is contained in the interior of R. Let $L_1, \ldots, L_{2g+5}$ be the first $2g+5$ such paths, enumerated in the order that they appear along R. Let w_i be the endpoint of L_i in Z_{r-1}. We apply Lemma 7.11 to Z_{r-1}, $\mathcal{T}_{r-1}$, G_{r-1}, and $\{w_1, \ldots, w_{2g+5}\}$ to obtain F, G^*, $\mathcal{T}^*$, Z^*, w_l, $\mathcal{P}$ and Q. We set $Z_r = Z^*$, $G_r = G^*$, $\mathcal{T}_r = \mathcal{T}^*$, $H_r = H_{r-1} \cup F$, we concatenate $P_{r-1,j}$ with an element of $\mathcal{P}$ to obtain $P_{r,j}$, and we use T_{r-1} and Q to create T_r. To this end, let R_1 and R_2 be the paths of $\mathcal{P}_r$ containing the endpoints of Q with $L_l \subset R_1$. Note that there is a path Q' in $Q \cup R_1 \cup R_2$ between a vertex of $L_l \cap T_{r-1}$ and $R_2 \cap T_{r-1}$ with no internal vertex in T_{r-1} which intersects both R_1 and R_2 in a path. Let C be the unique cycle of $T_{r-1} \cup Q'$. We know that C contains either a subpath of R between L_{l-2} and L_{l-1} or a subpath of R between L_{l+1} and L_{l+2}. By symmetry, we may assume the second possibility occurs. Let w be the endpoint of Q' not in L_l. Assume first that w is not in L_{l+2}. Then letting Q^* be the maximal subpath of R with one endpoint x in L_{l+1} and the other y in L_{l+2} we see that letting T_r be the tree $T_{r-1}+Q'-(Q^*-x-y)$ yields the desired result (as x and y are leaves of T_r which are not leaves of T_{r-1} and every leaf of T_{r-1} with the possible exception of w is a leaf of T_r). Otherwise, w is in $R \cap L_{l+2}$ and hence is not a leaf of T_{r-1}. So, letting Q^* be a maximal path of R with one endpoint w and the other x in L_{l+1} and setting $T_r = T_{r-1}+Q-(Q^*-w-x)$ yields the desired result. ∎

Proof of Lemma 7.9 The proof of Lemma 7.9 relies on the following easy corollary of the Closest Cutter Lemma.

> **Lemma 7.12 (The Relinking Lemma)** *Let $\mathcal{T}$ be a tangle of order at least $h+2$ in a graph G. Let Z be a $\mathcal{T}$-linked set in G with h vertices. Let z be a vertex of Z. Then there is some neighbour y of z such that $Z-z+y$ is a $\mathcal{T}/z$-linked set in $G-z$.*
>
> **Proof** Since Z is $\mathcal{T}$-linked, $Z-z$ is $\mathcal{T}/z$-linked in $G-z$. So, by the Closest Cutter Lemma, there is a set X^* of $h-1$ vertices of $G-z$ such that X^* separates $Z-z$ from $\mathcal{T}/z$ and such that any set Y of $h-1$ vertices of $G-z$ separating $Z-z$ from $\mathcal{T}/z$ is disjoint from $f_{\mathcal{T}/z}(X^*)$. Note that $f_{\mathcal{T}/z}(X^*) = f_{\mathcal{T}}(X^*)$. Thus, because Z is $\mathcal{T}$-linked, there is a neighbour y of z in $f_{\mathcal{T}/z}(X^*)$. By our choice of X^*, $Z-z+y$ is $\mathcal{T}/z$-linked. ∎

Now, repeated applications of the Relinking Lemma yield:

> **Lemma 7.13 (The Linked Set Moving Lemma)** *Let $\mathcal{T}$ be a tangle of order at least $2h+1$ in a graph G. Let $Z=\{z_1,\dots,z_h\}$ be a $\mathcal{T}$-linked set in G with h vertices. Then there is a set $Y=\{y_1,y_2,\dots,y_h\}$ of vertices of $G-Z$ such that y_i is a neighbour of z_i and Y is a $\mathcal{T}/Z$-linked set in $G-Z$.*

With these lemmas in hand we are ready to prove Lemma 7.9. So, let h, G, $\mathcal{T}$, Z, and C be as in the statement of the lemma. By induction, we can assume the statement holds for any minor of G. By the Closest Cutter Lemma, there is a $\mathcal{T}$-linked set X^* such that X^* is a set of h vertices separating Z from $\mathcal{T}$, any set of h vertices separating Z from $\mathcal{T}$ is disjoint from $f_{\mathcal{T}}(X^*)$, and there are h vertex disjoint paths between Z and X^*.

We consider two cases.

Case 1: $C \cap f_{\mathcal{T}}(X^*) = \emptyset$.

In this case, by the Linked Set Moving Lemma, there exists a $\mathcal{T}/X^*$ linked set Y of h vertices in $G - X^*$ such that there is a matching between Y and X^*. Obviously, $Y \subseteq f_{\mathcal{T}}(X^*) \subseteq G - C$. Furthermore, the paths from Z to X^* are disjoint from $f_{\mathcal{T}}(X^*)$ and $\mathcal{T}/X^*$ is contained in $f_{\mathcal{T}}(X^*)$. So, setting $G' = f_{\mathcal{T}}(X^*)$, H equal to the subgraph of G induced by $G - (f_{\mathcal{T}}(X^*) - Y)$, $Z' = Y$, and $\mathcal{T}' = \mathcal{T}/X^*$ we see that Lemma 7.9 holds.

Case 2: $C \cap f_{\mathcal{T}}(X^*) \neq \emptyset$.

In this case, since C is connected and $C \cap Z \neq \emptyset$, there is an edge e of C with at least one endpoint in $f_{\mathcal{T}}(X^*)$. Let G^* be the graph obtained by contracting e. Let Z^* consist of those vertices of Z in G^* and the vertex of G^* not in G if one of the endpoints of e was in Z. Clearly $\mathcal{T}$ is a tangle in $G^* - C = G - C$. We claim that Z^* is a $\mathcal{T}$-linked set in G^*. Otherwise, there is a set Y of at most $h - 1$ vertices of G^* separating Z^* from $\mathcal{T}$. Now, Y corresponds to a set of vertices of G which separate Z from $\mathcal{T}$. This set is either Y or is a set of h vertices containing both endpoints of e. In the first case, we contradict the fact that Z is $\mathcal{T}$-linked. In the second case, we contradict the choice of X^*. So, by the induction hypothesis, we can find subgraphs H and G' of G^* as described in Lemma 7.9. By "uncontracting" the edge e, we convert these to subgraphs which show that Lemma 7.9 holds for G. ■

Proof of Lemma 7.8 We begin with the following lemma.

Lemma 7.14 *Let $h \geq 2$ be an integer, let G be a graph, let Z be a well-linked set of $24h+6$ vertices of G, let (S, T) be a partition of Z with $|S| = |T| = 12h + 3$. Then there is a path P with one endpoint in S, the other in T, and otherwise disjoint from Z such that for any set Y of fewer than $h + 2$ vertices of $G - P$, there is a component of $G - P - Y$ which contains at least $16h + 3$ vertices of Z.*

Proof Let $\mathcal{T}$ be the tangle consisting of those connected subgraphs of G containing more than two thirds of the vertices of Z. Because Z is well-linked, $\mathcal{T}$ must have order at least $8h + 2$. To begin, we show:

(1) If there is a path P from S to T and a set X of at most $7h$ vertices such that X separates P from $\mathcal{T}$, then P shows that the lemma is true.

Proof of (1): To see this note that for any set Y of less than $h + 2$ vertices of G none of which are on P, $f_{\mathcal{T}}(X \cup Y) \subseteq f_{\mathcal{T}}(X)$ and hence is disjoint from P. So, $f_{\mathcal{T}}(X \cup Y)$ is contained in a component of $G - P - Y$ which therefore contains at least $16h + 3$ vertices of Z. □

Now, let $\mathcal{P}$ be a set of $12h+3$ vertex-disjoint paths from S to T in G. Then, every element of $\mathcal{P}$ is internally vertex-disjoint from Z and we shall prove that one of them satisfies the theorem. We say a path P_1 in $\mathcal{P}$ splits a pair (P_2, P_3) of paths in $\mathcal{P}$, if there do not exist $h+2$ vertex disjoint paths from P_2 to P_3 in $G - P_1$. We may assume that:

(2) For any three distinct paths P_1, P_2, and P_3 of $\mathcal{P}$, if P_1 splits (P_2, P_3) then P_2 does not split (P_1, P_3).

Proof of (2): If there do not exist $7h$ vertex disjoint paths from P_3 to P_1 then (1) implies that one of P_1 or P_3 shows that the lemma is true. So, we can assume such paths exist. Each of these paths either hits a vertex of P_2 first or a vertex of P_1 first. The result follows. □

By (2), there are at most $\binom{12h+3}{3}$ pairs $\{P_1, (P_2, P_3)\}$ such that P_1, P_2, and P_3 are distinct paths of $\mathcal{P}$ and P_1 splits (P_2, P_3). It follows that there is some path P of $\mathcal{P}$ which splits at most $(12h+2)(12h+1)/6$ unordered pairs of paths of $\mathcal{P} - P$. We claim that P shows the lemma is true. For, if not then there is a set Y of less than $h+2$ vertices of $G - P$ such that every component of $G - Y - P$ contains fewer than $16h+3$ vertices of Z. It follows that there is a separation (A, B) of $G - P$ with $V(A \cap B) = Y$ such that A and B both contain at least $8h+2$ vertices of $Z - P$. Since $|Y| < h+2$, there are at least $11h+1$ paths of $\mathcal{P} - P$ which are either wholly within A or wholly within B. Furthermore, there must be at least $7h/2$ paths of $\mathcal{P} - P$ wholly within A and $7h/2$ paths of $\mathcal{P} - P$ wholly within B. Also, P splits any pair of paths, one of which is wholly within A and the other of which is wholly within B. It follows that P splits at least $7h(15h+2)/4$ paths, a contradiction. ■

We turn now to the proof of Lemma 7.8. So, we let θ, G, $\mathcal{T}$, $\{s,t\}$ be as in the statement of the lemma. We choose a set Z of at most $24\theta+6$ vertices of G, a partition (S, T) of Z, and two disjoint connected subgraphs J and K of $G - f_{\mathcal{T}}(Z)$ such that

(1) Z is $\mathcal{T}$-linked,

(2) $|S|, |T| \leq 12\theta + 3$,

(3) $s \in V(J)$, $t \in V(K)$, $J \cap Z$ is a vertex of S, $K \cap Z$ is a vertex of T, and

(4) every vertex of $S - J$ is adjacent to a vertex of J, every vertex of $T - K$ is adjacent to a vertex of K.

Furthermore, subject to (1)–(4) we choose Z so as to minimize $f_{\mathcal{T}}(Z)$. Such a choice is possible because setting $Z = \{s,t\}$, $J = S = s$, $K = T = t$, satisfies (1)–(4).

We claim that:

(5) there is no set Y of $|Z|$ vertices which separates Z from $\mathcal{T}$.

Proof of (5): If such a Y exists then by the Closest Cutter Lemma, we can choose Y so that there are $r = |Z|$ vertex disjoint paths $P_1, \ldots, P_r$ from Z to Y. We enumerate Z as $v_1, \ldots, v_r$ and Y as $w_1, \ldots, w_r$ so that P_i has endpoints v_i and w_i. Let J' be the union of J and

$$\bigcup_{v_i \in S} P_i - w_i + \{w_i \mid v_i \text{ is the unique vertex of } J \cap S\},$$

then J' is connected by (4). Let K' be the union of K and

$$\bigcup_{v_i \in T} P_i - w_i + \{w_i \mid v_i \text{ is the unique vertex of } K \cap T\},$$

then K' is also connected by (4). Let $S' = \{w_i \mid v_i \in S\}$ and let $T' = \{w_i \mid v_i \in T\}$. Then, Y, S', T', J', K' contradicts the minimality of $f_{\mathcal{T}}(Z)$. □

From (5), we obtain easily

(6) $|S| = |T| = 12\theta + 3$.

Proof of (6): Suppose S contains fewer than $12\theta+3$ vertices. Then let v be the unique vertex in $J \cap S$. Since Z is $\mathcal{T}$-linked, v must have a neighbour w in $f_{\mathcal{T}}(Z)$. Furthermore, by (5), $Z+v$ must be $\mathcal{T}$-linked. But now, $Z+v$, $S+v$, T, J, K contradicts the minimality of $f_{\mathcal{T}}(Z)$. So $|S| = 12\theta + 3$. Similarily, we obtain $|T| = 12\theta + 3$, proving (6). □

We are now in a position to define a tangle in $f_{\mathcal{T}}(Z) \cup Z$, using:

(7) Z is a well-linked set in the graph G' induced by $Z \cup f_{\mathcal{T}}(Z)$.

Proof of (7): Suppose not, then there is a separation (A, B) of G' such that setting $Y = V(A \cap B)$, we have $|Y \cup (A \cap Z)|, |Y \cup (B \cap Z)| \le 24\theta + 5$. Let $Y' = (A \cap Z) \cup Y$. Let $Y'' = (B \cap Z) \cup Y$.

Note that, by (1), $f_{\mathcal{T}}(Y') \not\subset f_{\mathcal{T}}(Z)$ and $f_{\mathcal{T}}(Y'') \not\subset f_{\mathcal{T}}(Z)$. However, clearly if $f_{\mathcal{T}}(Y')$ intersects A then it is contained in $A - Z \subset f_{\mathcal{T}}(Z)$, a contradiction. So, $f_{\mathcal{T}}(Y') \cap A = \emptyset$. Similarly, $f_{\mathcal{T}}(Y'') \cap B = \emptyset$. So, $f_{\mathcal{T}}(Y') \cap f_{\mathcal{T}}(Y'') \cap f_{\mathcal{T}}(Z) = \emptyset$ and further any edge with an endpoint in $f_{\mathcal{T}}(Y') \cap f_{\mathcal{T}}(Y'')$ is disjoint from $f_{\mathcal{T}}(Z)$. But this contradicts the fact that $\mathcal{T}$ is a tangle (consider the tangle elements in $f_{\mathcal{T}}(Y')$, $f_{\mathcal{T}}(Y'')$, and $f_{\mathcal{T}}(Z)$). □

Now, from (7), (6), and Lemma 7.14 we obtain:

(8) There is a path P' of G' from S to T, with no internal vertex in Z such that for every set Y of less than $\theta + 2$ vertices of G' some component of $G' - Y - P'$ contains more than $16\theta + 2$ vertices of $Z - P$.

Now, let P be a path of G from s to t whose vertex set is contained in $J \cup K \cup P'$ (such a path exists by (4)). By (3), $|Z - P - J - K| \geq 24\theta + 2$ and $|(V(P) - V(P')) \cap G'| \leq 2$. So, by (8), for any set Y of less than θ vertices of G' there is a component of $G' - Y - P$ containing at least $16\theta + 3$ vertices of $Z - P$. It follows that the tangle $\mathcal{T}' = \{B \mid B$ is a connected subgraph of $G' - P$ containing more than two thirds of the vertices of $Z - P\}$ has order at least θ. Furthermore, this tangle has order at most $8\theta + 2$ as any set of $8\theta + 2$ vertices of $Z - P$ is a hitting set for it. To complete the proof, we need to show that $\mathcal{T}'$ is indistinguishable from $\mathcal{T}$ in G. To this end, consider any set Y of less than $\mathrm{ord}(\mathcal{T}') \leq 8\theta + 2$ vertices of G. Clearly, $Y \cup (f_{\mathcal{T}}(Y) \cap Z)$ separates Z from $\mathcal{T}$. Since Z is $\mathcal{T}$-linked, it follows that $Z \cap f_{\mathcal{T}}(Z) \geq |Z| - (8\theta + 2) > 16\theta + 3$. Thus $f_{\mathcal{T}'}(Y) = f_{\mathcal{T}}(Y)$, as required. This completes the proof of Lemma 7.8 and hence of Theorem 1.2 ∎

This proof can easily be strengthened to obtain Theorem 4.2. We omit the details because even the most doughty reader must be suffering from fatigue.

Acknowledgements

I would like to thank Colin McDiarmid for his valuable suggestions on an earlier draft. His long-term support for my project to write a survey on this topic has been even more valuable. I would also like to thank Rosemary Bailey for her painstaking work as editor. Her contributions include, but are not limited to, drawing the figures and verifying and amending the reference list.

References

[1] N. Alon, P. Seymour & R. Thomas, A separator theorem for graphs with an excluded minor and its applications, in *Proceedings of the 22nd Annual Association for Computing Machinery Symposium on Theory of Computing*, ACM Press, New York (1990), pp. 293–299.

[2] N. Alon, P. D. Seymour & R. Thomas, Planar separators, *SIAM Journal on Discrete Mathematics*, **7** (1994), 184–193.

[3] D. Archdeacon & P. Huneke, A Kuratowski theorem for nonorientable surfaces, *Journal of Combinatorial Theory, Series B*, **46** (1989), 173–231.

[4] S. Arnborg, D. G. Corneil & A. Proskurowski, Complexity of finding embeddings in a k-tree, *SIAM Journal on Algebraic and Discrete Methods*, **8** (1987), 277–284.

[5] S. Arnborg, B. Courcelle, A. Proskurowski & D. Seese, An algebraic theory of graph reduction, Technical report LaBRI-90-02, Université de Bordeaux, 1990.

[6] S. Arnborg, J. Lagergren & D. Seese, Easy problems for tree-decomposable graphs, *Journal of Algorithms*, **12** (1991), 308–340.

[7] S. Arnborg & A. Proskurowski, Linear time algorithms for NP-hard problems restricted to partial k-trees, *Discrete Applied Mathematics*, **23** (1989), 11–24.

[8] H. L. Bodlaender, Dynamic programming on graphs of bounded treewidth, in *Proceedings of the 15th International Colloquium on Automata, Languages and Programming* (eds. T. Lepistö & A. Salomaa), *Lecture Notes in Computer Science*, 317, Springer Verlag, Berlin (1988), pp. 105–118.

[9] H. L. Bodlaender, Polynomial algorithms for graph isomorphism and chromatic index on partial k-trees, *Journal of Algorithms*, **11** (1990), 631–643.

[10] H. L. Bodlaender, A linear time algorithm for finding tree decompositions of small treewidth, in *Proceedings of the 25th Annual Association for Computing Machinery Symposium on Theory of Computing*, ACM Press, New York (1993), 226–234.

[11] H. Bodlaender & T. Kloks, Better algorithms for pathwidth and treewidth of graphs, in *Proceedings of the 18th International Colloquium on Automata, Languages and Programming* (eds. J. Leach Albert, B. Monien & M. Rodríguez Artalejo), *Lecture Notes in Computer Science*, 510, Springer Verlag, Berlin (1991), 544–555.

[12] F. R. K. Chung, *Spectral Graph Theory*, American Mathematical Sociey, Providence, Rhode Island (1997).

[13] B. Courcelle, The monadic second order logic of graphs. I. Recognizable sets of finite graphs, *Information and Computation*, **85** (1990), 12–75.

[14] I. Dejter & V. Neumann-Lara, Unboundedness for generalized odd cycle traversability and a Gallai conjecture, paper presented at the Fourth Caribbean Conference on Computing, Puerto Rico, 1985.

[15] P. Erdős & L. Pósa, On independent circuits contained in a graph, *Canadian Journal of Mathematics*, **17** (1965), 347–352.

[16] M. R. Fellows & M. A. Langston, Nonconstructive advances in polynomial-time complexity, *Information Processing Letters*, **26** (1987), 157–162.

[17] M. R. Fellows & M. A. Langston, Nonconstructive tools for proving polynomial-time decidability, *Journal of the Association for Computing Machinery*, **35** (1988), 727–739.

[18] I. S. Filotti, G. L. Miller & J. Reif, On determining the genus of a graph in $O(|V|^{O(g)})$ steps, in *Proceedings of the 11th Annual Association for Computing Machinery Symposium on Theory of Computing*, ACM Press, New York (1979), pp. 27–37.

[19] T. Gallai, Problem 6, in *Proceedings Colloquium Tihany*, Academic Press, (1966), page 362.

[20] A. Gyarfas, Fruit salad, manuscript.

[21] T. C. Hu, *Integer Programming and Network Flows*, Addison-Wesley, Don Mills, Ontario (1969).

[22] R. M. Karp, On the complexity of combinatorial problems, *Networks*, **5** (1975), 45–68.

[23] J. B. Kruskal, Well-quasi-ordering, the tree theorem, and Vázsonyi's conjecture, *Transactions of the American Mathematical Society*, **95** (1960), 210–225.

[24] K. Kuratowski, Sur le problème des courbes gauches en topologie, *Fundamenta Mathematicae*, **15** (1930), 271–283.

[25] J. Lagergren, Upper bounds on the sizes of obstructions and intertwines, manuscript.

[26] R. J. Lipton & R. E. Tarjan, A separator theorem for planar graphs, *SIAM Journal on Applied Mathematics*, **36** (1979), 177–189.

[27] L. Lovász & A. Schrijver, personal communication.

[28] J. Lynch, The equivalence of theorem proving and the interconnection problem, *Association for Computing Machinery's Special Interest Group on Design Automation Newsletter*, **5** 1976.

[29] W. McCuaig, Intercyclic digraphs, in *Graph Structure Theory (Proceedings of the AMS-IMS-SIAM Joint Summer Research Conference on Graph Minors, Seattle, 1991)* (eds. N. Robertson & P. Seymour), *Contemporary Mathematics*, 147, American Mathematical Society, Providence, Rhode Island (1993), 203–247.

[30] K. Menger, Zur allgemeinen Kurventheorie, *Fundamenta Mathematicae*, **10** (1927), 96–115.

[31] C. St.J. A. Nash-Williams, On well-quasi-ordering infinite trees, *Proceedings of the Cambridge Philosophical Society*, **61** (1965), 697–720.

[32] B. A. Reed, Finding approximate separators and computing tree width quickly, in *Proceedings of the 24th Annual Association for Computing Machinery Symposium on Theory of Computing*, ACM Press, New York (1992), 221–228.

[33] B. Reed, Mangoes and blueberries, manuscript.

[34] B. Reed, Disjoint connected paths: faster algorithms and shorter proofs, manuscript.

[35] B. Reed, N. Robertson, P. Seymour & R. Thomas, On packing directed circuits, *Combinatorica*, in press.

[36] N. Robertson & P. D. Seymour, Graph Minors. II. Algorithmic aspects of tree-width, *Journal of Algorithms*, **7** (1986), 309–322.

[37] N. Robertson & P. D. Seymour, Graph Minors. IV. Tree-width and well-quasi-ordering, *Journal of Combinatorial Theory, Series B*, **48** (1990), 227–254.

[38] N. Robertson & P. D. Seymour, Graph Minors. V. Excluding a planar graph, *Journal of Combinatorial Theory, Series B*, **41** (1986), 92–114.

[39] N. Robertson & P. D. Seymour, Graph Minors. VII. Disjoint paths on a surface, *Journal of Combinatorial Theory, Series B*, **45** (1988), 212–254.

[40] N. Robertson & P. D. Seymour, Graph Minors. VIII. A Kuratowski theorem for general surfaces, *Journal of Combinatorial Theory, Series B*, **48** (1990), 255–288.

[41] N. Robertson & P. D. Seymour, Graph Minors. X. Obstructions to tree-decomposition, *Journal of Combinatorial Theory, Series B*, **52** (1991), 153–190.

[42] N. Robertson & P. D. Seymour, Graph Minors. XIII. The disjoint paths problem, *Journal of Combinatorial Theory, Series B*, **63** (1995), 65–110.

[43] N. Robertson & P. D. Seymour, Graph Minors. XVI. Excluding a non-planar graph, manuscript.

[44] N. Robertson & P. D. Seymour, Graph Minors. XX. Wagner's Conjecture, manuscript, 1988.

[45] N. Robertson, P. D. Seymour & R. Thomas, A survey of linkless embeddings, in *Graph Structure Theory (Proceedings of the AMS-IMS-SIAM Joint Summer Research Conference on Graph Minors, Seattle, 1991)* (eds. N. Robertson & P. Seymour), *Contemporary Mathematics*, 147, American Mathematical Society, Providence, Rhode Island (1993), 125–136.

[46] N. Robertson, P. Seymour & R. Thomas, Quickly excluding a planar graph, *Journal of Combinatorial Theory, Series B*, **62** (1994), 323–348.

[47] P. Seymour, A bound on the excluded minors for a surface, manuscript.

[48] P. D. Seymour & R. Thomas, Graph searching and a min-max theorem for tree-width, *Journal of Combinatorial Theory, Series B*, **58** (1993), 22–33.

[49] R. Thomas, A Menger-like property of tree-width: the finite case, *Journal of Combinatorial Theory, Series B*, **48** (1990), 67–76.

[50] C. Thomassen, On the presence of disjoint subgraphs of a specified type, *Journal of Graph Theory*, **12** (1988), 101–111.

[51] W. Tutte, Algebraic Theory of Graphs, Ph. D. Thesis, Cambridge, 1948.

[52] V. G. Vizing, On the estimate of the chromatic class of a p-graph, *Metody Diskretnogo Analiza*, **3** (1964), 25–30.

[53] D. Younger, Graphs with interlinked directed circuits, in *Proceedings of the Midwest Symposiom on Circuit Theory* (1973), pp. XVI 2.1-2.7.

Équipe Combinatoire
CNRS
Case 189
Université de Paris VI
4 place Jussieu
Paris 75005
France

Minor-monotone Graph Invariants

Alexander Schrijver

Summary A graph parameter $\phi(G)$ is called *minor-monotone* if $\phi(H) \leq \phi(G)$ for any minor H of G. We survey recent work on minor-monotone graph parameters motivated by the parameter $\mu(G)$ introduced by Colin de Verdière.

1 Introduction

A function $\phi(G)$ defined for any undirected graph G is called *minor-monotone* if for any graph G and any minor H of G one has

$$\phi(H) \leq \phi(G). \tag{1}$$

In this paper, all graphs are undirected, loopless and without multiple edges. A *minor* of a graph arises by a series of deletions and contractions of edges and deletions of isolated vertices, suppressing any multiple edges and loops that may arise.

The interest in minor-monotone graph parameters is activated because the Robertson–Seymour theory of graph minors can be applied to them. Recently a number of minor-monotone parameters have been studied, motivated in particular by the graph parameter $\mu(G)$ introduced by Colin de Verdière [5] (cf. [6]). The parameter $\mu(G)$ can be described in terms of properties of matrices related to G. It was motivated by the study of the maximum multiplicity of the second eigenvalue of certain Schrödinger operators. When such an operator is defined on a Riemann surface, one can approximate the surface by a densely enough embedded graph G, in such a way that $\mu(G)$ is the maximum multiplicity of the second eigenvalue of the operator.

The interest raised by Colin de Verdière's parameter can be explained not only by its background in differential geometry, but also by the facts that it is minor-monotone (so that the Robertson–Seymour graph minors theory applies to it), and that it characterizes planarity of graphs. Indeed, one has that $\mu(G) \leq 3$ if and only if G is planar. Moreover, as follows from the results in [19] and [15], $\mu(G) \leq 4$ if and only if G is linklessly embeddable in $\mathbb{R}^3$. (A graph G is *linklessly embeddable* if it can be embedded in $\mathbb{R}^3$ in such a way that the images of any two disjoint circuits in G are unlinked.) So with the help of μ, topological properties of a graph can be characterized in terms of spectral properties of matrices associated to the graph.

In this paper we give a survey of the graph parameter $\mu(G)$, and some related parameters, in particular the parameter $\lambda(G)$ introduced in [12]. We first give an overview of $\mu(G)$ and $\lambda(G)$, after which we give proofs of a number of results. Finally, we consider the parameters $\lambda'(G)$ (defined by oriented matroids) and $\kappa(G)$.

For more information we refer to the thesis by van der Holst [11], where in addition a few other minor-monotone parameters are studied.

2 Overview of $\mu(G)$

Let $G = (V, E)$ be an undirected graph, which we assume without loss of generality to have vertex set $V = \{1, \ldots, n\}$. Then $\mu(G)$ is the largest corank of any symmetric real-valued $n \times n$ matrix $M = (m_{i,j})$ such that:

(i) M has exactly one negative eigenvalue, of multiplicity 1, (2)

(ii) for all i, j with $i \neq j$: $m_{i,j} < 0$ if i and j are adjacent, and $m_{i,j} = 0$ if i and j are nonadjacent,

(iii) there is no nonzero symmetric $n \times n$ matrix $X = (x_{i,j})$ such that $MX = 0$ and such that $x_{i,j} = 0$ whenever $i = j$ or $m_{i,j} \neq 0$.

There is no condition on the diagonal entries $m_{i,i}$. The *corank* corank(M) of a matrix M is the dimension of its kernel (= null space).

Note that for each graph $G = (V, E)$ a matrix M satisfying (2) exists. If G is connected, let A be the adjacency matrix of G. Then we can choose λ in such a way that $\lambda I - A$ has exactly one negative eigenvalue and is nonsingular. If G is disconnected, we can choose such a λ for each component separately and obtain again a nonsingular matrix with exactly one negative eigenvalue.

Condition (iii) is called the *Strong Arnol'd Hypothesis* (or *Strong Arnol'd Property*). There are a number of equivalent formulations of the Strong Arnol'd Hypothesis, amounting to the fact that M is in a certain general position. Let $M = (m_{i,j})$ be a symmetric $n \times n$ matrix. Let R_M be the set of all symmetric $n \times n$ matrices A with rank(A) = rank(M). Let S_M be the set of all symmetric $n \times n$ matrices $A = (a_{i,j})$ such that $a_{i,j} = 0$ whenever $i \neq j$ and $m_{i,j} = 0$.

Then M fulfils the Strong Arnol'd Hypothesis (2)(iii) if and only if

$$R_M \text{ intersects } S_M \text{ at } M \text{ `transversally'}; \tag{3}$$

that is, if the tangent space of R_M at M and the tangent space of S_M at M together span the space of all symmetric $n \times n$ matrices. In other words, if the intersection of the normal spaces at M of R_M and of S_M only consists of the all-zero matrix.

It is elementary linear algebra to show that the tangent space of R_M at M consists of all symmetric $n \times n$ matrices N such that $x^T N x = 0$ for each $x \in \ker(M)$. Thus the normal space of R_M at M is equal to the space generated by all matrices xx^T with $x \in \ker(M)$. (We assume that our underlying space is the space of real-valued symmetric $n \times n$ matrices.) This space is equal to the space of all symmetric $n \times n$ matrices X satisfying $MX = 0$. Trivially, the normal space of S_M at M consists of all symmetric $n \times n$ matrices $X = (x_{i,j})$ such that $x_{i,j} = 0$ whenever $i = j$ or $m_{i,j} \neq 0$. Therefore, (3) is equivalent to (2)(iii).

An important property of $\mu(G)$ proved by Colin de Verdière [5] is that it is monotone under taking minors:

$$\text{the graph parameter } \mu(G) \text{ is minor-monotone.} \tag{4}$$

Proving this is nontrivial, and the Strong Arnol'd Hypothesis is needed. We give the elementary proof as given in van der Holst [11] in Section 4.

The minor-monotonicity of $\mu(G)$ is especially interesting in the light of the Robertson–Seymour theory of graph minors [16], which has as principal result that if $\mathcal{C}$ is a collection of graphs so that no graph in $\mathcal{C}$ is a minor of another graph in $\mathcal{C}$, then $\mathcal{C}$ is finite. This can be equivalently formulated as follows. For any graph property $\mathcal{P}$ closed under taking minors, call a graph G a *forbidden minor* for $\mathcal{P}$ if G does not have property $\mathcal{P}$, but each proper minor of G does have property $\mathcal{P}$. Note that a minor-closed property $\mathcal{P}$ is completely characterized by the collection of its forbidden minors. Now Robertson and Seymour's theorem states that each graph property that is closed under taking minors, has only finitely many forbidden minors. (See Reed's paper elsewhere in this volume.)

Since

$$\mu(K_n) = n - 1 \tag{5}$$

for each n (cf. Section 5), Hadwiger's conjecture implies that $\gamma(G) \leq \mu(G) + 1$ (where $\gamma(G)$ denotes the colouring number of G); this last inequality is conjectured by Colin de Verdière [5]. Since Hadwiger's conjecture holds for graphs not containing any K_6-minor (Robertson, Seymour, and Thomas [18]), we know that $\gamma(G) \leq \mu(G) + 1$ holds if $\mu(G) \leq 4$.

In studying $\mu(G)$, we can restrict ourselves to considering connected graphs, since if G has at least one edge, then $\mu(G)$ is equal to the maximum of $\mu(K)$ taken over all components K of G.

The following characterizations show that with the help of $\mu(G)$, topological properties of a graph can be characterized algebraically:

(i) $\mu(G) \leq 1 \iff G$ is a disjoint union of paths. (6)

(ii) $\mu(G) \leq 2 \iff G$ is outerplanar.

(iii) $\mu(G) \leq 3 \iff G$ is planar.

(iv) $\mu(G) \leq 4 \iff G$ is linklessly embeddable.

Here (i), (ii), and (iii) are due to Colin de Verdière [5]. In (iv), $\Longrightarrow$ is due to Robertson, Seymour, and Thomas [17] (based on the hard theorem of [19] that the Petersen family (Figure 2 on page 188) is the collection of forbidden minors for linkless embeddability), and $\Longleftarrow$ to Lovász and Schrijver [15]. In fact, in (6) each $\Longrightarrow$ follows from a forbidden minor characterization of the right-hand statement.

We give a proof of (i), (ii), and (iii) in Sections 9, 11, and 12, respectively. In Section 15, we indicate how $\Longleftarrow$ in (iv) can be proved, with the help of a certain Borsuk-type theorem on the existence of 'antipodal links'.

Interestingly, Kotlov, Lovász, and Vempala [14] showed that with the value $n-\mu(G)$ (for a graph G with n vertices) one is close to characterizing that the complementary graph $\overline{G}$ of G is outerplanar or planar. In fact they showed:

$$\begin{aligned}&\text{if } \overline{G} \text{ is a disjoint union of paths then } \mu(G) \geq n-3;\\ &\text{if } \overline{G} \text{ is outerplanar then } \mu(G) \geq n-4;\\ &\text{if } \overline{G} \text{ is planar then } \mu(G) \geq n-5.\end{aligned} \tag{7}$$

Conversely, one has, if G does not have 'twin vertices' (two (adjacent or non-adjacent) vertices u, v that have the same neighbours $\neq u, v$), then:

$$\begin{aligned}&\text{if } \mu(G) \geq n-3 \text{ then } \overline{G} \text{ is outerplanar;}\\ &\text{if } \mu(G) \geq n-4 \text{ then } \overline{G} \text{ is planar.}\end{aligned} \tag{8}$$

The proof by Colin de Verdière [5] of the planarity characterization (6)(iii) uses a result of Cheng [4] on the maximum multiplicity of the second eigenvalue of Schrödinger operators defined on the sphere. A short direct proof was given by van der Holst [10], based on the following lemma. For any vector x, let $\operatorname{supp}(x)$ denote the support of x (i.e., the set $\{i \mid x_i \neq 0\}$). Moreover, denote $\operatorname{supp}^+(x) := \{i \mid x_i > 0\}$ and $\operatorname{supp}^-(x) := \{i \mid x_i < 0\}$. We say that a vector $x \in \ker(M)$ *has minimal support* if x is nonzero and for each nonzero vector $y \in \ker(M)$ with $\operatorname{supp}(y) \subseteq \operatorname{supp}(x)$ one has $\operatorname{supp}(y) = \operatorname{supp}(x)$. For any subset U of V, let $G|U$ denote the subgraph of G induced by U.

Then Van der Holst's lemma states:

> Let M satisfy (2) and let $x \in \ker(M)$ have minimal support. Then $G|\operatorname{supp}^+(x)$ and $G|\operatorname{supp}^-(x)$ are connected. (9)

We give the proof in Section 10.

3 Overview of $\lambda(G)$

Van der Holst's lemma motivated van der Holst, Laurent, and Schrijver [12] to introduce a related graph parameter $\lambda(G)$, defined as follows. Let $G = (V, E)$ be a graph. Call a subspace X of $\mathbb{R}^V$ *representative* for G if

> for each nonzero vector $x \in X$, $\operatorname{supp}^+(x)$ is nonempty and $G|\operatorname{supp}^+(x)$ is connected. (10)

Then $\lambda(G)$ is defined as the maximum dimension of any representative subspace X of $\mathbb{R}^V$.

Clearly, (10) implies that also $\mathrm{supp}^-(x)$ is nonempty and induces a connected subgraph of G for each nonzero $x \in X$.

The results characterizing μ and λ for small values, suggest that λ is close to μ. In fact, recently Rudi Pendavingh showed that $\mu(G) \leq \lambda(G)+2$ for each graph G. Conversely, it might be that $\lambda(G) \leq \mu(G)$ holds.

There is a direct equivalent characterization of $\lambda(G)$. Let $G=(V,E)$ be a graph and let $d \in \mathbb{N}$. Call a function $\phi: V \to \mathbb{R}^d$ *representative* for G if

$$\text{for each halfspace } H \text{ of } \mathbb{R}^d\text{, the set } \phi^{-1}(H) \text{ is nonempty and induces a connected subgraph of } G. \tag{11}$$

(Here $\phi^{-1}(H) := \{v \in V \mid \phi(v) \in H\}$.) A subset H of $\mathbb{R}^d$ is called a *halfspace* if $H = \{x \in \mathbb{R}^d \mid c^T x > 0\}$ for some nonzero $c \in \mathbb{R}^d$. Note that if $\phi: V \to \mathbb{R}^d$ is representative, then the vectors $\phi(v)$ $(v \in V)$ span $\mathbb{R}^d$ (since otherwise there would exist a halfspace H with $\phi^{-1}(H) = \emptyset$).

Now $\lambda(G)$ is equal to the largest d for which there is a representative function $\phi: V \to \mathbb{R}^d$. This is easy to see. Suppose X is a d-dimensional subspace of $\mathbb{R}^V$ representative for G. Let vectors $x_1, \ldots, x_d$ form a basis of X. Define $\phi(v) := (x_1(v), \ldots, x_d(v))$ for each $v \in V$. Then ϕ is a representative function for G. Conversely, let $\phi: V \to \mathbb{R}^d$ be representative. Define for any $c \in \mathbb{R}^d$ the function $x_c \in \mathbb{R}^V$ by: $x_c(v) := c^T\phi(v)$ for $v \in V$. Then $X := \{x_c \mid c \in \mathbb{R}^d\}$ is a representative space for G.

It is easy to show that the function $\lambda(G)$ is minor-monotone (much easier than for $\mu(G)$):

Theorem 3.1 *If H is a minor of G then $\lambda(H) \leq \lambda(G)$.*

Proof Let $H=(V',E')$. If H arises from G by deleting an isolated vertex v_0, the inequality $\lambda(H) \leq \lambda(G)$ is easy: if $\phi': V' \to \mathbb{R}^d$ is representative for H with $d = \lambda(H)$, then defining $\phi(v_0) := 0$ and $\phi(v) := \phi'(v)$ for all other vertices v of G, gives a representative function for G.

So we may assume that $H=(V',E')$ arises from $G=(V,E)$ by deleting or contracting one edge $e = uw$. Let $\phi': V' \to \mathbb{R}^d$ be representative for H with $d = \lambda(H)$. If H arises from G by deleting e, then $V = V'$, and ϕ' is also representative for G. Hence $\lambda(G) \geq d = \lambda(H)$.

If H arises from G by contracting e, let v_0 be the vertex of H which arose by contracting e. Define $\phi(u) := \phi(w) := \phi'(v_0)$, and define $\phi(v) := \phi'(v)$ for all other vertices v of G. Then ϕ is representative of G. ∎

One easily shows that

$$\lambda(K_n) = n-1 \tag{12}$$

(cf. Section 5). Hence, Hadwiger's conjecture implies that $\gamma(G) \leq \lambda(G)+1$ (where $\gamma(G)$ denotes the colouring number of G). So by the truth of Hadwiger's conjecture for K_6-free graphs (Robertson, Seymour, and Thomas [18]), the inequality $\gamma(G) \le \lambda(G) + 1$ holds if $\lambda(G) \le 4$.

As for the colouring number, also the function $\lambda(G)$ cannot be increased by 'clique sums'. Graph $G = (V, E)$ is a *clique sum* of graphs $G_1 = (V_1, E_1)$ and $G_2 = (V_2, E_2)$ if $V = V_1 \cup V_2$ and $E = E_1 \cup E_2$, where $V_1 \cap V_2$ is a clique both in G_1 and in G_2. Then $\gamma(G) = \max\{\gamma(G_1), \gamma(G_2)\}$ if G is a clique sum of G_1 and G_2. A similar relation holds for the size of the largest clique minor in G. Now in Section 6 we shall show:

If G has at least one edge and is a clique sum of G_1 and G_2, then $\lambda(G) = \max\{\lambda(G_1), \lambda(G_2)\}$. (13)

This directly gives with (12):

(i) $\lambda(G) \le 1$ if and only if G is a forest; (14)

(ii) $\lambda(G) \le 2$ if and only if G is a series-parallel graph.

Indeed, forests can be characterized as the graphs not having a K_3-minor and also as the graphs obtainable from K_2 by taking clique sums and subgraphs. Similarly, series-parallel graphs can be characterized as the graphs not having a K_4-minor and also as the graphs obtainable from K_3 by taking clique sums and subgraphs.

In Section 13 we show that

$\lambda(G) \le 3$ if and only if G can be obtained from planar graphs by taking clique sums and subgraphs. (15)

The kernel of the proof here is to show that $\lambda(G) \le 3$ for any planar graph G. Having this, a fundamental decomposition theorem of Wagner [20] then implies the full characterization. Indeed, let V_8 be the graph with vertices $v_1, \ldots, v_8$,

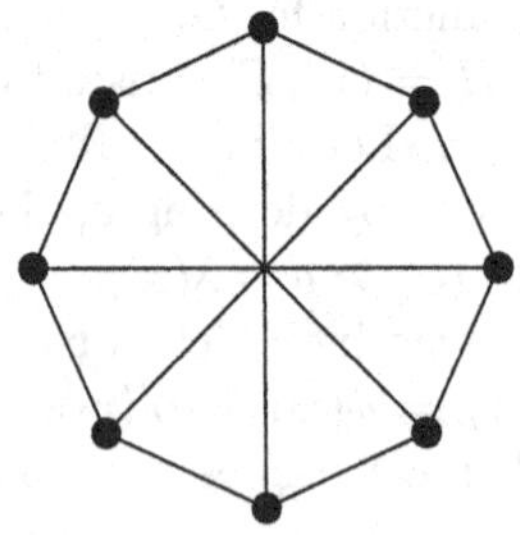

Figure 1: The graph V_8

where v_i and v_j are adjacent if and only if $|i - j| \in \{1, 4, 7\}$. Then Wagner showed:

> G can be obtained from planar graphs by taking clique sums and subgraphs $\iff$ G does not have a K_5- or V_8-minor. (16)

Since $\lambda(K_5) = 4$ and since $\lambda(V_8) = 4$ (as we show in Section 13), we obtain (15).

In Section 16 we give a few observations concerning the class of graphs G with $\lambda(G) \leq 4$. In particular, we show the result of [15] that $\lambda(G) \leq 4$ for any linklessly embeddable graph G. This implies with (13):

> if G is obtainable from linklessly embeddable graphs by taking clique sums and subgraphs, then $\lambda(G) \leq 4$. (17)

As mentioned, an open question is if there is any direct relation between $\lambda(G)$ and $\mu(G)$. It might be the case that $\lambda(G) \leq \mu(G)$ for each graph G. That is, for any subspace X of $\mathbb{R}^V$ representative for G there is a matrix M satisfying (2) with $\dim(X) \leq \text{corank}(M)$. This is true if $\mu(G) \leq 4$.

In fact, a tempting, more general speculation is that for any natural number t:

> (???) a graph G satisfies $\lambda(G) \leq t$ if and only if G is obtainable from graphs H satisfying $\mu(H) \leq t$ by taking clique sums and subgraphs (???) (18)

This has been proved for $t \leq 3$, and the 'if' part for $t \leq 4$.

4 Some basic facts on $\mu(G)$

We first prove a number of elementary facts on the parameter $\mu(G)$. We use the following notation. If M is a matrix and I is a set of rows of M and J is a set of columns of M, then $M_{I\times J}$ is the submatrix induced by the rows in I and the columns in J. If $I = J$ we write M_I for $M_{I\times I}$.

First we have the following important property due to Colin de Verdière [6], which we prove with the method described by van der Holst [11]:

Theorem 4.1 *For any edge e of any graph G one has $\mu(G - e) \leq \mu(G)$.*

Proof For any smooth manifold $\mathcal{M}$, any smooth submanifold $\mathcal{A}$ of $\mathbb{R}^d$, any smooth function $f\colon \mathcal{M} \to \mathbb{R}^d$, and any $x \in \mathcal{M}$ with $f(x) \in \mathcal{A}$, we say that f intersects $\mathcal{A}$ *transversally* at $x \in \mathcal{M}$, in notation: $f\#_x\mathcal{A}$, if

$$T_{f(x)}\mathcal{A} + Df_x(T_x\mathcal{M}) = \mathbb{R}^d. \tag{19}$$

Here $T_y\mathcal{N}$ denotes the tangent space of $\mathcal{N}$ at y, and Df_x the differential of f at x.

A basic property of transversality is:

> If $f\#_x\mathcal{A}$, then there is a neighbourhood U of x in $\mathcal{M}\cap f^{-1}(\mathcal{A})$ such that $f\#_y\mathcal{A}$ for each $y\in U$, and such that U has the same codimension in $\mathcal{M}$ as $\mathcal{A}$ has in $\mathbb{R}^d$. (20)

Let $\mathcal{S}_n$ denote the collection of real-valued symmetric $n\times n$ matrices, and $\mathcal{S}_{n,k}$ the collection of matrices in $\mathcal{S}_n$ of corank k. For any graph $G=(V,E)$ let $\mathcal{O}_G$ be the collection of real-valued symmetric $V\times V$ matrices M satisfying (2)(ii).

First assume that graph H arises from graph G by deleting an edge $e=uw$. We may assume that G has vertex set $V=\{1,\ldots,n\}$, and that $u=1$ and $w=2$. Let $W:=\{3,\ldots,n\}$. Let $f:\mathbb{R}\times\mathcal{O}_H\to\mathcal{S}_n$ be defined by

$$f(h,K):=\begin{pmatrix} k_{1,1} & h & K_{\{1\}\times W}\\ h & k_{2,2} & K_{\{2\}\times W}\\ K_{W\times\{1\}} & K_{W\times\{2\}} & K_{W\times W}\end{pmatrix}, \tag{21}$$

where $K=(k_{i,j})\in\mathcal{O}_H$. Let $f_0(K):=f(0,K)$.

Let $M'=(m'_{i,j})$ satisfy (2), with corank $k=\mu(H)$. By (2)(iii),

$$f_0\#_{M'}\mathcal{S}_{n,k}, \tag{22}$$

which implies

$$f\#_{(0,M')}\mathcal{S}_{n,k}. \tag{23}$$

Then by (20), there is a neighbourhood U of $(0,M')$ in $\mathbb{R}\times\mathcal{O}_H$ such that for all $x\in U$

$$f\#_x\mathcal{S}_{n,k}. \tag{24}$$

Also by (20), $U\cap(\{0\}\times f_0^{-1}(\mathcal{S}_{n,k}))$ is a submanifold of $U\cap(\{0\}\times\mathcal{O}_H)$ of codimension $\frac{1}{2}k(k+1)$ (since the codimension of $\mathcal{S}_{n,k}$ in $\mathcal{S}_n$ is $\frac{1}{2}k(k+1)$). Moreover, $f^{-1}(\mathcal{S}_{n,k})\cap U$ is a submanifold of U of codimension $\frac{1}{2}k(k+1)$. Hence there exists a $(h,L)\in U$ with $h<0$ such that $M:=f(h,L)\in\mathcal{S}_{n,k}$. By taking (h,L) close to $(0,M')$ we may assume that M has exactly one negative eigenvalue. Since $f\#_{(h,L)}\mathcal{S}_{n,k}$, M fulfils the Strong Arnol'd Hypothesis ((3)). Hence M satisfies (2), and therefore $\mu(G)\geq\mu(H)$. ∎

This theorem implies:

Theorem 4.2 *For any subgraph H of any graph G one has*

$$\mu(H)\leq\mu(G). \tag{25}$$

Proof By Theorem 4.1 we can assume that H arises from $G=(V,E)$ by deleting an isolated vertex v. Let M' be a matrix satisfying (2) with respect to H, with $\text{corank}(M')=\mu(H)$, and let M be the $V\times V$ matrix arising from M' by adding 0's, except in position (v,v), where $M_{v,v}=1$. Then trivially $\text{corank}(M)=\text{corank}(M')$ and M satisfies (2) with respect to G. This shows (25). ∎

This implies:

Theorem 4.3 *If G has at least one edge, then*

$$\mu(G) = \max_{K} \mu(K), \tag{26}$$

where K extends over the components of G.

Proof By Theorem 4.2 we know that $\geq$ holds in (26). To see equality, let M be a matrix satisfying (2). Since G has at least one edge, we know $\mu(G) > 0$ (since trivially $\mu(K_2) = 1$), and hence $\text{corank}(M) > 0$. Then there is exactly one component L of G with $\text{corank}(M_L) > 0$. For suppose that there are two such components, K and L. Choose nonzero vectors $x \in \ker(M_K)$ and $y \in \ker(M_L)$. Extend x and y by zeros on the positions not in K and L, respectively. Then the matrix $X := xy^T + yx^T$ is nonzero and symmetric, has zeros in positions corresponding to edges of G, and satisfies $MX = 0$. This contradicts the Strong Arnol'd Hypothesis.

So $\text{corank}(M) = \text{corank}(M_L)$. Suppose now that M_L has no negative eigenvalue. Then 0 is the smallest eigenvalue of M_L, and hence, by the connectivity of L and the Perron-Frobenius theorem, $\text{corank}(M_L) = 1$. So $\mu(G) = 1$. Let L' be a component of G with at least one edge. Then $\mu(L') \geq 1$, proving (26).

One easily shows that M_L satisfies the Strong Arnol'd Hypothesis, implying $\mu(G) = \mu(L)$, thus proving (26). ∎

Next we have:

Theorem 4.4 *Let $G = (V, E)$ be a graph and let $v \in V$ such that $G - v$ has at least one edge. Then*

$$\mu(G) \leq \mu(G - v) + 1. \tag{27}$$

Proof Let M be a matrix satisfying (2) with $\text{corank}(M) = \mu(G)$. Let $M' := M_{V\setminus\{v\}}$. Clearly, $\text{corank}(M') \geq \text{corank}(M) - 1$, since $\text{rank}(M') \leq \text{rank}(M)$. So it suffices to show that M' satisfies (2) with respect to G'.

Trivially, M' satisfies (2)(ii). To see that M' satisfies (2)(i), it suffices to show that M' has at least one negative eigenvalue. If M' has no negative eigenvalue, then M' is positive semidefinite, and 0 is an eigenvalue of multiplicity at least $\mu(G) + 1$. Hence (by the Perron-Frobenius theorem) for each component K of $G - v$, if the matrix M_K has eigenvalue 0, then it has multiplicity 1. As the theorem trivially holds if $\mu(G) \leq 2$ (since $\mu(G - v) \geq 1$ as $G - v$ has at least one edge), we can assume that $\mu(G) \geq 3$. Hence $G - v$ has at least $\mu(G) + 1 \geq 4$ components K with M_K singular. Let $K_1, \ldots, K_4$ be four such components. For $i = 1, \ldots, 4$, let x_i be a nonzero vector with $M_{K_i} x_i = 0$. By the Perron-Frobenius theorem we know that we can assume $x_i > 0$ for each i. Extend x_i to a vector in $\mathbb{R}^V$ by adding components 0.

Let z be an eigenvector of M belonging to the smallest eigenvalue of M. By scaling the x_i we can assume that $z^T x_i = 1$ for each i. Now define

$$X := (x_1 - x_2)(x_3 - x_4)^T + (x_3 - x_4)(x_1 - x_2)^T. \qquad (28)$$

Then $MX = 0$, since $M(x_1 - x_2) = 0$ (as $(x_1 - x_2)^T M(x_1 - x_2) = 0$ and as $x_1 - x_2$ is orthogonal to z), and similarly $M(x_3 - x_4) = 0$. This contradicts the fact that M satisfies (2)(iii). So M' satisfies (2)(i).

To see that M' satisfies the Strong Arnol'd Hypothesis (2)(iii), let X' be a $(V \setminus v) \times (V \setminus v)$ matrix with 0's in positions (i, j) where $i = j$ or i and j are adjacent, and satisfying $M'X' = 0$. We must show that $X' = 0$. Let X be the $V \times V$ matrix obtained from X' by adding 0's.

Since M' has exactly one negative eigenvalue, we know by interlacing that $\text{corank}(M') \leq \text{corank}(M)$. If $MX = 0$ we know by (2)(iii) that $X = 0$ and hence $X' = 0$. So we can assume that $MX \neq 0$. As $\text{corank}(M') \leq \text{corank}(M)$, it follows that there is a vector $x \in \ker(M)$ with $x_v \neq 0$. Hence the first column of M is a linear combination of the other columns of M. Therefore $MX = 0$, a contradiction. ∎

On the other hand we have, where $S(G)$ arises from G by adding one new vertex v adjacent to all other vertices of G:

Theorem 4.5 *For any graph G with at least one edge, one has*

$$\mu(S(G)) = \mu(G) + 1. \qquad (29)$$

Proof By Theorem 4.4 it suffices to show that $\mu(S(G)) \geq \mu(G) + 1$, and by Theorem 4.3 we can assume that G is connected. Let M be a matrix satisfying (2) with $\text{corank}(M) = \mu(G)$. Let z be an eigenvector of M belonging to the smallest eigenvalue λ_1 of M. We can assume that $z < 0$ and that $\|z\| = 1$. Let M' be the matrix

$$M' := \begin{pmatrix} \lambda_1^{-1} & z^T \\ z & M \end{pmatrix}. \qquad (30)$$

Since $(0, x)^T \in \ker(M')$ for each $x \in \ker(M)$ and since $(-\lambda_1, z)^T \in \ker(M')$, we know that $\text{corank}(M') \geq \text{corank}(M) + 1$. By interlacing it follows that M' has exactly one negative eigenvalue. One similarly easily checks that M' satisfies the Strong Arnol'd Hypothesis (2)(iii). ∎

Above we gave a proof that $\mu(G)$ is monotone under taking subgraphs. More strongly, as Colin de Verdière [5] proved, $\mu(G)$ is minor-monotone. Again we give the elementary proof due to van der Holst [11] of this fact.

Theorem 4.6 *$\mu(G)$ is minor-monotone.*

Proof By Theorem 4.2 it suffices to show that $\mu(H) \leq \mu(G)$ if H arises from G by contracting edge $e = uw$. Let the new vertex of H be v. Let $n := |V|$ and $n' := |V'|$. So $n = n' + 1$. We may assume that $u = 1$ and $w = 2$. Let $W := \{3, \ldots, n\}$.

Let $\mathcal{Z}$ be the set of all matrices $K = (k_{i,j}) \in \mathcal{O}_G$ with $k_{1,1} = 0 = k_{1,2}$. Define a function

$$f: \mathbb{R} \times \mathcal{Z} \to \mathcal{S}_{n'} \tag{31}$$

by

$$f(h, K) = \begin{pmatrix} k_{2,2} & K_{\{1\}\times W} + K_{\{2\}\times W} \\ K_{W\times\{1\}} + K_{W\times\{2\}} & K_{W\times W} - hK_{W\times\{1\}}K_{\{1\}\times W} \end{pmatrix}, \tag{32}$$

and let $f_0(K) = f(0, K)$.

Let $M' = (m'_{i,j})$ satisfy (2) with respect to H, with corank $k = \mu(H)$. Trivially there is a $P \in \mathcal{Z}$ such that $f(0, P) = M'$. Since the tangent space of $\mathcal{O}_H$ at M' is a subspace of the space of all vectors $Df_{(0,P)}(A)$ with $A \in T_{(0,P)}(\mathbb{R} \times \mathcal{Z})$ we know that

$$f \#_{(0,P)} \mathcal{S}_{n',k}. \tag{33}$$

Again by (20), there is a neighbourhood U of $(0, P)$ such that for all $x \in U$

$$f \#_x \mathcal{S}_{n',k}. \tag{34}$$

Also by (20), $(\{0\} \times f_0^{-1}(\mathcal{S}_{n',k})) \cap U$ is a submanifold of $U \cap (\{0\} \times \mathcal{O}_H)$ of codimension $\frac{1}{2}k(k+1)$ (since the codimension of $\mathcal{S}_{n',k}$ in $\mathcal{S}_{n'}$ is $\frac{1}{2}k(k+1)$). Moreover, $f^{-1}(\mathcal{S}_{n',k}) \cap U$ is a submanifold of U of codimension $\frac{1}{2}k(k+1)$. Hence there is an $(h, L) \in U$ with $h > 0$ such that $f(h, L) \in \mathcal{S}_{n',k}$ and

$$f \#_{(h,L)} \mathcal{S}_{n',k}. \tag{35}$$

Taking (h, L) close to $(0, P)$ we may assume that $f(h, L)$ has exactly one negative eigenvalue.

Define

$$M := \begin{pmatrix} \frac{1}{h} & -\frac{1}{h} & L_{\{1\}\times W} \\ -\frac{1}{h} & L_{\{2\}\times\{2\}} + \frac{1}{h} & L_{\{2\}\times W} \\ L_{W\times\{1\}} & L_{W\times\{2\}} & L_{W\times W} \end{pmatrix}. \tag{36}$$

Clearly $M \in \mathcal{O}_G$. We show that M satisfies (2) and has corank k. Let

$$P := \begin{pmatrix} 1 & 1 & -hL_{\{1\}\times W} \\ 0 & 1 & 0 \\ 0 & 0 & I \end{pmatrix}. \tag{37}$$

Then

$$P^T M P = \begin{pmatrix} \frac{1}{h} & 0 \\ 0 & f(h,L) \end{pmatrix}. \tag{38}$$

Therefore, by Sylvester's law of inertia and since $\frac{1}{h} > 0$, $f(h,L)$ has the same number of negative eigenvalues and the same corank as M. It remains to show that M fulfils the Strong Arnol'd Hypothesis ((2)(iii)).

Choose $F \in \mathcal{S}_n$. We must show that there exists an $N \in T_M \mathcal{O}_G$ such that $x^T F x = x^T N x$ for all $x \in \ker(M)$. Define

$$Q := \begin{pmatrix} 1 & -hL_{\{1\}\times W} \\ 1 & 0 \\ 0 & I \end{pmatrix} \tag{39}$$

and $F' := Q^T F Q$.

Since $f \#_{(h,L)} \mathcal{S}_{n',k}$,

$$Df_{(h,L)}(T_{(h,L)}(\mathbb{R} \times \mathcal{Z})) + T_{f(h,L)}(\mathcal{S}_{n',k}) = \mathcal{S}_{n'}. \tag{40}$$

The tangent space of $\mathcal{S}_{n',k}$ at $f(h,L)$ is the set of all real-valued symmetric matrices C for which $x'^T C x' = 0$ for all $x' \in \ker(f(h,L))$. Hence there is an $(a,B) \in T_{(h,L)}(\mathbb{R} \times \mathcal{Z})$ such that

$$x'^T Df_{(h,L)}(a,B) x' = x'^T F' x' \tag{41}$$

for all $x' \in \ker(f(h,L))$.

Now let

$$N := \begin{pmatrix} \frac{a}{h^2} & -\frac{a}{h^2} & B_{\{1\}\times W} \\ -\frac{a}{h^2} & B_{\{2\}\times\{2\}} + \frac{a}{h^2} & B_{\{2\}\times W} \\ B_{W\times\{1\}} & B_{W\times\{2\}} & B_{W\times W} \end{pmatrix}. \tag{42}$$

So $N \in T_M \mathcal{O}_N$. A calculation shows

$$Df_{(h,L)}(a,B) = Q^T N Q. \tag{43}$$

For each vector $x \in \ker(M)$, the vector

$$x' = \begin{pmatrix} x_2 \\ x_W \end{pmatrix} \tag{44}$$

belongs to $\ker(f(h,L))$ and satisfies $Qx' = x$. Hence

$$x^T F x = x'^T Q^T F Q x' = x'^T Df_{(h,L)}(a,B) x' = x^T N x. \quad \blacksquare \tag{45}$$

5 $\mu(G)$ and $\lambda(G)$ for complete graphs

It is easy to see that for each graph G with n vertices one has

$$\mu(G) \leq n-1 \quad \text{and} \quad \lambda(G) \leq n-1. \tag{46}$$

This follows from the fact that any matrix M satisfying (2) has a negative eigenvalue, and that the all-one vector does not belong to any representative subspace X of $\mathbb{R}^V$.

Moreover:

Theorem 5.1 *For any graph G with n vertices, $\mu(G) = n-1$ if and only if G is complete or $n \leq 2$.*

Proof Let G have n vertices. To see sufficiency, first note that trivially $\mu(G) = n-1$ if $n \leq 2$. Moreover $\mu(K_n) = n-1$ follows from the fact that the all -1 matrix satisfies (2) and has corank $n-1$.

To see necessity, let $n \geq 3$ and $\mu(G) = n-1$. Let M be a matrix satisfying (2) with corank $n-1$. So M has rank 1.

Suppose that M has an all-zero row, say row 1. Then 1 is an isolated vertex of G. Since M has rank 1, the dimension of the kernel of M is at least 2, and hence there is a nonzero vector $x \in \mathbb{R}^n$ with $x_1 = 0$ and $Mx = 0$. Let $y \in \mathbb{R}^n$ be given by $y_1 := 1$ and $y_i := 0$ for $i > 1$. Let $X := xy^T + yx^T$. Then $MX = 0$, and hence by (2)(iii), $X = 0$, a contradiction.

So M does not have any all-zero row, and hence (as M has rank 1), all entries in M are nonzero. So G is complete. ■

It follows that for each $t \geq 1$, the graph K_{t+2} is a forbidden minor for the property $\mu(G) \leq t$.

Similarly, one has:

Theorem 5.2 *For any graph G with n vertices, $\lambda(G) = n-1$ if and only if G is complete or $n \leq 2$.*

Proof To see sufficiency, if $n \leq 2$, then trivially $\lambda(G) = n-1$. If $G = K_n$, then $\lambda(G) \geq n-1$, since the set X of functions $x \in \mathbb{R}^V$ with $\sum_{v \in V} x(v) = 0$ is representative for K_n.

To see necessity, let $n \geq 3$ and $\lambda(G) = n-1$. Suppose that G is not complete, and let vertices u and u' be nonadjacent. Let X be a subspace of dimension $n-1$ representative for G. So there is a nonzero vector $c \in \mathbb{R}^V$ such that X consists of all vectors $x \in \mathbb{R}^V$ with $c^T x = 0$. We can assume that $c_u = 1$. Then each entry of c is positive. For suppose that $c_w \leq 0$. Then the vector x with $x_w = 1$, $x_u = -c_w$, and $x_v = 0$ for all other vertices v, belongs to X. So $\text{supp}^-(x) = \emptyset$, contradicting (10).

Now by scaling we can assume that each entry in c is 1. Let v be any vertex different from u and u'. Then for the vector x with $x_u = 1$, $x_{u'} = 1$, $x_w = -2$, and $x_v = 0$ for all other vertices v, the graph $G|\text{supp}^+(x)$ is disconnected, contradicting (10). ■

Hence, for each $t \geq 1$, the graph K_{t+2} is a forbidden minor for the property $\lambda(G) \leq t$.

6 Clique sums

As mentioned, Colin de Verdière conjectures that $\gamma(G) \leq \mu(G) + 1$, where $\gamma(G)$ is the colouring number of G. This conjecture would follow from Hadwiger's conjecture (as $\mu(K_n) = n - 1$), and is true for $\mu(G) \leq 4$. A similar relation holds for the size of the largest clique minor in a graph. We therefore are interested in studying the behaviour of $\mu(G)$ and $\lambda(G)$ under clique sums.

To study this for $\lambda(G)$, we first give an auxiliary result. For any finite subset Z of $\mathbb{R}^d$ let $\text{cone}(Z)$ denote the smallest nonempty convex cone containing Z; that is, it is the intersection of all closed halfspaces $\{x \in \mathbb{R}^d \mid c^T x \geq 0\}$ containing Z. (Thus $\text{cone}(\emptyset) = \{0\}$, while $\text{cone}(Z) = \mathbb{R}^d$ if there are no halfspaces containing Z.)

For any graph $G = (V, E)$ and $U \subseteq V$, let $G - U$ denote the graph obtained from G by deleting the vertices of U. (So $G - U = G|(V \setminus U)$.)

Theorem 6.1 *Let $\phi: V \to \mathbb{R}^d$ be representative for a graph $G = (V, E)$ and let $U \subseteq V$. Assume that $\text{cone}(\phi(U))$ is not a hyperplane in $\mathbb{R}^d$. Then there is at most one component K of $G - U$ for which the inclusion $\phi(K) \subseteq \text{cone}(\phi(U))$ does not hold.*

Proof We may assume that $\text{cone}(\phi(U)) \neq \mathbb{R}^d$. Since $\text{cone}(\phi(U))$ is not a hyperplane in $\mathbb{R}^d$, the set

$$C := \{c \in \mathbb{R}^d \mid c \neq 0,\ c^T \phi(v) \leq 0 \text{ for each } v \in U\}, \tag{47}$$

is nonempty and topologically connected (because the polar cone $C \cup \{0\}$ of $\text{cone}(\phi(U))$ is not a line). For $c \in \mathbb{R}^d$, let $H_c := \{x \in \mathbb{R}^d \mid c^T x > 0\}$. Let $K_1, \ldots, K_t$ be the components of $G - U$. Let C_i be the set of vectors $c \in C$ for which H_c intersects $\phi(K_i)$. So if $i \neq j$ then $C_i \cap C_j = \emptyset$, since if $c \in C$ then $\phi^{-1}(H_c)$ is connected and is disjoint from U. As $C_1 \cup \cdots \cup C_t = C$ and since each C_i is an open subset of C, it follows that $C_i = \emptyset$ for all but one i. Hence $\phi(K_i) \subseteq \text{cone}(\phi(U))$ for all but one i. ∎

This implies ([12]):

Theorem 6.2 *If G has at least one edge and is a clique sum of G_1 and G_2, then*

$$\lambda(G) = \max\{\lambda(G_1), \lambda(G_2)\}. \tag{48}$$

Proof We have $\lambda(G) \geq \max\{\lambda(G_1), \lambda(G_2)\}$, since G_1 and G_2 are subgraphs of G, So it suffices to show that $\lambda(G) = \lambda(G_i)$ for some $i = 1, 2$. Assume that $\lambda(G) > \max\{\lambda(G_1), \lambda(G_2)\}$. Let $d := \lambda(G)$, $G = (V, E)$, and $G_i = (V_i, E_i)$ for $i = 1, 2$.

Let $\phi: V \to \mathbb{R}^d$ be representative for G. As $d > \lambda(G_i)$, $\phi|V_i$ is not representative for G_i, for $i = 1$ and $i = 2$. Let $K := V_1 \cap V_2$ and $t := |K|$. We may assume that we have chosen the counterexample so that $|K|$ is as small as possible.

Then $G|(V_1 \setminus K)$ has a component L such that each vertex in K is adjacent to at least one vertex in L. Otherwise G would be a repeated clique sum of subgraphs of G_1 and G_2 with common clique being smaller than K. In that case $\lambda(G) = \max\{\lambda(G_1), \lambda(G_2)\}$ would follow by the minimality of K.

So G_1 has a K_{t+1}-minor. So $\lambda(G_1) \geq t$, and hence $\lambda(G) > t = |K|$. Therefore, cone($\phi(K)$) is not a hyperplane in $\mathbb{R}^d$. (Here we use that it is not the case that $K = \emptyset$ and $d = 1$.) So by Theorem 6.1, we may assume that $\phi(V_1) \subseteq \text{cone}(\phi(K))$.

As $d > \lambda(G_2)$, there exists a halfspace H of $\mathbb{R}^d$ such that $G|(\phi^{-1}(H) \cap V_2)$ is empty or disconnected. If it is empty, then $\phi(v) \in H$ for some $v \in V_1 \setminus K$, contradicting the facts that $\phi(v) \in \text{cone}(\phi(K))$ and that $\phi(K) \cap H = \emptyset$. So it is disconnected. But then also $\phi^{-1}(H)$ would induce a disconnected subgraph of G, as K is a clique. This is a contradiction. ∎

Hence we have that for each $t \geq 1$:

the class of graphs G with $\lambda(G) \leq t$ is closed under taking clique sums. (49)

A statement like this for μ does not hold. A critical example is the graph $K_{t+3} \setminus \Delta$ (the graph obtained from the complete graph K_{t+3} by deleting the edges of a triangle). One has $\mu(K_{t+3} \setminus \Delta) = t + 1$ (since the star $K_4 \setminus \Delta$ has $\mu(K_4 \setminus \Delta) = 2$ (see Theorem 8.2 below), and since adding a new vertex adjacent to all existing vertices increases μ by 1).

However, $K_{t+3} \setminus \Delta$ is a clique sum of K_{t+1} and $K_{t+2} \setminus e$ (the graph obtained from K_{t+2} by deleting an edge), with common clique of size t. Both K_{t+1} and $K_{t+2} \setminus e$ have $\mu = t$. So, generally one does not have that, for fixed t, the property $\mu(G) \leq t$ is maintained under clique sums. Similarly, $K_{t+3} \setminus \Delta$ is a clique sum of two copies of $K_{t+2} \setminus e$, with common clique of size $t + 1$.

These examples where μ increases by taking a clique sum are in a sense the only cases, as shown in [13]:

Theorem 6.3 *If G has at least one edge and is a clique sum of G_1 and G_2, with common clique S, then $\mu(G) > t := \max\{\mu(G_1), \mu(G_2)\}$ if and only if:*

either (i) $|S| = t$ and $G - S$ has three components the contraction of which makes with S a $K_{t+3} \setminus \Delta$, (50)

or (ii) $|S| = t+1$ and $G - S$ has two components the contraction of which makes with S a $K_{t+3} \setminus \Delta$.

Moreover, if $\mu(G) > t$ then $\mu(G) = t + 1$, $\mu(G_1) = \mu(G_2) = t$, and we can contract two or three components of $G - S$ so that the contracted vertices together with S form a $K_{t+3} \setminus \Delta$.

7 Behaviour of $\mu(G)$ and $\lambda(G)$ under YΔ and ΔY

The results on clique sums can be applied to study the behaviour of $\mu(G)$ and $\lambda(G)$ under applying the YΔ- and ΔY-operations. The YΔ-operation works as follows, on a graph G: choose a vertex v of degree 3, make its three neighbours pairwise adjacent, and delete v and the three edges incident with v. The ΔY-operation is the reverse operation, starting with a triangle and adding a new vertex.

Note that the if H arises by a ΔY from G, then H is a subgraph of a clique sum of G and K_4. Then Theorem 6.3 implies that $\mu(H) \leq \mu(G)$ if $\mu(G) \geq 4$, and Theorem 6.2 that $\lambda(H) \leq \lambda(G)$ if $\lambda(G) \geq 3$.

In fact, Bacher and Colin de Verdière [1] proved:

Let H arise by a ΔY operation from G. Then $\mu(H) \leq \mu(G)$. (51)
If moreover $\mu(G) \geq 4$, then $\mu(H) = \mu(G)$.

8 $\mu(G)$ and $\lambda(G)$ for complete bipartite graphs

Since complete bipartite graphs are often candidates for forbidden minors, in this section we give formulas for $\mu(K_{m,n})$ and $\lambda(K_{m,n})$. This also exhibits a difference between $\mu(G)$ and $\lambda(G)$. First we consider $\lambda(G)$:

Theorem 8.1 *For $n \geq m \geq 1$, $\lambda(K_{m,n}) = m$.*

Proof On the one hand, K_{m+1} is a minor of $K_{m,n}$, and on the other hand, $K_{m,n}$ is a subgraph of a clique sum of K_{m+1}'s. So by Theorem 6.2 $\lambda(K_{m,n}) = \lambda(K_{m+1}) = m$. ∎

Characterizing $\mu(G)$ for complete bipartite graphs is a little more complicated:

Theorem 8.2 *For $n \geq m \geq 1$ we have*

$$\mu(K_{m,n}) = \begin{cases} m & \text{if } n \leq 2, \\ m+1 & \text{if } n \geq 3. \end{cases} \tag{52}$$

Proof Note that $\mu(K_{m,n}) \leq m+1$ by Theorem 6.3, since $K_{m,n}$ is a subgraph of a clique sum of K_{m+1}'s. It is not hard to see that $\mu(K_{1,1}) = \mu(K_{1,2}) = 1$ and $\mu(K_{2,2}) = 2$. Hence $\mu(K_{m,n}) = m$ if $n \leq 2$.

So let $n \geq 3$. If $m \leq 3$ we can assume that $n = 3$. Let $K_{m,3}$ have vertices $1, \ldots, m+3$, with colour classes $\{1, \ldots, m\}$ and $\{m+1, m+2, m+3\}$. Let M be the $(m+3) \times (m+3)$ matrix with $m_{i,j} = -1$ if $i \leq m < j$ or $j \leq m < i$, and $m_{i,j} = 0$ otherwise. Then M has rank 2 and hence corank $m+1$. Moreover M satisfies (2). Indeed, (2)(ii) is trivial. Moreover, (2)(i) follows directly from the fact that neither M nor $-M$ is positive semi-definite. Finally,

M satisfies the Strong Arnol'd Hypothesis ((2)(iii)). Otherwise there is a nonzero symmetric matrix X with $MX = 0$ and $x_{i,j} = 0$ if $i = j$ or $i \leq m < j$, which can be seen to be impossible using the fact that $m \leq n = 3$.

If $m \geq 4$, we can assume that $n = m$. Choose two adjacent vertices u and v of $K_{m,m}$. Delete the edge uv, and delete $m-4$ other edges incident with u and $m-4$ other edges incident with v. So in the new graph, u and v have degree 3. Applying $\mathrm{Y}\Delta$ to u and to v we obtain a $K_{m-1,m-1}$ with a triangle added to each of the colour classes. The $2(m-4)$ vertices not covered by these triangles span a matching of size $m-4$. Contracting each edge of this matching, we obtain a K_{m+2}. Since $\mu(K_{m+2}) = m+1$, we obtain $\mu(K_{m,m}) = m+1$ (using (51)). ∎

9 Characterizing $\lambda(G) \leq 1$ and $\mu(G) \leq 1$

Note that one trivially has:

$$\mu(G) = 0 \iff \lambda(G) = 0 \iff G \text{ has exactly one vertex.} \tag{53}$$

We next describe the collections of graphs G satisfying $\mu(G) \leq 1$ and $\lambda(G) \leq 1$.

For $\mu(G)$ it is ([5]):

Theorem 9.1 $\mu(G) \leq 1$ *if and only if G is a vertex-disjoint union of paths; that is, if G does not have a K_3 or $K_{1,3}$-minor.*

Proof Since $\mu(K_3) = 2$ by Theorem 5.1 and $\mu(K_{1,3}) = 2$ by Theorem 8.2, the minor-monotonicity of μ gives the 'only if' part.

To see the 'if' part, we can assume, by the minor-monotonicity of $\mu(G)$, that G is a path. Then trivially any matrix M satisfying (2) has rank at least $n-1$, and hence corank at most 1. So $\mu(G) \leq 1$. ∎

The class of graphs G with $\lambda(G) \leq 1$ is a little larger ([12]):

Theorem 9.2 $\lambda(G) \leq 1$ *if and only if G is a forest; that is, if and only if G does not have a K_3-minor.*

Proof If $\lambda(G) \leq 1$ then G has no K_3-minor, as $\lambda(K_3) = 2$. Conversely, if G is a forest, then G arises by taking clique sums and subgraphs from the graph K_2. As $\lambda(K_2) = 1$, Theorem 6.2 gives the corollary. ∎

10 Van der Holst's lemma

In characterizing $\mu(G) \leq 2$ and $\mu(G) \leq 3$ the lemma due to van der Holst [10] turns out to be very helpful.

If $x \in \mathbb{R}^n$ and $I \subseteq \{1, \ldots, n\}$, then x_I denotes the subvector of x induced by the indices in I.

Recall that a vector $x \in \ker(M)$ *has minimal support* if x is nonzero and for each nonzero vector $y \in \ker(M)$ with $\mathrm{supp}(y) \subseteq \mathrm{supp}(x)$ one has $\mathrm{supp}(y) = \mathrm{supp}(x)$.

Theorem 10.1 (Van der Holst's lemma) *Let G be a connected graph and let M satisfy (2). Let $x \in \ker(M)$ have minimal support. Then $G|\operatorname{supp}^+(x)$ and $G|\operatorname{supp}^-(x)$ are both connected.*

Proof Suppose that (say) $G|\operatorname{supp}^+(x)$ is disconnected. Let I and J be two of the components of $G|\operatorname{supp}^+(x)$. Let $K := \operatorname{supp}^-(x)$. Since $m_{i,j} = 0$ if $i \in I$, $j \in J$, we have:

$$\begin{aligned} M_{I\times I}x_I + M_{I\times K}x_K &= 0, \\ M_{J\times J}x_J + M_{J\times K}x_K &= 0. \end{aligned} \tag{54}$$

Let z be an eigenvector of M with negative eigenvalue. By the Perron-Frobenius theorem we may assume $z > 0$. (Strictly speaking, we apply the Perron-Frobenius theorem to the (nonnegative and indecomposable) matrix $\lambda I - M$ choosing λ large enough.)

Let

$$\lambda := \frac{z_I^T x_I}{z_J^T x_J}. \tag{55}$$

Define $y \in \mathbb{R}^n$ by: $y_i := x_i$ if $i \in I$, $y_i := -\lambda x_i$ if $i \in J$, and $x_i := 0$ if $i \notin I \cup J$. By (55), $z^T y = z_I^T x_I - \lambda z_J^T x_J = 0$. Moreover, one has (since $m_{i,j} = 0$ if $i \in I$ and $j \in J$):

$$\begin{aligned} y^T M y &= y_I^T M_{I\times I} y_I + y_J^T M_{J\times J} y_J \\ &= x_I^T M_{I\times I} x_I + \lambda^2 x_J^T M_{J\times J} x_J \\ &= -x_I^T M_{I\times K} x_K - \lambda^2 x_J^T M_{J\times K} x_K \\ &\le 0, \end{aligned} \tag{56}$$

(using (54)) since $M_{I\times K}$ and $M_{J\times K}$ are nonpositive, and since $x_I > 0$, $x_J > 0$ and $x_K < 0$.

Now $z^T y = 0$ and $y^T M y \le 0$ imply that $My = 0$ (as M is symmetric and has exactly one negative eigenvalue, with eigenvector z). Therefore, $y \in \ker(M)$. This contradicts the fact that x has minimal support. ∎

We note that if M satisfies (2), then each vertex $v \notin \operatorname{supp}(x)$ adjacent to some vertex in $\operatorname{supp}^+(x)$ is also adjacent to some vertex in $\operatorname{supp}^-(x)$, and conversely; that is,

$$\text{for each } x \in \ker(M): \quad N(\operatorname{supp}^+(x)) \setminus \operatorname{supp}(x) = N(\operatorname{supp}^-(x)) \setminus \operatorname{supp}(x). \tag{57}$$

Here $N(U)$ is the set of vertices in $V \setminus U$ that are adjacent to at least one vertex in U.

11 Characterizing $\mu(G) \leq 2$ and $\lambda(G) \leq 2$

We can now derive the following result of Colin de Verdière [5]:

Theorem 11.1 *$\mu(G) \leq 2$ if and only if G is outerplanar; that is, if and only if G does not have a K_4- or $K_{2,3}$-minor.*

Proof Since $\mu(K_4) = 3$ by Theorem 5.1 and $\mu(K_{2,3}) = 3$ by Theorem 8.2, the minor-monotonicity of μ gives the 'only if' part (using the forbidden minor characterization of outerplanarity).

To see the 'if' part, we may assume that G is maximally outerplanar. Suppose that $\mu(G) > 2$, and let M be a matrix satisfying (2) of corank more than 2. Let uv be a boundary edge of G. Then there exists a nonzero vector $x \in \ker(M)$ with $x_u = x_v = 0$. We can assume that x has minimal support. By Van der Holst's lemma (Theorem 10.1), $G|\,\mathrm{supp}^+(x)$ and $G|\,\mathrm{supp}^-(x)$ are nonempty and connected. As G is maximally outerplanar, G is 2-connected. Hence there exist two vertex-disjoint paths P_1 and P_2 from $\mathrm{supp}(x)$ to $\{u, v\}$. Let P_1' and P_2' be the parts outside $\mathrm{supp}(x)$. Then the first vertices of P_1' and P_2' both belong to $N(\mathrm{supp}(x))$, and hence (by(57)) to both $N(\mathrm{supp}^+(x))$ and $N(\mathrm{supp}^-(x))$. Contracting each of $\mathrm{supp}^+(x)$, $\mathrm{supp}^-(x)$, P_1', and P_2' to one point, gives an embedded outerplanar graph with uv on the boundary and u and v connected by two paths of length two. This is not possible. ■

The corresponding characterization for $\lambda(G)$ is easier, and was given in [12]:

Theorem 11.2 *$\lambda(G) \leq 2$ if and only if G is a series-parallel graph; that is, if and only if G does not have a K_4-minor.*

Proof If $\lambda(G) \leq 2$ then G has no K_4-minor, as $\lambda(K_4) = 3$.

Conversely, if G is a series-parallel graph, then G arises by taking clique sums and subgraphs from the graph K_3. As $\lambda(K_3) = 2$, Theorem 6.2 gives the corollary. ■

12 Characterizing $\mu(G) \leq 3$

We apply Van der Holst's lemma (Theorem 10.1) similarly to the case $\mu \leq 3$, a main result of Colin de Verdière [5]:

Theorem 12.1 *$\mu(G) \leq 3$ if and only if G is planar; that is, if and only if G does not have a K_5- or $K_{3,3}$-minor.*

Proof Since $\mu(K_5) = 4$ by Theorem 5.1 and $\mu(K_{3,3}) = 4$ by Theorem 8.2, the minor-monotonicity of μ gives the 'only if' part (using Kuratowski's forbidden minor characterization of planarity).

To see the 'if' part, we may assume that G is maximally planar (triangulated). Suppose that $\mu(G) > 3$ and let M be a matrix satisfying (2) of corank more than 3. Let uvw be a face of G. Then there exists a nonzero vector $x \in \ker(M)$ with $x_u = x_v = x_w = 0$. We can assume that x has minimal support. By Van der Holst's lemma (Theorem 10.1), $G|\operatorname{supp}^+(x)$ and $G|\operatorname{supp}^-(x)$ are nonempty and connected. As G is maximally planar, G is 3-connected. Hence there exist three vertex-disjoint paths P_1, P_2, P_3 from $\operatorname{supp}(x)$ to $\{u, v, w\}$. Let P_1', P_2', P_3' be the parts outside $\operatorname{supp}(x)$. Then the first vertices of the P_i' belong to $N(\operatorname{supp}(x))$, and hence (by (57)) to both $N(\operatorname{supp}^+(x))$ and $N(\operatorname{supp}^-(x))$. Contracting each of $\operatorname{supp}^+(x)$, $\operatorname{supp}^-(x)$, P_1', P_2', P_3' to one point, would give an embedded outerplanar graph with uvw forming a face and u, v, and w having two common neighbours. This is not possible. ■

13 Characterizing $\lambda(G) \leq 3$

We characterize in this section the graphs G satisfying $\lambda(G) \leq 3$, a result of [12]. The main ingredient is:

Theorem 13.1 *If G is planar then $\lambda(G) \leq 3$.*

Proof Suppose $G = (V, E)$ is a planar graph with $\lambda(G) \geq 4$ and $|V|$ minimal. We assume that we have an embedding of G in the sphere. For each face f of G let V_f be the set of vertices incident with f. Note that G is 4-connected, since otherwise it would be a subgraph of clique sums of smaller planar graphs, and hence we would have $\lambda(G) \leq 3$ by Theorem 6.2.

Let $\phi: V \to \mathbb{R}^4$ be representative for G. Then $\phi(v) \neq 0$ for each $v \in V$, since otherwise we can delete v, contradicting the minimality of G. So we can assume that $\|\phi(v)\| = 1$ for each $v \in V$.

We may assume that, for each edge uv, $\phi(u) \neq \pm\phi(v)$, since otherwise, either $\phi(u) = \phi(v)$, in which case we can contract the edge $\{u, v\}$ in G, or $\phi(u) = -\phi(v)$, in which case we can delete the edge $\{u, v\}$ from G. In either case we obtain a contradiction with the minimality of G.

Observe that if f and f' are faces with $\dim(\phi(V_f)) = \dim(\phi(V_{f'})) = 2$ and having a common edge, e say, then $\text{lin.hull}(\phi(V_f)) = \text{lin.hull}(\phi(V_{f'}))$, as it is equal to $\text{lin.hull}(\phi(e))$. Similarly, $\text{lin.hull}(\phi(V_f)) \subseteq \text{lin.hull}(\phi(V_{f'}))$ if $\dim(\phi(V_f)) = 2$, $\dim(\phi(V_{f'})) = 3$ and f, f' share a common edge.

Fixing V, we choose E maximal under the condition that $\phi(u) \neq \pm\phi(v)$ for each edge $\{u, v\}$. Then $\dim(\phi(V_f)) \in \{2, 3\}$ for each face f. Indeed, $\dim(\phi(V_f)) \geq 2$, as each edge $e = uv$ has $\dim(\phi(\{u, v\})) \geq 2$. Moreover, if $\dim(\phi(V_f)) = 4$, then V_f contains at least two nonadjacent vertices u, v with $\dim(\phi(\{u, v\})) = 2$. As we can add the edge uv, this contradicts the maximality of E.

For $c \in \mathbb{R}^4$ let

$$c^+ := \{v \in V \mid c^T\phi(v) > 0\},$$
$$c^- := \{v \in V \mid c^T\phi(v) < 0\}, \tag{58}$$

and let $\mathcal{F}_c$ be the set of faces f for which V_f intersects both c^+ and c^-. Then:

> Let f and f' be two faces with $\dim(\phi(V_f \cup V_{f'})) = 4$. Then there is a $c \in \mathbb{R}^4$ with $f, f' \in \mathcal{F}_c$. (59)

To see this, we note that because $\dim(\phi(V_f)) \geq 2$, $\dim(\phi(V_{f'})) \geq 2$, and $\dim(\phi(V_f \cup V_{f'})) = 4$, there exist vertices $u, v \in V_f$ and $u', v' \in V_{f'}$ with $\dim(\phi(\{u, v, u', v'\})) = 4$. Therefore, we can find a $c \in \mathbb{R}^4$ such that $u, u' \in c^+$ and $v, v' \in c^-$. So $f, f' \in \mathcal{F}_c$, proving (59).

For $c \in \mathbb{R}^4$, let $W_c := \bigcup\{V_f | f \in \mathcal{F}_c\}$. To finish the proof of the theorem, it suffices to show:

$$\dim(\phi(W_c)) \leq 3 \text{ for each } c \in \mathbb{R}^4. \tag{60}$$

This is sufficient, since (60) implies an immediate contradiction with (59), as there exist faces f and f' with $\dim(V_f \cup V_{f'}) = 4$, since $\dim(\phi(V)) = 4$ and as there is a face f with $\dim(\phi(V_f)) = 3$ (since if $\dim(\phi(V_f)) = 2$ for each face f then $\dim(\phi(V)) = 2$, since $\text{lin.hull}(\phi(V_f)) = \text{lin.hull}(\phi(V_{f'}))$ for any two adjacent faces f, f').

We show that (60) holds. It suffices to show the result for those c with W_c inclusionwise maximal, and hence with $c^T\phi(v) \neq 0$ for each vertex v.

Let such a c be given. As both $G|c^+$ and $G|c^-$ are connected, the cut $\delta(c^+)$ corresponds in the dual graph of G to a circuit C which traverses exactly two edges in each face $f \in \mathcal{F}_c$.

Suppose, to obtain a contradiction, that $\dim(\phi(W_c)) = 4$. Then there exist faces $f, f' \in \mathcal{F}_c$ with $\dim(\phi(V_f)) = \dim(\phi(V_{f'})) = 3$ and such that $\text{lin.hull}(\phi(V_f)) \neq \text{lin.hull}(\phi(V_{f'}))$ (as otherwise $\text{lin.hull}(\phi(V_f)) = \text{lin.hull}(\phi(V_{f'}))$ for all $f, f' \in \mathcal{F}_c$ with $\dim(\phi(V_f)) = 3$ and $\dim(\phi(V_{f'})) = 3$, which implies that $\dim(\phi(W_c)) = 3$). They correspond to two vertices on C. Denote by $f_1, \ldots, f_t$ the faces between f and f' when travelling from f to f' along C (in a given direction). Set $f_0 := f$ and $f_t := f'$. Then we may assume that $\dim(\phi(V_{f_i})) = 2$ for all $i = 1, \ldots, t$. (Otherwise we can make t smaller.)

For $i = 0, 1, \ldots, t$, let u_iv_i be the edge common to the faces f_i and f_{i+1}. So each u_iv_i belongs to $\delta(c^+)$ (as G is 4-connected). We may assume that $u_i \in c^+$ and $v_i \in c^-$ for each i.

Now choose $w \in V_f$ so that $\phi(w) \notin \text{lin.hull}(\phi(V_{f'}))$ and $w' \in V_{f'}$ so that $\phi(w') \notin \text{lin.hull}(\phi(V_f))$. Then the set $\phi(\{u_0, v_0, w, w'\})$ has dimension 4. Hence, there exists a $d \in \mathbb{R}^4$ such that $d^T\phi(w) > 0$, $d^T\phi(w') > 0$, $d^T\phi(u_0) = 0$, and $d^T\phi(v_0) = 0$. Then the set $d^+ \cup d^-$ contains none of the vertices on the faces $f_1, \ldots, f_t$ (since $V_{f_i} \subseteq \text{lin.hull}(\phi(\{u_0, v_0\}))$ for all $i = 1, \ldots, t$). In particular, $u_i, v_i \notin d^+ \cup d^-$ for $i = 1, \ldots, t$. By the connectivity of $G|d^+$ there exists a path P from w to w' which is entirely contained in d^+.

Consider the region $R := \bigcup_{i=0}^{t+1} \overline{f}_i$ (where $\overline{f}_i$ is the topological closure of f_i). As P joins two vertices on the boundary of R, $R \cup P$ partitions the rest of the sphere into two regions R_1 and R_2. We choose indices such that R_1 has the vertices $u_0, \ldots, u_t$ on its boundary, while R_2 has the vertices $v_0, \ldots, v_t$ on its boundary.

By the connectivity of $G|d^-$, d^- is contained either in $\overline{R}_1$ or in $\overline{R}_2$. Suppose first that d^- is contained in $\overline{R}_1$. Consider the vector $\tilde{d} = d + \varepsilon c$, with $\varepsilon > 0$ small enough such that $d^+ \subseteq \tilde{d}^+$ and $d^- \subseteq \tilde{d}^-$. Then, $\tilde{d}^- \supseteq \{v_0, \ldots, v_t\} \cup d^-$, while $u_0, \ldots, u_t \in \tilde{d}^+$. Then there is no path joining v_0 and d^- which is entirely contained in $\tilde{d}^-$, contradicting the connectivity of $G|\tilde{d}^-$.

If d^- is contained in $\overline{R}_2$, we arrive similarly at a contradiction, by considering $\tilde{d} = d - \varepsilon c$. ∎

We can now characterize the graphs G satisfying $\lambda(G) \leq 3$. Having Theorem 13.1, Theorem 6.2 gives that $\lambda(G) \leq 3$ also holds for graphs G obtained from planar graphs by taking clique sums and subgraphs. This characterizes the graphs G with $\lambda(G) \leq 3$, as follows from the following two theorems.

Theorem 13.2 *If G has no K_5- or V_8-minor, then G can be obtained by taking clique sums and subgraphs from planar graphs.*

Proof Suppose G is not planar. If G is not 3-connected, then it is easy to see that G is a subgraph of a clique sum of two smaller graphs not having a K_5- or V_8-minor. So we may assume that G is 3-connected.

Then by Wagner's theorem [20], G can be obtained as a subgraph of a 3-clique sum of two smaller graphs G_1 and G_2 both with no K_5-minor. Let K be the clique.

It suffices to show that G_1 and G_2 have no V_8-minor. Suppose to the contrary that G_1, say, has a V_8-minor. As V_8 does not contain any triangle, the V_8-minor in G_1 does not need all three edges of K. So $G_1 - e$ has a V_8-minor for some edge e in K. However, $G_1 - e$ is a minor of G (by the 3-connectedness of G), contradicting the fact that G does not have a V_8-minor. ∎

In [12] also the following was shown (we thank Andries Brouwer for communicating the proof below to us):

Theorem 13.3 $\lambda(V_8) = 4$.

Proof The inequality $\lambda(V_8) \leq 4$ follows from the fact that for any vertex v of V_8, the graph $V_8 - v$ is planar. Hence $\lambda(V_8) \leq \lambda(V_8 - v) + 1 \leq 4$ by Theorem 13.1.

We next show $\lambda(V_8) \geq 4$. Represent V_8 as the graph G with vertex set $V = \{0, \ldots, 7\}$. Let $M = (m_{i,j})$ be the 8×8 matrix defined by

$$m_{i,j} = \begin{cases} 1 & \text{if } i = j, \\ -\sqrt{2} & \text{if } |i-j| = 1 \text{ or } 7, \\ -1 & \text{if } |i-j| = 4, \\ = 0 & \text{otherwise,} \end{cases} \tag{61}$$

where we assume that the rows and columns are labelled $0, \ldots, 7$.

One can show that M has rank at most 4 as follows. For $\alpha = \frac{1}{4}\pi\mathrm{i}$ and $\alpha = \frac{1}{2}\pi\mathrm{i}$, let x^α be the vector in $\mathbb{C}^8$ defined by

$$x_j^\alpha := e^{\alpha j} \tag{62}$$

for $j = 0, \ldots, 7$. These two vectors are linearly independent and both satisfy $Mx^\alpha = 0$. Indeed, by symmetry it suffices to show that for both choices of α one has $(Mx^\alpha)_0 = 0$. Now

$$(Mx^\alpha)_0 = x_0^\alpha - x_{-1}^\alpha\sqrt{2} - x_1^\alpha\sqrt{2} - x_4^\alpha = 1 - e^{-\alpha}\sqrt{2} - e^\alpha\sqrt{2} - e^{4\alpha} \tag{63}$$

taking subscripts mod 8. If $\alpha = \frac{1}{4}\pi\mathrm{i}$, then $e^\alpha + e^{-\alpha} = 2\cos\frac{1}{4}\pi = \sqrt{2}$, while $e^{4\alpha} = \cos\pi + \mathrm{i}\sin\pi = -1$, and hence $(Mx^\alpha)_0 = 0$. If $\alpha = \frac{1}{2}\mathrm{i}\pi$, then $e^\alpha + e^{-\alpha} = 2\cos\frac{1}{2}\pi = 0$, while $e^{4\alpha} = \cos 2\pi + \mathrm{i}\sin 2\pi = 1$, and hence again $(Mx^\alpha)_0 = 0$. Since the real and imaginary parts of the two vectors x^α give four vectors linearly independent over $\mathbb{R}$, we know that $\text{corank}(M) \geq 4$.

Let X be the kernel (null space) of M. We show that X is representative for G. Choose a nonzero $x \in X$. So

$$x_j = x_{j-1}\sqrt{2} + x_{j+1}\sqrt{2} + x_{j+4}. \tag{64}$$

Let $W := \text{supp}^+(x)$. Then $W \neq \emptyset$, since otherwise for any j with $x_{j+1} < 0$, the value of x_j would be strictly smaller than x_{j+1} by (64).

Assume that W induces a disconnected subgraph of V_8. Let $U := V \setminus W$, and let K_1 and K_2 be two of the components of $G|W$. Then $|K_i| \geq 2$, since otherwise K_i would consist of one vertex, contradicting (64). So $|U| \leq 4$. Since V_8 is 3-connected, since each cutset of size 3 consists of the set of vertices adjacent to one vertex, and since U separates K_1 and K_2, it follows that $|U| = 4$, and that the subgraph induced by W consists of two disjoint edges.

Now for each edge $e = \{j, j+1\}$ of V_8, each other edge e' of V_8 disjoint from e contains at least one vertex that is adjacent to at least one vertex in e. It follows that $W = \{1, 3, 5, 7\}$ or $W = \{0, 2, 4, 6\}$. Then (64) implies that $x_j \leq x_{j+4}$ for each $j \in W$, and hence $x_j = x_{j+4}$ for each $j \in W$. But then $x_j = 0$ for each $j \in U$, contradicting the fact that $\text{supp}^-(x) \neq \emptyset$. ∎

Thus we have the following theorem:

Theorem 13.4 *Let G be a graph. Then $\lambda(G) \leq 3$ if and only if G arises by taking clique sums and subgraphs from planar graphs; that is, if and only if G has no K_5- or V_8-minor.*

Proof Directly from Theorems 5.2, 6.2, 13.1, 13.2, and 13.3. ∎

14 A Borsuk theorem for antipodal links

We next come to studying $\mu(G) \leq 4$ and $\lambda(G) \leq 4$. The following Borsuk-type theorem on the existence of certain antipodal links is essential in the proof. This theorem, for general dimension, is proved in [15].

Let P be a convex polytope in $\mathbb{R}^n$. We say that two faces F and F' are *antipodal* if there exists a nonzero vector c in $\mathbb{R}^n$ such that the linear function $c^T x$ is maximized by every point of F and minimized by every point of F'. Let $(P)_1$ denote the 1-skeleton of P. For any face F of P, let ∂F be its boundary.

Theorem 14.1 *Let P be a full-dimensional convex polytope in $\mathbb{R}^5$ and let ϕ be an embedding of $(P)_1$ into $\mathbb{R}^3$. Then there exists a pair of antipodal 2-faces F and F' such that $\phi(\partial F)$ and $\phi(\partial F')$ are linked.*

In [15] this is derived (for general dimension) from a Borsuk-type theorem on the existence of antipodal intersections, extending a result of Bajmóczy and Bárány [2] slightly. A direct proof of Theorem 14.1 can be sketched as follows. First:

> We can assume that if F and F' are antipodal 2-faces of P, then $F - F$ and $F' - F'$ do not have any nonzero vector in common. (65)

This can be shown by applying a small projective perturbation to P.

For any two disjoint closed curves C and C' in $\mathbb{R}^3$, let $\text{lk}(C, C')$ denote their *linking number*, which is the number mod 2 of crossings in any link diagram where C is over C'. (This is a topological invariant.) Then:

> There exists an embedding $\psi\colon (P)_1 \to \mathbb{R}^3$ with the property that there is exactly one pair of antipodal 2-faces F, F' for which $\text{lk}(\psi(\partial F), \psi(\partial F')) = 1$. (66)

Indeed, we can assume that by maximizing the last coordinate x_5 we obtain some 2-face F_0 and by minimizing x_5 we obtain some 2-face F_0' antipodal to F_0. Moreover, we can assume that $(0, 0, 0, 0, 1)$ belongs to the relative interior of F_0 and that $(0, 0, 0, 0, -1)$ belongs to the relative interior of F_0'. For any vector $x = (x_1, \ldots, x_5)$ in $\mathbb{R}^5$, let $\tilde{x} := (x_1, \ldots, x_4)$. Then define $\psi\colon (P)_1 \to S^3$ by:

$$\psi(x) := \frac{\tilde{x}}{\|\tilde{x}\|}. \qquad (67)$$

We can assume that ψ is an embedding of $(P)_1$ into S^3 (by moving P slightly). Then F_0, F_0' is the only pair of antipodal 2-faces F, F' for which we have $\text{lk}(\psi(\partial F), \psi(\partial F')) = 1$.

Finally:

We can deform ψ to ϕ while only edges are moved through each other; at each such operation, the quantity $\sum_{F,F'} \text{lk}(\partial F, \partial F')$ remains invariant. (68)

This follows from the fact that for any two edges e, e' of $(P)_1$, the number of pairs of antipodal 2-faces F, F' with $e \subset \partial F$ and $e' \subset \partial F'$, is even.

15 Characterizing $\mu(G) \leq 4$

We first give a brief introduction to the work of Robertson, Seymour, and Thomas on linklessly embeddable graphs. An embedding of a graph G into $\mathbb{R}^3$ is called *linkless* if any two disjoint circuits in G have unlinked images in $\mathbb{R}^3$. A graph G is *linklessly embeddable* (in $\mathbb{R}^3$) if it has a linkless embedding in $\mathbb{R}^3$.

There are a number of equivalent characterizations of linklessly embeddable graphs. Call an embedding of G *flat* if for each circuit C in G there is a disk D (a 'panel') disjoint from (the embedding of) G and having boundary equal to C. Clearly, each flat embedding is linkless, but the reverse does not hold. (For instance, if G is just a circuit C, then any embedding of G is linkless, but only the unknotted embeddings are flat.) However, if G has a linkless embedding, it also has a flat embedding. So the collections of linklessly embeddable graphs and of flatly embeddable graphs are the same. This was shown by Robertson, Seymour, and Thomas [19], as a byproduct of a proof of an even deeper forbidden-minor characterization of linklessly embeddable graphs.

To understand this forbidden-minor characterization, it is important to note that the class of linklessly embeddable graphs is closed under the YΔ- and ΔY-operations. It implies that also the class of forbidden minors for linkless embeddability is closed under applying YΔ and ΔY. Now Robertson, Seymour, and Thomas [19] showed:

the Petersen family is the collection of forbidden minors for linkless embeddability. (69)

Here the *Petersen family* is the class of graphs arising from the Petersen graph by any series of ΔY- and YΔ-operations. The Petersen family consists of seven graphs, and includes the graph K_6 (see Figure 2).

It turns out not to be difficult to prove that $\mu(G) = 5$ for each graph in the Petersen family. In fact, by result (51) of Bacher and Colin de Verdière [1], the class of graphs G with $\mu(G) = 5$ is closed under ΔY and YΔ. Since moreover $\mu(K_6) = 5$, we know $\mu(G) = 5$ for each graph G in the Petersen family.

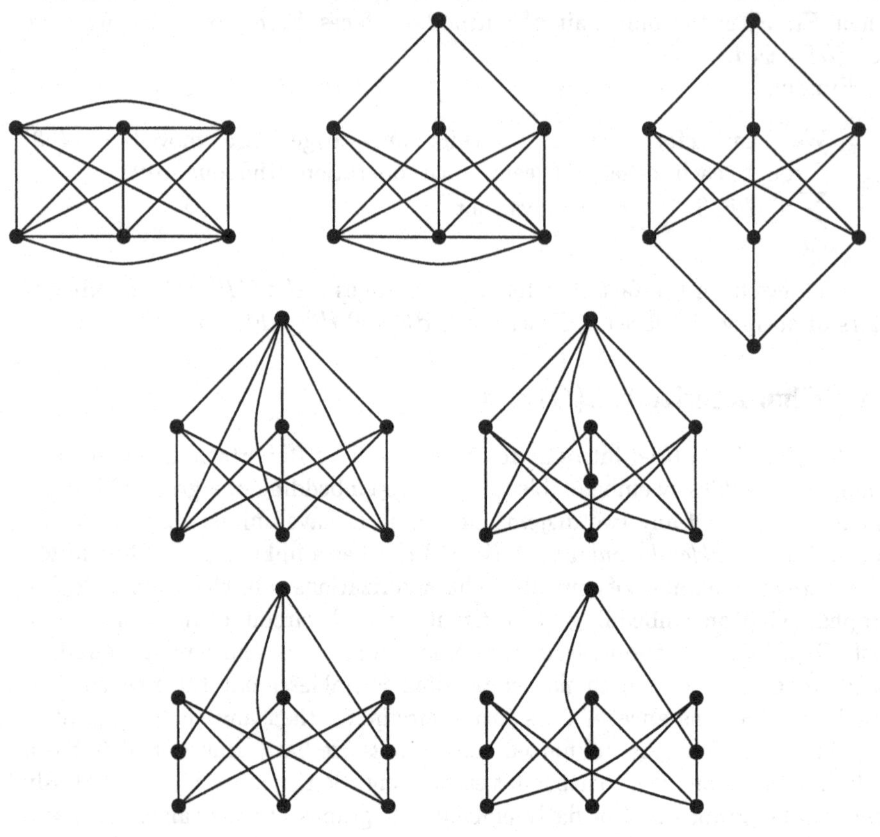

Figure 2: The Petersen family

So with the forbidden minor characterization of Robertson, Seymour, and Thomas [19], we know that if $\mu(G) \leq 4$ then G is linklessly embeddable. The reverse implication was conjectured by Robertson, Seymour, and Thomas [17] and proved in [15], and thus we have:

Theorem 15.1 *$\mu(G) \leq 4$ if and only if G is linklessly embeddable; that is, if and only if G does not have a minor in the Petersen family.*

We will not give the full proof of this here, but rather give an indication of the proof by showing that $\lambda(G) \leq 4$ for each linklessly embeddable graph G. The proof that also $\mu(G) \leq 4$ for linklessly embeddable graphs G is similar, but requires a few more technicalities, and we do not give it in this paper.

16 Towards characterizing $\lambda(G) \leq 4$

We do not know a complete characterization of the class of graphs G satisfying $\lambda(G) \leq 4$. However, we have ([15]):

Theorem 16.1 *If G is linklessly embeddable, then $\lambda(G) \leq 4$.*

Proof Let G be linklessly embedded in $\mathbb{R}^3$, and suppose that $\lambda(G) \geq 5$. Then there is a 5-dimensional subspace L of $\mathbb{R}^V$ such that $G|\operatorname{supp}^+(x)$ is nonempty and connected for each nonzero $x \in L$.

Call two elements x and x' of L *equivalent* if $\operatorname{supp}^+(x) = \operatorname{supp}^+(x')$ and $\operatorname{supp}^-(x) = \operatorname{supp}^-(x')$. The equivalence classes decompose L into a centrally symmetric complex $\mathcal{P}$ of pointed polyhedral cones. Choose a sufficiently dense set of vectors of unit length from every cone in $\mathcal{P}$, in a centrally symmetric fashion, and let P be the convex hull of these vectors. Then P is a 5-dimensional centrally symmetric convex polytope such that every face of P is contained in a cone of $\mathcal{P}$.

We define an embedding ϕ of $(P)_1$ in $\mathbb{R}^3$. For each vertex v of P, we choose a vertex v' of G in $\operatorname{supp}^+(v)$, and we let $\phi(x)$ be a point in $\mathbb{R}^3$ very near v'. For each edge $e = uv$ of P, we choose a path e' connecting u' and v' in $G|\operatorname{supp}^+(x)$, where x is an interior point of e. (By our construction, $\operatorname{supp}^+(x)$ is independent of the choice of x, and contains both $\operatorname{supp}^+(u)$ and $\operatorname{supp}^+(v)$.) Then we map e onto a Jordan curve connecting $\phi(u)$ and $\phi(v)$ very near e'. Clearly we can choose the images of the vertices and edges so that this map ϕ is one-to-one.

Then by Theorem 14.1, P has two antipodal 2-faces F and F' such that the images of their boundaries are linked. Since P is centrally symmetric, there is a facet D of P such that $F \subseteq \overline{D}$ and $F' \subseteq -\overline{D}$. Let y be a vector in the interior of D. Then the images of ∂F and $\partial F'$ are very near subgraphs spanned by $\operatorname{supp}^+(y)$ and $\operatorname{supp}^-(y)$, respectively, and hence some circuit of G spanned by $\operatorname{supp}^+(y)$ must be linked with some circuit in $\operatorname{supp}^-(y)$, a contradiction. ■

Corollary 16.2 *If G is obtained from linklessly embedded graphs by taking clique sums and subgraphs, then $\lambda(G) \leq 4$.*

Proof Directly from Theorems 6.2 and 16.1. ■

By Theorem 5.2, $G = K_6$ is a forbidden minor for the class of graphs G with $\lambda(G) \leq 4$. Any other graph G in the Petersen family of graphs however satisfies $\lambda(G) \leq 4$, since:

Theorem 16.3 *Let G be in the Petersen family with $G \neq K_6$. Then G is obtainable by taking clique sums and subgraphs from K_5.*

Proof Inspection of the Petersen family (Figure 2) shows that G is either a subgraph of the graph obtained from K_7 by deleting the edges of a triangle, and this graph is a clique sum of three K_5's, or G arises from such a subgraph by one or more ΔY-transformations, that is, it is a subgraph of a clique sum with K_4's. ∎

This immediately implies that $\lambda(G) \leq 4$ for each graph $G \neq K_6$ in the Petersen family. Moreover, it follows that each such graph is obtainable by taking clique sums and subgraphs from linklessly embeddable graphs.

Note that the graph G obtained from V_8 by adding a new vertex adjacent to all vertices of V_8, cannot be obtained from linklessly embeddable graphs by taking clique sums and subgraphs; but G does not have a K_6-minor. In fact, it satisfies $\lambda(G) = 5$. However it is not minor-minimal for the property $\lambda(G) \geq 5$.

Let V_9' arise from V_8 by adding an extra vertex v_0, adjacent to v_2, v_4, v_6, v_7, v_8 (see Figure 3). Similarly, let V_9'' arise from V_8 by adding an extra vertex v_0

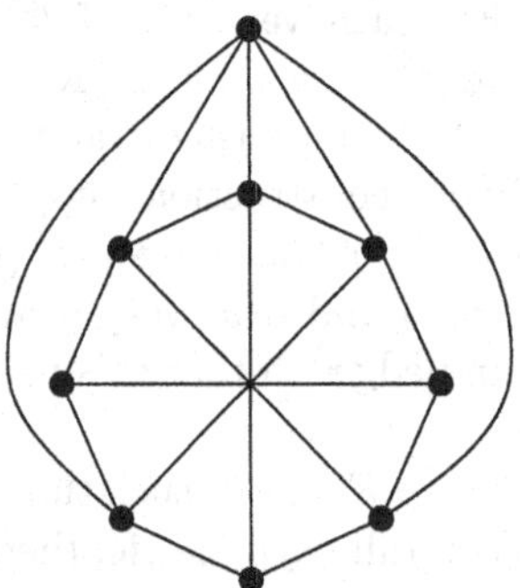

Figure 3: The graph V_9'

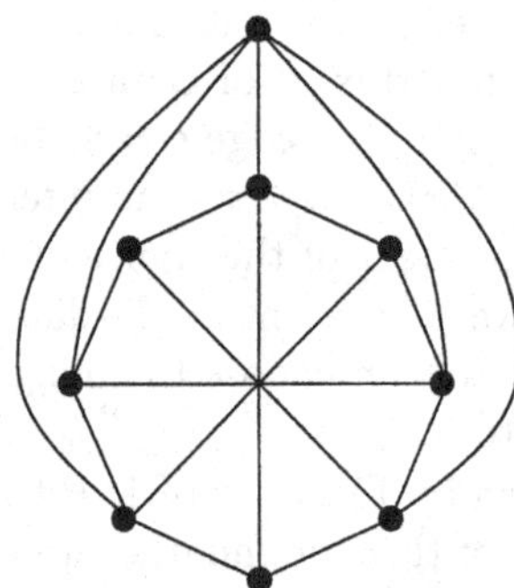

Figure 4: The graph V_9''

adjacent to v_2, v_3, v_5, v_7, v_8 (see Figure 4). It is shown in [12] that V_9' and V_9'' are minor-minimal graphs G with $\lambda(G) \geq 5$.

The graphs V_9' and V_9'' are also minor-minimal graphs not obtainable from linklessly embeddable graphs by taking clique sums and subgraphs. This can be seen as follows. Since $\lambda(V_9') = \lambda(V_9'') = 5$, it follows from Corollary 16.2 that these two graphs indeed are not obtainable in such a way. Moreover, to see that they are minor-minimal, observe that deleting or contracting any edge of V_9' or V_9'', produces a graph that has a vertex whose deletion makes the graph a clique sum of planar graphs.

Since the class of graphs G with $\lambda(G) \leq 4$ is closed under taking ΔY operations (not under YΔ), we can obtain other graphs with $\lambda(G) \geq 5$ by applying a YΔ operation to V_9' or V_9''. Any of them contains a K_6-minor, except if we apply YΔ to vertex v_1 (or equivalently, to v_5) of V_9'.

17 An extension to oriented matroids

It turns out that the results described above for $\lambda(G)$ can be extended to oriented matroids, as is shown in [7]. Before describing this, we first give the definition of and a little further background on oriented matroids (see Björner, Las Vergnas, Sturmfels, White, and Ziegler [3] for more information).

It is convenient to introduce, for any ordered pair $x = (a, b)$, the notation $x^+ := a$ and $x^- := b$, and $-x = (b, a)$.

An *oriented matroid* (V, X) consists of a finite set V and a collection X of ordered pairs $x = (x^+, x^-)$ of subsets of V such that:

(i) for each $x \in X$, $x^+ \cap x^- = \emptyset$; (70)

(ii) $\mathbf{0} := (\emptyset, \emptyset) \in X$;

(iii) if $x \in X$ then $-x \in X$;

(iv) if $x, y \in X$, then $x \cdot y \in X$ where $x \cdot y$ is defined by $x \cdot y := (x^+ \cup (y^+ \setminus x^-), x^- \cup (y^- \setminus x^+))$;

(v) if $x, y \in X$ and $u \in x^+ \cap y^-$, then there exists a $z \in X$ such that $u \notin z^+ \cup z^-$, $(x^+ \setminus y^-) \cup (y^+ \setminus x^-) \subseteq z^+ \subseteq x^+ \cup y^+$, and $(x^- \setminus y^+) \cup (y^- \setminus x^+) \subseteq z^- \subseteq x^- \cup y^-$.

The elements of X are called the *vectors* of the oriented matroid. ($\mathbf{0}$ is the *zero*.) Any linear subspace Y of $\mathbb{R}^V$ gives an oriented matroid (V, X), by taking

$$X := \{(\text{supp}^+(x),\ \text{supp}^-(x)) \mid x \in Y\}. \tag{71}$$

For any oriented matroid $M = (V, X)$, the minimal nonempty subsets of $\{x^+ \cup x^- \mid x \in X\}$ form the circuit collection of a matroid, again denoted by M. Thus matroid terminology applies to oriented matroids, and we can speak of the *rank* rank(M) of an oriented matroid M: it is the maximum size of a subset of V not containing any circuit as a subset. The *corank* corank(M) of M is equal to $|V| - \text{rank}(M)$. It is not difficult to prove that if M is given by (71), then

$$\text{corank}(M) = \dim(Y). \tag{72}$$

Now the graph parameter $\lambda'(G)$ is defined as follows. Let $G = (V, E)$ be an undirected graph. An oriented matroid $M = (V, X)$ is called *representative* for G if

for each nonzero $x \in X$, x^+ is nonempty and induces a connected subgraph of G. (73)

Then $\lambda'(G)$ is the largest corank of an oriented matroid representative for G.

From (72) one derives that for each graph G:

$$\lambda(G) \le \lambda'(G). \tag{74}$$

One of the consequences of the results described below is that there are no graphs G with $\lambda(G) \leq 3$ and $\lambda(G) < \lambda'(G)$. In fact, we do not know any graph G with strict inequality in (74).

Any result we know for $\lambda(G)$, also holds for $\lambda'(G)$. First of all, $\lambda'(G)$ is minor-monotone:

if G is a minor of H then $\lambda'(G) \leq \lambda'(H)$. (75)

Moreover one has:

$$\lambda'(K_n) = n - 1. \quad (76)$$

So again Hadwiger's conjecture implies the conjecture that $\gamma(G) \leq \lambda'(G) + 1$ for each graph G, where $\gamma(G)$ is the colouring number of G.

Moreover:

For any graph G and vertex v of G one has $\lambda'(G - v) \geq \lambda'(G) - 1$. (77)

Again for each $t \geq 1$ the class of graphs G with $\lambda(G) \leq t$ is closed under taking clique sums, since:

If G has at least one edge and is a clique sum of G_1 and G_2, then $\lambda'(G) = \max\{\lambda'(G_1), \lambda'(G_2)\}$. (78)

This directly implies characterizations of those graphs G satisfying $\lambda'(G) \leq 1$ and $\lambda'(G) \leq 2$:

$\lambda'(G) \leq 1$ if and only if G is a forest, (79)

and

$\lambda'(G) \leq 2$ if and only if G is a series-parallel graph. (80)

Moreover, it can be proved that

a graph G satisfies $\lambda'(G) \leq 3$ if and only if G can be obtained from planar graphs by taking clique sums and subgraphs. (81)

Recently, Rudi Pendavingh showed:

if G is obtainable from linklessly embeddable graphs by taking subgraphs and clique sums, then $\lambda'(G) \leq 4$. (82)

18 The related graph invariant $\kappa(G)$

We finally describe a graph invariant related to $\lambda(G)$ (introduced in [12]), for which the set of forbidden minors can be precisely characterized. For any connected graph $G = (V, E)$, define $\kappa(G)$ to be the largest d for which there exists a function $\phi : V \to \mathbb{R}^d$ such that:

(i) $\phi(V)$ affinely spans a d-dimensional affine space; (83)

(ii) for each affine halfspace H of $\mathbb{R}^d$, $\phi^{-1}(H)$ induces a connected subgraph of G (possibly empty).

(An *affine halfspace* is a set of the form $\{x \mid c^T x > \delta\}$ for some nonzero vector c.) Note that such a function ϕ does not exist for disconnected graphs; so $\kappa(G)$ is undefined if G is disconnected.

Observe that if G is the 1-skeleton of a full-dimensional polytope in $\mathbb{R}^d$, then $\kappa(G) \geq d$, as the polytope gives the embedding in $\mathbb{R}^d$.

By similar arguments as used in the proof of Theorem 3.1 one shows that if H is a connected minor of G then $\kappa(H) \leq \kappa(G)$. So again for each d there is a finite collection of forbidden minors for the collection of graphs satisfying $\kappa(G) \leq d$. This collection of graphs is equal to $\{K_{d+2}\}$, as is shown in the next theorem.

First observe that

$$\kappa(G) \leq \lambda(G) \tag{84}$$

holds for each connected graph G, since if $\phi: V \to \mathbb{R}^d$ satisfies (83), then we may assume that the origin belongs to the interior of the convex hull of $\phi(V)$. But then trivially ϕ is representative for G.

Basic in the characterization is the following observation (Grünbaum and Motzkin [9], Grünbaum [8]):

Theorem 18.1 *If G is the 1-skeleton of a d-dimensional polytope P, then G has a K_{d+1}-minor.*

Proof By induction on d, the case $d = 0$ being trivial. If $d > 0$, let F be a facet of P. By the induction hypothesis, the 1-skeleton of F can be contracted to K_d. Moreover, the vertices of P not on F induce a connected subgraph of G, and hence can be contracted to one vertex. This yields a contraction of G to K_{d+1}, as each vertex of F is adjacent to at least one vertex of P not on F. ■

This gives:

Theorem 18.2 *For each connected graph G and each d, $\kappa(G) \geq d$ if and only if G has a K_{d+1}-minor.*

Proof *Sufficiency.* One has $\kappa(K_{d+1}) = d$ since the vertices of a simplex in $\mathbb{R}^d$ give a function ϕ satisfying (83). So if G has a K_{d+1}-minor, then $\kappa(G) \geq d$.

Necessity. Let $G = (V, E)$ be a connected graph and let $d := \kappa(G)$, such that for each proper connected minor H one has $\kappa(H) < d$. By Theorem 18.1 it suffices to show that G is the 1-skeleton of a d-dimensional polytope.

Let $\phi: V \to \mathbb{R}^d$ satisfy (83). Let P be the convex hull of $\phi(V)$. So P is a d-dimensional polytope in $\mathbb{R}^d$. We show that G is the 1-skeleton of P.

First observe that for each vertex x of P, the set $\phi^{-1}(x)$ induces a connected subgraph of G, as it is equal to $\phi^{-1}(H)$ for some affine halfspace H of $\mathbb{R}^d$. Hence if $\phi^{-1}(x)$ consists of more than one vertex of G, then we can contract this subgraph to one vertex, contradicting the minimality of G.

Similarly, for each edge xy of P, the set $\phi^{-1}(xy)$ induces a connected subgraph of G. Hence it contains a path from $\phi^{-1}(x)$ to $\phi^{-1}(y)$.

As this is true for each edge, G contains a subdivision of the 1-skeleton of P as a subgraph. By the minimality of G this implies that G is equal to the 1-skeleton of P. ∎

So Hadwiger's conjecture is equivalent to $\gamma(G) \leq \kappa(G) + 1$ for each connected graph G.

Acknowledgements

We thank the referee for several suggestions improving the presentation.

References

[1] R. Bacher & Y. Colin de Verdière, Multiplicités des valeurs propres et transformations étoile-triangle des graphes, *Bulletin de la Société Mathématique de France*, **123** (1995), 101–117.

[2] E. G. Bajmóczy & I. Bárány, On a common generalization of Borsuk's and Radon's theorem, *Acta Mathematica Academiae Scientiarum Hungaricae*, **34** (1979), 347–350.

[3] A. Björner, M. Las Vergnas, B. Sturmfels, N. White & G. Ziegler, *Oriented Matroids*, Cambridge University Press, Cambridge (1993).

[4] S. Y. Cheng, Eigenfunctions and nodal sets, *Commentarii Mathematici Helvetici*, **51** (1976), 43–55.

[5] Y. Colin de Verdière, Sur un nouvel invariant des graphes et un critère de planarité, *Journal of Combinatorial Theory, Series B*, **50** (1990), 11–21.

[6] Y. Colin de Verdière, On a new graph invariant and a criterion for planarity, in *Graph Structure Theory (Proceedings of the AMS-IMS-SIAM Joint Summer Research Conference on Graph Minors, Seattle, 1991)* (eds. N. Robertson & P. Seymour), *Contemporary Mathematics*, 147, American Mathematical Society, Providence, Rhode Island (1993), 137–147.

[7] J. Edmonds, M. Laurent & A. Schrijver, A minor-monotone graph parameter based on oriented matroids, *Discrete Mathematics*, in press.

[8] B. Grünbaum, On the facial structure of convex polytopes, *Bulletin of the American Mathematical Society*, **71** (1965), 559–560.

[9] B. Grünbaum & T S. Motzkin, On polyhedral graphs, in *Convexity* (ed. V. Klee), *Proceedings of Symposia in Pure Mathematics*, 7, American Mathematical Society, Providence, Rhode Island (1963), pp. 285–290.

[10] H. van der Holst, A short proof of the planarity characterization of Colin de Verdière, *Journal of Combinatorial Theory, Series B*, **65** (1995), 269–272.

[11] H. van der Holst, Topological and Spectral Graph Characterizations, Ph.D. Thesis, University of Amsterdam, 1996.

[12] H. van der Holst, M. Laurent & A. Schrijver, On a minor-monotone graph invariant, *Journal of Combinatorial Theory, Series B*, **65** (1995), 291–304.

[13] H. van der Holst, L. Lovász & A. Schrijver, On the invariance of Colin de Verdière's graph parameter under clique sums, *Linear Algebra and its Applications*, **226** (1995), 509–517.

[14] A. Kotlov, L. Lovász & S. Vempala, The Colin de Verdière number and sphere representations of a graph, preprint, 1996.

[15] L. Lovász & A. Schrijver, A Borsuk theorem for antipodal links and a spectral characterization of linklessly embeddable graphs, *Proceedings of the American Mathematical Society*, in press.

[16] N. Robertson & P. D. Seymour, Graph minors. XX. Wagner's conjecture, preprint, 1988.

[17] N. Robertson, P. D. Seymour & R. Thomas, A survey of linkless embeddings, in *Graph Structure Theory (Proceedings of the AMS-IMS-SIAM Joint Summer Research Conference on Graph Minors, Seattle, 1991)* (eds. N. Robertson & P. Seymour), *Contemporary Mathematics*, 147, American Mathematical Society, Providence, Rhode Island (1993), pp. 125–136.

[18] N. Robertson, P. Seymour & R. Thomas, Hadwiger's conjecture for K_6-free graphs, *Combinatorica*, **13** (1993), 279–361.

[19] N. Robertson, P. Seymour & R. Thomas, Sachs' linkless embedding conjecture, *Journal of Combinatorial Theory, Series B*, **64** (1995), 185–227.

[20] K. Wagner, Über eine Eigenschaft der ebene Komplexe, *Mathematische Annalen*, **114** (1937), 570–590.

CWI
Kruislaan 413
1098 SJ Amsterdam
The Netherlands
and
Department of Mathematics
University of Amsterdam
Plantage Muidergracht 24
1018 TV Amsterdam
The Netherlands
lex@cwi.nl

Some Applications of Algebraic Curves in Finite Geometry and Combinatorics

T. Szőnyi

Summary Various applications of Weil's theorem in finite geometry and combinatorics are surveyed. Several illustrative proofs are sketched. As a by-product, we give an up-to-date account on what is known about complete arcs, minimal blocking sets and (k, n)-arcs in the desarguesian plane $\mathrm{PG}(2, q)$.

1 Introduction

Four years ago at the British Combinatorial Conference, Aart Blokhuis [16] gave a lecture on some applications of polynomials in finite geometry and combinatorics. The aim of this survey is to collect those applications of polynomials where the polynomial is considered as a curve, or more generally as an algebraic variety over a finite field. Most of the applications depend on estimates for the number of $\mathrm{GF}(q)$-rational points of curves. In some of the applications Bézout's theorem is enough, but typically the deep theorem of Weil, or a variant or consequence of it, is used. Weil's theorem gives a very strong bound on the number of $\mathrm{GF}(q)$-rational points of a curve:

Let f be an absolutely irreducible curve of degree d, defined over $\mathrm{GF}(q)$, *and denote by N the number of* $\mathrm{GF}(q)$*-rational points on it. Then*

$$q - (d-1)(d-2)\sqrt{q} + 1 \le N \le q + (d-1)(d-2)\sqrt{q} + 1. \qquad (1)$$

Some refinements of the theorem are discussed in Section 2. To decide whether a given curve is absolutely irreducible or not is not always easy, so in Section 3 we collect some applications where a simpler condition has to be verified (for example, one criterion is to decide if the polynomial is a square of another or not). There are several other bounds for the number of $\mathrm{GF}(q)$-rational points on a curve, but they mainly yield upper bounds only. Some of them will be discussed in Section 5. The advantage of Weil's theorem is that it gives also a lower bound, that is one can use it to show that a certain curve does have points over $\mathrm{GF}(q)$. If the curve has low degree, then it actually gives the asymptotics for the number of rational points.

There are two different philosophies in using Weil's theorem. The first one is to use it for explicitly given or constructed curves, when we want to construct a combinatorial or geometric object or verify certain properties of given algebraically defined structures. In this case either the asymptotics, or just the existence of a $\mathrm{GF}(q)$-rational point is needed.

The second philosophy applies for problems when we want to say something about the size or structure of a geometric object defined by combinatorial properties. So, for a given geometric object in the plane, a curve reflecting the geometric properties of the object is to be associated. This also means that in this case only the existence of a curve is known, but typically not the equation of the curve. In most cases an upper bound for the number of rational points is used.

In Section 3 applications of Weil's theorem in combinatorics are collected. They are applications according to the first philosophy. In this section we do not restrict ourselves to Weil's theorem, but also collect various methods (or just tricks) to estimate the number of solutions of a system of equations. An almost classical illustration of this type of result is due to Graham and Spencer [52] who used the character sum version of Weil's theorem to show that *the Paley graphs provide a solution to a dominance problem of Erdős and Schütte.*

Then finite geometry applications following the first philosophy are collected in Section 4. In this section the definitions of the basic geometric objects: arcs, (k, n)-arcs, blocking sets are given. The first occurrence of this philosophy for solving a problem in finite geometry is a *construction of complete arcs*, due to Tallini Scafati [104]. The method will be illustrated here by a result of V. Abatangelo [1].

The last two sections are devoted to applications in finite geometry following the second philosophy. In this case the weaponry is bigger, there are several improvements on Weil's theorem due to Stöhr–Voloch [92], Hirschfeld–Korchmáros [62, 63], but sometimes also elementary upper bounds are sufficient. This is a typical and very attractive method in Galois geometry, the classical example being Segre's [87] theorem: *a set of $q+1$ points in* $\mathrm{PG}(2, q)$, *q odd, not containing three collinear points must be a conic.* The main point here is to construct a curve that somehow reflects the properties of the geometric object. The curve-making methods of Segre [90] based on a generalization of Menelaus' theorem in classical geometry (Section 5) and recent work of the author [100, 101] (see Section 6) are discussed with some examples. The various methods will be illustrated by sketching two proofs of the following theorem of Tallini [103]: *a set of q points in* $\mathrm{PG}(2, q)$, *q even, not containing three collinear points is contained in a set of $q+2$ points with the same property.*

To sum up, we present a survey of applications of algebraic curves in combinatorics and finite geometry. One important theory, the theory of codes and curves, is completely omitted, since there are textbooks available on the subject [53, 54]. This also applies to MDS codes, see [93, 64, 109]. Finite geometry exclusively means finite geometry over a field, and only the planes $\mathrm{PG}(2, q)$ are considered. So we had to omit important methods of the same flavour, like the approach of Bruen, Thas and Blokhuis [39], which can be regarded as a natural higher dimensional generalization of Segre's method. On the other hand, as a by-product, an up-to-date account on what is known about complete arcs,

(k,n)-arcs and minimal blocking sets in $\mathrm{PG}(2,q)$, is given. The combinatorial applications are connected to graph properties that mimic properties of the random graphs. They are all related to the problem of estimating the number of common neighbours of a small set of vertices in some algebraically defined graphs.

2 Weil's theorem and its variants

A *plane curve* over a field K is the equivalence class of homogeneous polynomials in three variables, where two polynomials are equivalent if they are constant multiples of each other. This means that multiple components are allowed. The degree of the curve is just the degree of the polynomial. A curve is called *absolutely irreducible* if it is irreducible over the algebraic closure of K. A point of the curve f is called *K-rational*, if its coordinates are in K. A point of the curve f is *singular* if all three partial derivatives of f vanish at this point. Weil's theorem gives a surprisingly strong bound for the number of $\mathrm{GF}(q)$-rational points of absolutely irreducible curves defined over $\mathrm{GF}(q)$. First we state it for non-singular plane curves. Note that non-singularity obviously implies absolute irreducibility.

Theorem 2.1 *Let C be a non-singular plane curve of degree d, defined over $\mathrm{GF}(q)$, and denote by N the number of its $\mathrm{GF}(q)$-rational points. Then*

$$|N-(q+1)| \leq (d-1)(d-2)\sqrt{q}.$$

The particular case $d=3$ was proved by Hasse in the thirties, so sometimes even the general result is called the theorem of Hasse–Weil. In this case the theorem says that the number of $\mathrm{GF}(q)$-rational points is between $q-2\sqrt{q}+1$ and $q+2\sqrt{q}+1$. Later Waterhouse [117] showed that the bound is sharp in the sense that essentially every N in the interval does occur (for the precise statement see [121] and also Schoof [86], Hirschfeld [60]).

In most of the applications, one cannot guarantee that the curve is non-singular, so an extension to curves having singular points is needed.

Theorem 2.2 *The number R of $\mathrm{GF}(q)$-rational points of a plane, absolutely irreducible curve C of degree d, genus g, defined over $\mathrm{GF}(q)$ satisfies*

$$q+1-2g\sqrt{q} \leq R \leq q+1+2g\sqrt{q}.$$

Non-singular $\mathrm{GF}(q)$-rational points are counted once, while a singular point of multiplicity m is counted m^* times in R for some $0 \leq m^* \leq m$.

The improvement is that the term $(d-1)(d-2)$ is replaced by $2g$, where g is the genus of the curve. The only thing we need to know about the invariant g is that for plane curves with singularities of multiplicity $r_1,\ldots,r_s$ it is at most

$$\binom{d-1}{2}-\sum_{i=1}^{s}\binom{r_i}{2}$$

(see any textbook on algebraic curves, or [58, equation (10.7)]); so this can be a substantial improvement, for example for curves having singular points with high multiplicities. The form of Weil's estimate mentioned in the introduction follows easily from the bound for the genus and the bounds for m^* in Theorem 2.2.

For a discussion of Weil's theorem with several examples, the reader is referred to Chapter 2 of the new edition of [58] and to other papers of Hirschfeld [60, 61]. In these papers the reader can find a detailed discussion of higher dimensional generalizations of Weil's theorem as well as more explicit formulas to count the singular points.

In order to apply Weil's theorem a technical difficulty still remains: one has to check whether the curve is absolutely irreducible or not. In most of the applications in finite geometry one can use a criterion due to Segre, which we discuss in Section 4, see Theorem 4.1. In most of the applications in combinatorics, the character sum version of Weil's theorem is used (see Weil [119]). This has the advantage that instead of absolute irreducibility a much simpler condition has to be checked. This theorem, due essentially to Burgess, can be found as Theorem 5.41 in the book [76] by Lidl and Niederreiter.

Theorem 2.3 *Let χ be a multiplicative character of* GF(q) *of order k and let $f(x)$ be a polynomial of degree d which cannot be written as $f(x) = cg(x)^k$. Then*

$$\left| \sum_{x \in \mathrm{GF}(q)} \chi(f(x)) \right| \leq (d-1)\sqrt{q}.$$

Let us consider the simplest (but probably most useful) particular case, when χ is the *quadratic character* ω. This means that $\omega(x) = +1$ if x is a non-zero square, $\omega(x) = -1$ if x is a non-square and $\omega(x) = 0$ if $x = 0$. In this case the previous sum essentially counts the number of x's for which $f(x)$ is a non-zero square. Indeed, if $f(x)$ has t zeroes in GF(q), and the sum is S then the number of x's for which $f(x)$ is a square is just $(q - t + S)/2$. Of course, this number can directly be estimated by Weil's theorem; there are t points on the curve $y^2 = f(x)$ with $y = 0$, and the number of the remaining points is two times the number of x's for which $f(x)$ is a square. Note that Theorem 2.3 gives a quite strong remainder term implicitly using the fact that the curve $y^2 = f(x)$ has genus at most $(d-1)/2$, if $d = \deg(f)$. Actually, for d even, the genus is at most $(d-2)/2$. Note that this bound is better than the one mentioned after Theorem 2.2. Indeed, the infinite point $(1, 0, 0)$ is a singular point with multiplicity $d-2$, hence the bound after Theorem 2.2 only gives $(d-2)$ as an upper bound for the genus. Roughly speaking this theorem says that the value of a polynomial "is a square with probability 1/2".

The biggest advantage of the character sum version of Weil's theorem is that it can easily be extended to a system of equations. We wish to prescribe that certain one-variable polynomials $f_i(x)$, $(i = 1, \ldots, m)$ take square or non-square values. (This can be considered as a system of equations by putting $(f_i(x))^{(q-1)/2} = \pm 1$.) Under some light and natural conditions one can show that such a system of equations has approximately $q/2^m$ solutions, where m is the number of equations. This supports the intuitive idea that being a square is a "random event with probability 1/2".

Theorem 2.4 ([98]) *Let $f_1(x), \ldots, f_m(x) \in \mathrm{GF}(q)[x]$ be given polynomials. Suppose that no partial product $f_{i_1}(x) \cdots f_{i_j}(x)$ $(1 \leq i_1 < i_2 < \cdots < i_j;$ $j \leq m)$ can be written as a constant multiple of a square of a polynomial. If $2^{m-1} \sum_{i=0}^{m} \deg(f_i) \leq \sqrt{q} - 1$, then there is an $x_0 \in \mathrm{GF}(q)$ such that $f_i(x_0)$ is a non-square for every $i = 1, \ldots, m$. More precisely if we denote the number of these x_0's by N, then*

$$\left| N - \frac{q}{2^m} \right| \leq \sum_{i=1}^{m} \deg(f_i) \frac{\sqrt{q}+1}{2}.$$

Multiplying the polynomials by an appropriate constant, the theorem also guarantees that $\omega(f_i(x))$ $(= \pm 1)$ can be prescribed. Note that the number of equations may depend on q. There are several extensions of the character sum version of Weil's theorem, for example for polynomials in more than one variable, additive characters, and so on. They can be found in Chapter 5 of [76], some of them in the bibliographical notes.

3 Applications of Weil's theorem in graph theory and combinatorics

There are several important classes of graphs which were defined to mimic some properties of the random graphs. In the random graph $G(n,p)$ every edge is chosen independently with probability p. In such a graph the point-degrees are roughly np, the number of common neighbours of two points is roughly np^2. One of the definitions of a (class of) *quasi-random graphs* (when $p = 1/2$, see [43]) is that

$$\sum_{v,v'} \left| |G(v) \cap G(v')| - \frac{n}{4} \right| = o(n^3),$$

where $G(v)$ is the set of neighbours of v and n denotes the number of vertices in G. This means that quasi-randomness is related to the average number of common neighbours of a pair of vertices. A similar class of graphs, modelling random graphs with arbitrary p, is the class of (p,α)-jumbled graphs, see Thomason [110]. If $0 < p < 1 \leq \alpha$, then the graph G is called (p,α)*-jumbled* if every induced subgraph H satisfies

$$\left| e(H) - p\binom{|H|}{2} \right| \leq \alpha |H|,$$

where $e(H)$ denotes the number of edges in H. One of the key theorems in [110] is the following: if the minimum degree in G is at least pn and the number of common neighbours of a pair of points is at most p^2n+l, then the graph G is $(p, \sqrt{(p+l)n})$-jumbled. Again, a bound for the number of common neighbours of a pair of vertices is needed. There are other interesting classes of graphs that are related to a bound for the number of common neighbours, such as expanders; so we tried to collect results which use Weil's theorem (or rather its character sum version) to bound the number of common neighbours in various algebraically defined graphs. Even the graphs considered here serve only as illustrations; for more examples and results the reader is referred to Thomason [110] and Chung, Graham and Wilson [43]. Besides these papers, a good general reference for this chapter is Alon [3].

In the first part of this section we shall be interested in the Paley graphs, their bipartite versions and generalizations. For a prime-power q the *Paley graph* P_q is defined as follows: the vertices of P_q are the elements of the field $\mathrm{GF}(q)$ and (a,b) is a directed edge of the graph if and only if $b-a$ is a square. If $q \equiv 3 \pmod 4$, then -1 is a non-square; so in this case precisely one of (a,b) and (b,a) is an edge. In other words, the graph P_q is a tournament. For $q \equiv 1 \pmod 4$, -1 is a square, hence (a,b) is an edge if and only if (b,a) is an edge. This means that P_q can simply be considered as an undirected graph. It is in fact a strongly regular graph (actually a conference graph) with parameters $v=q$, $k=(q-1)/2$, $\lambda=(q-5)/4$, $\mu=(q-1)/4$. In some sense, these graphs behave like random graphs with $p=1/2$; so they are quasi-random. This will follow from the next theorems.

The classical paper of Graham and Spencer [52] used the particular case of Theorem 2.4 when the polynomials f_i are linear. A vertex v *dominates* a set S of points in a directed graph if the edges joining v to a point of S are directed from v to the point of S.

Theorem 3.1 (Graham, Spencer [52]) *If $q \equiv 3 \pmod 4$, then in the Paley graph P_q, to every set S of $k \le (\frac{1}{2}-\varepsilon)\log_2 q$ vertices there is a vertex v that dominates them if $q \ge q_0(\varepsilon)$.*

If we wish to find a tournament in which to every set of k points there is a vertex dominating them, then the order of the graph cannot be too small. E. and G. Szekeres [102] proved that it must have at least $2^k(k+1)$ vertices. On the other hand, using the probabilistic method, Erdős (see [47] and [49, p. 40]) proved that there is such a tournament on at most $2^kk^2(\log 2+o(1))$ vertices. The above result of Graham and Spencer shows that the Paley graphs provide a constructive solution with at most ck^24^k vertices.

For undirected Paley graphs the corresponding result is due to Bollobás and Thomason [29]. Intuitively it says that the Paley graphs contain all small graphs as induced subgraphs.

Theorem 3.2 *Let H be an arbitrary graph on m vertices, and ε be a fixed positive constant. If $m \le (\frac{1}{2} - \varepsilon)\log_2 q$, and $q \equiv 1 \pmod 4$, then the Paley graph P_q contains H as an induced subgraph, if q is sufficiently large.*

A graph containing all graphs of order at most r was called *r-full* by Bollobás and Thomason [29]. The original question of Rosenfeld was to find a strongly regular r-full graph. The probabilistic method shows that an r-full graph must have at least $cr^2 2^{r/2}$ vertices. For a Paley graph to be r-full one has to choose its order at least $cr^2 2^{2r}$. Vu [112] constructs a strongly regular r-full graph on 2^{r+2} vertices.

A result in the spirit of the previous two theorems, due to Bollobás [27], is the following. Note that it implies immediately that Paley graphs are quasi-random.

Theorem 3.3 *The number of edges between a set of k vertices of the Paley graph P_q and another disjoint set of l vertices lies between $\frac{1}{2}kl - \frac{1}{2}\sqrt{klq}$ and $\frac{1}{2}kl + \frac{1}{2}\sqrt{klq}$.*

The three proofs are essentially the same and we sketch it for the undirected case (that is when $q \equiv 1 \pmod 4$). Let the graph H have k vertices: $v_1, \ldots, v_k$. We build up a copy of H in P_q recursively. Corresponding to v_i we wish to find an element $a_i \in \mathrm{GF}(q)$. The first element a_1 can be arbitrary. Suppose that the elements $a_1, \ldots, a_m$ have already been chosen. Consider v_{m+1} and denote by x the element corresponding to it. Let the polynomial $f_i(x) = x - a_i$ ($i = 1, \ldots, m$) and prescribe $\omega(f_i(x))$ in the following way: it will be $+1$ if (v_{m+1}, v_i) is an edge in H, -1 otherwise. Since the elements $a_1, \ldots, a_m$ are pairwise distinct, these polynomials satisfy the conditions of Theorem 2.4. Thus we can find an x if $m2^{m-1} \le \sqrt{q} - 1$. Since $m \le k \le (1/2 - \varepsilon)\log_2 q$, such an x exists.

A special case of Theorem 3.2 is to bound the size of the largest independent set and the largest clique of the Paley graph. Using the fact that these are the same, since multiplying a clique by a non-square yields an independent set and vice versa, the classical Erdős–Szekeres upper bound $R(k,k) < 4^k$ for the Ramsey number $R(k,k)$ gives something slightly better than the previous argument. In general, it is not true that the largest independent set has size $c\log q$, it can be $c\log q \log\log\log q$ infinitely often. For this and more information, see [43].

If one considers also Paley graphs of prime-power order q, then the largest clique can be substantially bigger than $\log q$. For example, when q is an odd square, then $\mathrm{GF}(\sqrt{q})$ consists of squares in $\mathrm{GF}(q)$, so we have a clique of size $\sqrt{q}$. An easy counting argument shows that this is the maximum possible size. Blokhuis [15] proved that a clique containing $0, 1$ must be $\mathrm{GF}(\sqrt{q})$. Modifying this clique, Baker, Ebert, Hemmeter and Woldar [7] obtained other maximal cliques of size $(\sqrt{q}+1)/2$ or $(\sqrt{q}+3)/2$, according as $\sqrt{q}$ is 1 or 3 modulo 4. A nice conjecture, due to Blokhuis [15] and Baker, Ebert, Hemmeter and Woldar [7], is that these cliques are the second largest cliques in the Paley graphs of square order. The conjecture has been verified for $\sqrt{q} \leq 25$. However, it is not true that all the maximal cliques of this size are equivalent, see [7].

Let us see how the maximality of the cliques constructed in [7] follows from Theorem 2.4. First of all note that the elements of $\mathrm{GF}(q)$ can be written as $a+bi$, with $a, b \in \mathrm{GF}(\sqrt{q})$, where $i^2 = k$ is a non-square in $\mathrm{GF}(\sqrt{q})$. By direct computation, one can verify that $c + di$ is a square in $\mathrm{GF}(q)$ if and only if $c^2 - kd^2$ is a square in $\mathrm{GF}(\sqrt{q})$. Let

$$C = \{i\} \cup \{x : x \in \mathrm{GF}(\sqrt{q}),\ x - i \text{ is a square in } \mathrm{GF}(\sqrt{q})\}.$$

Then C is obviously a clique. To find the size of C, it needs to be determined how often $x - i$ is a square, that is how often $x^2 - k$ is a square. Now $x^2 - k = y^2$ is the equation of a hyperbola, hence it has $\sqrt{q} - 1$ affine points. Since this hyperbola has no points on the x-axis, $x^2 - k$ is a square for half of the affine points, which means $|C| = (\sqrt{q}+1)/2$. For the maximality, let $a + bi$ be an element of $\mathrm{GF}(q) \setminus \mathrm{GF}(\sqrt{q})$. Consider the two quadratic polynomials $f_1(x) = x^2 - k$ and $f_2(x) = (a-x)^2 - kb$ describing that $i - x$ and $(a+bi) - x$ are both squares in $\mathrm{GF}(q)$. If f_1 and f_2 are not constant multiples of each other, then Theorem 2.4 gives that there is an x for which $f_1(x)$ and $f_2(x)$ are squares in $\mathrm{GF}(q)$ if $4 \cdot 2^1 \leq \sqrt[4]{q} - 1$. If f_1 and f_2 are constant multiples of each other, then $a = 0$ and $b = \pm 1$. Elementary properties of finite fields show that $-i$ can be added to C precisely when $\sqrt{q} \equiv 3 \pmod 4$.

In general, if adjacency in a graph can be described by the condition that the value of a certain polynomial is a square, then Theorem 2.4 can be used to bound the sizes of cliques and to show that certain cliques are indeed maximal. This applies for the class of graphs, defined using exterior and interior points of conics, considered in Baker, Ebert, Hemmeter and Woldar [7].

The case of general χ was also extended to a system of equations, at least when all the polynomials f_i are linear. This was done by Babai, Gál and Wigderson [4], who used it in computer science. Here only the combinatorial part of their result will be given.

Theorem 3.4 *Let $a_1, \ldots, a_t$ be distinct elements of the finite field* $\mathrm{GF}(q)$ *and k be a divisor of $q-1$ $(k, t \geq 2)$. Then the number of solutions to the system of equations $(a_i+x)^{(q-1)/k} = 1$ $(i = 1, \ldots, t)$ is between $q/k^t - t\sqrt{q}$ and $q/k^t + t\sqrt{q}$.*

They used this theorem to construct self-avoiding Sperner families. A system $\mathcal{F}$ of subsets of a set X is a *Sperner family* if it does not contain two elements $F \subset F'$, $F, F' \in \mathcal{F}$. A Sperner family $\mathcal{F}$ is *self-avoiding* if one can associate a set $D(H)$ with each $H \in \mathcal{F}$, called the *core* of H, such that

1. $D(H)$ determines H (with respect to $\mathcal{F}$);

2. for any $H \in \mathcal{F}$ and any subset $T \subset D(H)$, the set

$$S(T) := \bigcup_{G \in \mathcal{F}, G \cap T \neq \emptyset} G \setminus T,$$

does not contain any member of $\mathcal{F}$. The set $S(T)$ is called the *spread* of T.

Let $P(q,k)$ be the *bipartite Paley graph*, that is a graph Γ with bipartition $V(\Gamma) = V_1 \cup V_2$, where $V_1 = \mathrm{GF}(q)$ and also $V_2 = \mathrm{GF}(q)$ and with edge-set $E(\Gamma) = \{(x,y) : x \in V_1,\ y \in V_2,\ (x+y)^{(q-1)/k} = 1\}$. For a vertex $v \in V(\Gamma)$ the set of neighbours of v is denoted by $\Gamma(v)$. For a subset $A \subset V(G)$, $\Gamma(A)$ denotes the set of common neighbours, that is $\Gamma(A) = \bigcap_{v \in A} \Gamma(v)$. Also, let $\Delta(A) = A \cup \Gamma(A)$.

Theorem 3.5 *Fix an integer $t > 1$. If $k \geq 3t$ and $q > 4t^4k^{2t-2}$ then the system*

$$\mathcal{F} = \{\Delta(A) : A \subset V_1,\ |A| = t-1\}$$

is self-avoiding.

In a previous version of their paper, Babai, Gál and Wigderson [4] used different bipartite graphs to construct large self-avoiding Sperner families, based on a recent work by Kollár, Rónyai and Szabó [72] concerning Zarankiewicz' problem. Since the key ingredient in [72] is to estimate the number of solutions of a system of equations, we include their result here.

Let H be a fixed graph. The *Turán number* $ex(n,H)$ is the maximum number of edges in a graph on n vertices which does not contain a copy of H. Zarankiewicz' problem is to determine the Turán number for (complete) bipartite graphs. Let $K_{t,s}$ denote the complete bipartite graph on $t+s$ vertices and with ts edges. For a fixed t and $s \geq t$, Kővári, T. Sós and Turán [74] proved that

$$ex(n, K_{t,s}) \leq c_{t,s} n^{2-\frac{1}{t}},$$

where $c_{t,s}$ is a constant depending on s and t. The best general lower bound, obtained by the probabilistic method, yields only

$$c' n^{2-\frac{s+t-2}{st-1}} \leq ex(n, K_{t,s}),$$

where c' is a positive absolute constant. Using the so-called norm-graph, Kollár, Rónyai and Szabó showed that, for $s \geq t!+1$ the Kővári, T. Sós, Turán bound gives the right order of magnitude

$$ex(n, K_{t,s}) \geq 2^{-t} n^{2-\frac{1}{t}}. \tag{2}$$

Their construction is algebraic. The *norm-graph* $G_{q,t} = G$ is defined as follows: the set of vertices of G is $\mathrm{GF}(q^t)$. Recall that the norm (with respect to $\mathrm{GF}(q)$) of an element $a \in \mathrm{GF}(q^t)$ is $N(a) = a \cdot a^q \cdot \dots \cdot a^{q^{t-1}} = a^{(q^t-1)/(q-1)}$. It is well-known (see e.g. [76]) that $N(x) \in \mathrm{GF}(q)$ and for any non-zero $u \in \mathrm{GF}(q)$, the number of solutions of $N(x) = u$ is $(q^t - 1)/(q - 1)$. Two vertices a, b of G are connected by an edge if $N(a + b) = 1$. For this graph Kollár, Rónyai and Szabó proved the following.

Theorem 3.6 (Kollár, Rónyai, Szabó [72]) *The graph $G = G_{q,t}$ contains no subgraph isomorphic to $K_{t,t!+1}$.*

Let us sketch the proof: if $d_1, \ldots, d_t$ are t different elements of G, then the vertices adjacent to all of them are the solutions of the system of equations $N(x + d_i) = 1$, for $i = 1, \ldots, t$. Using

$$N(x + d_i) = (x + d_i)(x^q + d_i{}^q) \cdots (x^{q^{t-1}} + d_i^{q^{t-1}}),$$

replacing $x^{q^{i-1}}$ by x_i and $d_{i-1}{}^{q^{j-1}}$ by $-a_{ij}$ and putting arbitrary b_i's on the right-hand side one gets the following system of equations:

$$\begin{aligned}
(x_1 - a_{11})(x_2 - a_{12})\cdots(x_t - a_{1t}) &= b_1 \\
(x_1 - a_{21})(x_2 - a_{22})\cdots(x_t - a_{2t}) &= b_2 \\
&\cdots \\
(x_1 - a_{t1})(x_2 - a_{t2})\cdots(x_t - a_{tt}) &= b_t.
\end{aligned}$$

The key step in the proof of the theorem is to show that such a system of equations has at most $t!$ solutions if the b_i's are different. Note that the highest degree terms on the left-hand side are the same in each equation, so the equations as hypersurfaces meet in a variety of codimension 2 (contained in the hyperplane at infinity). The result says that their affine parts intersect in a finite number (at most $t!$) of points.

The previous results showed that among the values of a polynomial, that is not a constant multiple of a square of another polynomial, the squares and non-squares are distributed almost evenly. The classical inequality of Pólya and Vinogradov says that the same property holds for intervals in prime fields. It gives that for any h

$$\left|\sum_{i=0}^{h} \omega(i)\right| \leq \sqrt{p} \log p. \tag{3}$$

This can be used to study another interesting class of algebraically defined graphs: the *graphs* $B(n, t)$. These graphs are also (p, α)-jumbled, see [110]. The vertices of $B(n, t)$ are the elements of $\mathrm{GF}(n)$, where n is prime. Two vertices x, y are joined by an edge if and only if $(x - y)^2 \in \{1, \ldots, t\}$. These graphs are regular with degree $d = |\{x : x^2 \in \{1, \ldots, t\}\}|$. Using the Pólya–Vinogradov inequality one sees that $d/t \to 1$, if $t \geq n^{1/4} \log n$. Bollobás [28] has found a pleasing proof, based on the Pólya–Vinogradov inequality, of the following result.

Theorem 3.7 *No two vertices of the graph $B(n, t)$ have more than $t^2/n + \sqrt{n} \log^2 n$ common neighbours.*

Further properties of $B(n,t)$ can be found in Thomason [110], together with references to improvements of the Pólya–Vinogradov inequality. Applications of the Pólya–Vinogradov inequality in finite geometry, and a far reaching generalization due to Smith, can be found in Section 4.

4 Applications of Weil's theorem in finite geometry

Let us define some important objects in finite geometry.

Definition A (k,n)-*arc* in a projective plane of order q is a set of k points with some n but no $n+1$ points on a line. $(k,2)$-arcs are simply called k-*arcs*. A k-arc is *complete* if it is not contained in a $(k+1)$-arc, that is when it is maximal subject to inclusion. Similarly a (k,n)-arc is *complete* if it is not contained in a $(k+1,n)$-arc.

Definition An s-*fold blocking set* K in $\mathrm{PG}(2,q)$ (or $\mathrm{AG}(2,q)$) is a set of points such that every line of $\mathrm{PG}(2,q)$ (or $\mathrm{AG}(2,q)$) intersects K in at least s points.

A 1-fold blocking set is simply called a *blocking set.* We say that a (1-fold) blocking set is *trivial* if it contains a line of $\mathrm{PG}(2,q)$. An s-fold blocking set is called *minimal* or *irreducible* when no proper subset of it is an s-fold blocking set. For $s=2$ and 3 we also speak of a *double blocking set* and a *triple blocking set.*

Note that (k,n)-arcs and s-fold blocking sets with $n+s=q+1$ are in fact each other's complement. Typically, the (k,n)-arc terminology is used if n is small compared to q and we use the s-fold blocking set terminology if s is small.

The aim of this section is to collect some results, mainly constructions, for these objects. Several good survey papers are recommended: Hirschfeld [59] and Hirschfeld–Storme [64] are general ones, whereas Blokhuis [19] is mainly about blocking sets.

Throughout this paper we use the usual representation of $\mathrm{AG}(2,q)$ and $\mathrm{PG}(2,q)$. This means that the points of $\mathrm{AG}(2,q)$ have affine coordinates (x,y) where x,y are elements of $\mathrm{GF}(q)$. The lines of this affine plane have equation $mX+b-Y=0$ or $X=c$. The coefficient m is the *slope* of the line, and the *infinite points* of $\mathrm{PG}(2,q)$ can be identified with slopes. So (m) will denote the infinite point of lines with slope m. Similarly (∞) will be the infinite point of vertical lines, that is lines with equation $X=c$.

We shall also use the following standard terminology: if K is a set of points and ℓ is a line intersecting K in exactly s points, then we call ℓ an s-*secant of* K. Instead of 1-secant also the expression *tangent* will be used.

The first natural question for k-arcs is to determine the maximum value of k. This was done by Bose (see Chapter 9 of [58]), who proved that $k\le q+2$ and showed that $k=q+2$ is only possible if q is even. $(q+1)$-arcs are called *ovals*, $(q+2)$-arcs are called *hyperovals*. There are natural examples: a conic is always a $(q+1)$-arc. For q even the tangents of a conic pass through a point, called the *nucleus of the conic*. Adding this point results in a hyperoval. The next step is to go beyond Bose's bound. If a k-arc consists of slightly fewer than $q+1$ or $q+2$ points, then it can always be embedded in an arc having the maximum number of points. This beautiful theory is due to Beniamino Segre [90], and we discuss his results and some recent achievements in the next section. Note also that for q odd, Segre proved that $(q+1)$-arcs are in fact conics, so very large arcs are completely described. Let us introduce a notation: $m'(2,q)$ denotes the *size of the largest complete arc, which is not an oval or hyperoval.* This implies that a k-arc with $k > m'(2,q)$ can be embedded in a (hyper)oval.

From the other end of the spectrum, Lunelli and Sce [79] proved that a complete k-arc has to have at least $k \geq \sqrt{2q}$ points. Here the important question is to get close to this theoretical lower bound. The *size of the smallest complete arc* in $\mathrm{PG}(2,q)$ will be denoted by $n(2,q)$. Most of the early constructions of complete arcs other than ovals or hyperovals give k-arcs with $k \geq (q+3)/2$, but they all have $k \sim q/2$, when completeness is known. For a list of references, see [99]. The first example of a complete k-arc with $k \sim q/3$ was given by V. Abatangelo [1]. Before discussing his construction let us describe in general how the method based on Weil's theorem works for proving completeness of arcs.

Most of the constructions use the following general suggestion of Segre and Lombardo–Radice [75]: *the points of the arc are chosen (with some exceptions) among the points of a conic or a cubic.* So let us start with a conic or cubic $\mathcal{C}$. The steps of the general construction scheme are the following:

1. Choose an algebraically parametrized subset $K \subset \mathcal{C}$. Preferably the parametrization should have low degree.

2. Construct an algebraic curve describing the collinearity of two points of K and a point P outside $\mathcal{C}$.

3. Show that the curve is absolutely irreducible for most $P \notin \mathcal{C}$. Then Weil's theorem will guarantee the existence of two points of K such that the line joining them passes through P; in other words, P cannot be added to K.

4. Extend the arc with some exceptional points $P \notin \mathcal{C}$, and with some points of $\mathcal{C}$.

Let us illustrate this scheme by a result of V. Abatangelo [1].

1. Start with the parabola $\mathcal{C}: \ y = x^2$. Let the algebraically parametrized subset K be $K = \{(u^3, u^6) : 0 \neq u \in \mathrm{GF}(q)\}$, where 3 divides $q-1$. Note that $|K| = (q-1)/3$.

2. Take first an affine point $P : (a, b)$ with $b \neq a^2$. The collinearity of $P_1 : (x^3, x^6)$, $P_2 : (y^3, y^6)$ $(P_1 \neq P_2)$ and P means that $a(x^3+y^3)-x^3y^3-b=0$. This is the equation of a curve $\mathcal{F}$ of degree 6. Here we used that $P_1 \neq P_2$ implies $x^3 \neq y^3$.

3. To show that $\mathcal{F}$ is absolutely irreducible the following criterion of Segre [89] can be used.

Lemma 4.1 *Let $\mathcal{F}$ be a plane algebraic curve of degree k over a field E. $\mathcal{F}$ is irreducible over the algebraic closure of E if there is a point P on $\mathcal{F}$ such that there is no linear component of $\mathcal{F}$ through P and there is a tangent r at P which counts once among the tangents at P and intersects $\mathcal{F}$ with multiplicity k.*

Note that the conditions for the tangent r are easy to verify, since we only have to prove that $r \cap \mathcal{F} = \{P\}$ for the multiplicity condition and the tangents can be found explicitly. In our case, if $a \neq 0$, the conditions can be verified for the infinite points of $\mathcal{F}$. These are $(0, 1, 0)$ and $(1, 0, 0)$ and the tangents are the lines $x = c$ and $y = d$ with $c^3 = d^3 = a$. (For $a = 0$ the tangents would not be distinct.) Substituting $x = c$ for such a c in the equation of $\mathcal{F}$ gives $a^2 - b = 0$ and this is not zero. Hence $\mathcal{F}$ is indeed absolutely irreducible. The curve is of degree 6 and it has two ordinary singular points with multiplicity 3, so its genus is at most 4 and Weil's theorem shows that there are points with $x^3 \neq y^3$ if q is large enough. A similar argument works for infinite points. Doing this precisely, Abatangelo [1] proved the following theorem.

Theorem 4.2 *Let $q = 2^h$, $h \geq 8$ and $K = \{(u^3, u^6) : 0 \neq u \in \mathrm{GF}(q)\}$. If $K \cup \{P\}$ is an arc, then either P belongs to the line $x = 0$ or P is the nucleus $(1, 0, 0)$ of the parabola $\mathcal{C}$ or $P \in \mathcal{C} \setminus K$.*

4. We can add to K the nucleus $(1, 0, 0)$ and the points $(0, g)$ and $(0, g^2)$ of the line $x = 0$, where g is a generator of the multiplicative group of $\mathrm{GF}(q)$.

Abatangelo's idea was generalized to arbitrary multiplicative subgroups by Korchmáros [73].

The same idea was used by the present author [94] for cubic curves, to construct complete k-arcs with $k = o(q)$. To completely follow the above scheme one has to start with a rational curve, see [94, 95]. However, the method can be extended to non-singular cubics but instead of the parametrization, that is a morphism to the projective line, morphisms to the curve itself have to be considered. Let us first mention that for cubic curves one can always work in an abelian group, since there is an abelian group describing collinearity, see Schoof [86]. In particular, if this abelian group has even order, then the coset of a subgroup of index two will be an arc containing half the points of the cubic. Voloch [115] proved that if the j-invariant of the elliptic cubic is non-zero and

$q \geq 175$ (if q is odd), or $q \geq 256$ (if q is even), then these arcs are complete. This construction, without proving the completeness, is due to Zirilli [122]. A nice additional property of these arcs is that they intersect a conic in at most 6 points by Bézout's theorem; for a general combinatorial problem related to this fact, see Cameron [41]. In a second paper Voloch [116] extended his method to cosets of other subgroups. Under some mild conditions, he proved that if the characteristic is not 2 and the j-invariant is not zero, then the coset of a subgroup of index m covers all the points outside the cubic curve provided that

$$q > 98m^4 + 34m^2 + 2m(7m^2+1)\sqrt{49m^2+20}.$$

Using essentially this result, the present author [96] proved by a construction that $n(2,q) \leq cq^{3/4}$. Very recently Kim and Vu [71] used a sophisticated version of the probabilistic method, the so-called Rödl nibble, to show that $n(2,q) \leq d\sqrt{q}\log^c q$, where d and c are absolute constants.

A next natural question is to determine the spectrum of complete arcs, that is to decide which are the possible cardinalities of complete k-arcs. Combining the results of Voloch [115], Hadnagy [55] and Szőnyi [96] the following result can be obtained.

Theorem 4.3 (Hadnagy, Voloch, Szőnyi) *Let $p \geq 5^{10}$ be a prime number. For every integer k satisfying*

$$[2.46 \cdot \log p \cdot p^{3/4}] \leq k < (\sqrt{p}+1)^2/2$$

there exists a complete arc with k points in $\mathrm{PG}(2,p)$.

Historically, Voloch proved the part $(\sqrt{p}-1)^2 \leq k \leq (\sqrt{p}+1)^2$ by using the complete arcs consisting of half the points of a non-singular cubic curve. He only needed $p \geq 175$, and his result extends to planes of non-prime order. Taking points of a conic, Szőnyi (see [96]) showed that for every integer k satisfying $[p/3]+3 \leq k \leq [p/2]+1$ there exists a complete k-arc in $\mathrm{PG}(2,p)$, if p is a sufficiently large prime. Finally, Éva Hadnagy dealt with the longest part of the interval. She needs the bound $p \geq 5^{10}$, which guarantees for example that the intervals are non-empty.

For the proof of completeness, Hadnagy uses a theorem of Smith about small solutions of congruences, which is somewhat similar to the Pólya–Vinogradov inequality.

Theorem 4.4 (Theorem of Smith) *Let p be an odd prime and denote by C the set of points $\mathbf{x} = (x_1,\ldots,x_n)$ satisfying $0 \leq x_i < p \quad (i=1,\ldots,n)$. Let $C^* = C \setminus \{(0,\ldots,0)\}$. We define a box $\mathcal{B}$ in C as the set of all points $\mathbf{x} \in C$, which satisfy $0 \leq v_i \leq x_i < v_i + h_i \leq p \quad (i=1,\ldots,n)$. Let $f(\mathbf{X})$ be a polynomial in n variables ($\mathbf{X} = (X_1,\ldots,X_n)$, $0 \leq X_i < p$, $i=1,\ldots,n$). Denote by $N(\mathcal{B})$ the number of $\mathbf{X} \in \mathcal{B}$ for which $f(\mathbf{X}) = 0 \pmod p$. Then*

$$N(\mathcal{B}) = \frac{|\mathcal{B}|}{|C|}N(C) + \frac{1}{|C|}\sum_{\mathbf{c}\in C^*} S_n(f,\mathbf{c})\mathcal{E}_{\mathbf{c}}(\mathcal{B}) \tag{4}$$

where

$$\mathcal{E}_{\mathbf{c}}(\mathcal{B}) = \sum_{\mathbf{a}\in\mathcal{B}} e^{(2\pi i/p)\cdot(-\mathbf{a}\cdot\mathbf{c})} \quad \text{and} \quad S_n(f,\mathbf{c}) = \sum_{f(\mathbf{x})=0} e^{(2\pi i/p)\cdot(-\mathbf{c}\cdot\mathbf{x})},$$

where the sum extends over all $\mathbf{x}$ *satisfying* $f(\mathbf{x}) = 0$, *and* $\mathbf{c}\cdot\mathbf{x}$ *denotes the ordinary inner product of* $\mathbf{c}$ *and* $\mathbf{x}$.

Note that the Pólya–Vinogradov inequality can be deduced from Smith's theorem with a slightly weaker remainder term: taking the box $\mathcal{B} = [1, h] \times [1, p-1]$ and the polynomial $f(x, y) = x - y^2$ the number of quadratic residues in the interval $[1, h]$ will just be half of $N(\mathcal{B})$. The next lemmas clarify what the order of magnitude of $N(\mathcal{B})$ is.

The remainder term of (4) can be bounded using the following lemma.

Lemma 4.5

$$E(\mathcal{B}) = \sum_{\mathbf{c}\in C^*} |\mathcal{E}_{\mathbf{c}}(\mathcal{B})| \leq Bp^n \cdot \log^n p, \tag{5}$$

where B *is an absolute constant depending only on* n. *For example if* $n = 2$ *and* $p \geq 60$ *then* $B = 1$ *can be supposed (see Section 3 of [42]).*

Lemma 4.6 (Lemma 1 of [91]) *Let* $f(X)$ *and* $\psi(X, Y)$ *be polynomials over* $\mathrm{GF}(p)$, $\deg\psi = d_1$, $\deg f = d_2$, $1 \leq d_1, d_2 < p$ *and suppose* $\psi(X, Y)$ *has no linear factors. Let*

$$S(f,\psi) = \sum_{\psi(x,y)=0} e^{(2\pi i/p)\cdot f(x)},$$

where the summation is over all $(x, y) \in \mathrm{GF}(p) \times \mathrm{GF}(p)$. *Then*

$$|S(f,\psi)| \leq (d_1^2 + 2d_1d_2 - 3d_1)p^{1/2} + d_1^2. \tag{6}$$

So for the Pólya–Vinogradov case we have $d_2 = 1$, $d_1 = 2$, hence (4) gives $N(\mathcal{B}) = h + \log^2 p(2\sqrt{p} + 4)$, so it is only a $\log p$-factor weaker than the Pólya–Vinogradov inequality. Actually, the extra $\log p$-factor comes from Lemma 4.5. Using the fact that in our case one of the intervals of the box $\mathcal{B}$ is the complete residue system, the proof (see [42, Section 3, p. 293]) gives for $n = 2$ that $|E(\mathcal{B})| \leq 2p^2 \log p$. This means that Smith's theorem almost gives back the Pólya–Vinogradov inequality.

The previous constructions produced relatively small complete arcs in the sense that the largest arcs contain approximately $q/2$ points and the upper bound for the size of a complete arc is $q+1$ or $q+2$. Therefore it is surprising that in $\mathrm{PG}(2, q)$, q a fixed square, there is only one larger non-oval complete arc known. If q is not a square then the largest non-oval complete arc is the one containing half of the points of an elliptic cubic. If q is a square then $q^2+q+1 = (q+\sqrt{q}+1)(q-\sqrt{q}+1)$ and taking the orbit of a subgroup of order $q-\sqrt{q}+1$ of a Singer group, one gets a (cyclic) arc of size $q-\sqrt{q}+1$. The arcs are complete for $q \geq 9$. These arcs were first found by Kestenband [69] using a different construction, and later with this representation by Ebert [46], Fisher, Hirschfeld and Thas [50], and by Boros, Szőnyi [30], see also Kestenband [70]. For exact values of $m'(2,q)$ and $n(2,q)$ in planes of small order, as well as information on the spectrum of complete arcs, see [59, 64, 99].

For minimal blocking sets the situation is similar to the one for complete arcs. Bruen [33, 34] proved that the size of a non-trivial blocking set is at least $q + \sqrt{q} + 1$. On the other hand, Bruen and Thas [38] proved that a minimal blocking set contains at most $q\sqrt{q}+1$ points. Both bounds are sharp if q is a square; a subplane of order $\sqrt{q}$ (a Baer subplane) achieves the lower, a Hermitian arc (unital) achieves the upper bound. Recall that the *Hermitian curve* is the curve $X_1^{\sqrt{q}+1} + X_2^{\sqrt{q}+1} + X_3^{\sqrt{q}+1}$, if q is a square. It has exactly $q\sqrt{q}+1$ GF(q)-rational points and the lines of PG$(2,q)$ meet this set in either 1 or $\sqrt{q}+1$ points, see [58]. *Hermitian arcs* (or *unitals*) are combinatorial analogues of Hermitian curves, that is sets of $q\sqrt{q}+1$ points in PG$(2,q)$, q square, which intersect the lines in 1 or $\sqrt{q}+1$ points. In particular, there is a unique tangent at each point. For further properties of Hermitian arcs and characterizations of Hermitian curves see [58]. However, it is not clear whether the upper bound is any good if q is not a square. In general, it would be desirable to know which are the possible sizes of minimal blocking sets, that is to determine the spectrum. This problem was posed for example in Cameron [41]. The next section contains some results in this direction for small blocking sets, that is for blocking sets whose size is at most $3(q+1)/2$. In particular, we shall see some improvements on Bruen's lower bound. Regarding large minimal blocking sets not much is known. For $q \equiv 1 \pmod 4$ there is a nice connection between maximal cliques in the Paley graph P_q and a particular class of minimal blocking sets obtained as the union of parabolas in a pencil. Denote by $\mathcal{P}_a$ the parabola $\{(x, x^2+a) : x \in \text{GF}(q)\} \cup \{(\infty)\}$, where as usual (∞) is the infinite point of vertical lines.

Lemma 4.7 ([98]) *Let $A \subset$ GF(q), $q \equiv 1 \pmod 4$, and $B = \bigcup_{a \in A} \mathcal{P}_a$. Then B is a minimal blocking set if and only if A is a maximal independent set in the Paley graph P_q.*

This immediately shows that for $q \equiv 1 \pmod 4$ there exist minimal blocking sets of size at least $cq \log q$ in PG$(2,q)$, see [98], where the same result is obtained using Theorem 2.4 also for $q \equiv 3 \pmod 4$ with a more complicated construction of similar flavour. Actually, in both cases it is shown that there are minimal blocking sets of size $cq \log q$. The construction idea of considering unions of conics was also used by Abbott and Liu [2] and Ughi [111]. The above lemma also means that the results on the cliques of the Paley graphs can be translated to results on minimal blocking sets. For example, if q is an odd square then the independent set $A = \delta\,\text{GF}(\sqrt{q})$, ($\delta$ is a non-square in GF(q)) gives a minimal blocking set of size $q\sqrt{q}+1$. This set is a nice example of a Hermitian arc which is not a Hermitian curve. Such Hermitian arcs were constructed by Buekenhout [40] and Metz [83] using 4-dimensional representations of projective planes of square order. For more on explicit planar representation of non-classical unitals, the reader is referred to the papers by Baker and Ebert [5, 6].

One can also repeat the trick used by Baker, Ebert, Hemmeter and Woldar about halving a maximal clique and obtain various minimal blocking sets. This was done by Hirschfeld and Szőnyi [65], who proved the following density theorem for planes of square order.

Theorem 4.8 *For every λ with $\frac{1}{4} < \lambda \leq \frac{1}{2}$ there are constants c_1 and c_2 such that in* PG$(2, q)$, *q square, $q > q_0(\lambda)$ there are minimal blocking sets B with*

$$c_1 q^{1+\lambda} \leq |B| \leq c_2 q^{1+\lambda}. \tag{7}$$

The interval $[q + 1, 3(q + 1)/2)$, that is small blocking sets, will be the subject of the next section. In the interval $(3(q+1)/2, 2q)$ there are examples of minimal blocking sets of size $2q + 1 - d$ for every divisor d of q or $q - 1$. There are also other examples known, such as a minimal blocking set of size 19 in PG$(2, 11)$; for more results, particularly for infinite series, see Gács [51]. In the interval $k \in [2q, 3q - 3]$ we know almost everything: Innamorati and Maturo [67] and independently Illés, Szőnyi and Wettl [66] constructed minimal blocking sets of size k for every such k. For a discussion of small planes, see [19] and [10].

Beyond this, almost nothing is known. Blokhuis and Metsch [22] proved that for $q \geq 49$, q square, there are no minimal blocking sets of size $q\sqrt{q}$ (that is one less than the upper bound of Bruen and Thas). Some more (trivial) examples can be found in [66].

The situation is quite analogous for (k, n)-arcs. Barlotti [13] proved that for $1 \leq n \leq q+1$, $k \leq qn - q + n$ and that equality can only occur when $n \mid q$. For q even Denniston [45] constructed (k, n)-arcs with $k = nq - q + n$ in PG$(2, q)$ for every divisor n of q. A different construction was given by Thas [105]. Sometimes these arcs are called *maximal arcs.* Recently Ball, Blokhuis and Mazzocca [12] proved the long-standing conjecture that for q odd, equality in Barlotti's bound cannot occur; so there are no maximal arcs. A much shorter proof in the same style has been subsequently found by Ball and Blokhuis [11]. Previously this was only known for $n = 3$, see Cossu [44] and Thas [105]. Such a hypothetical $(2q + 3, 3)$-arc would be a Steiner triple system embedded in PG$(2, q)$. It is interesting to note that Thas' proof is based on the observation that in this Steiner triple system 3 non-collinear points generate an affine plane of order 3 (hence such a system is a so-called Hall triple system). This is proved using Segre's lemma of tangents (see the next section for Segre's lemma). Then Thas actually shows that the Steiner system must be an affine space over GF(3) and it gives the contradiction that both q and $2q + 3$ must be powers of 3. If n is not a divisor of q, then the above bound by Barlotti was improved several times. Namely, if n is not a divisor of q then Lunelli and

Sce [79] proved that $k \le (n-1)q+n-3$. Later they improved the bound to $k \le (n-1)q+8n/13$, if n does not divide q and q is large compared to n. They also conjectured $k \le (n-1)q+1$ if q is not divisible by n, but this was disproved by Hill and Mason [57]. On the other hand, if there is a line disjoint from the (k,n)-arc, then Blokhuis [18] verified the Lunelli–Sce conjecture; that is he proved $k \le (n-1)q+1$ for such (k,n)-arcs. For large $n > 2q/3$, Hill [56] further improved the above bounds.

On the constructive side not much is known. For small planes, we refer to Ball [9]. Besides the counterexamples by Hill and Mason, large (k,n)-arcs with $k = (q-p^h)(p^h-1)$, $n = q-p^h$ (and $q = p^n$) were constructed by Mason [80, 81]. Hirschfeld and Szőnyi [65] constructed (k,n)-arcs with $k \ge (1-\varepsilon)nq$, if $n \ge q^{1/2+\delta}$, which shows that Barlotti's bound is essentially correct. We used a pencil of conics and Theorem 2.4. Another approach in [65] used the Pólya–Vinogradov inequality for planes of prime order, and assumed $n \ge c\sqrt{q}\log q$. Let p be sufficiently large and let h be larger than $\sqrt{p}\log p$ regarding its order of magnitude. Consider the set

$$K = \{(x, x^2+a) : x \in \mathrm{GF}(p),\ 0 \le a \le h\}.$$

Elementary computation shows that it is enough to bound the intersection of K and horizontal lines; in other words, for a horizontal line $\ell_c : y = c$ to find the number of quadratic residues (squares) in the set $\{c-a : 0 \le a \le h\}$. The Pólya–Vinogradov inequality then gives that $|\ell_c \cap K| \le h + 2\sqrt{p}\log p$. This immediately gives the asymptotic sharpness of Barlotti's bound if n is substantially larger than $\sqrt{p}\log p$.

Proposition 4.9 *Let p be a prime and suppose that $n(p)/(\sqrt{p}\log p) \to \infty$. Then there is a (k,n')-arc in $\mathrm{PG}(2,p)$ with $n(p) \le n' \le n(p) + 2\sqrt{p}\log p$ and $k \sim pn'$.*

Choosing a smaller n and adding some points one can actually achieve that $n' = n$. Using the probabilistic method instead of the algebraic constructions, one can extend the result to $n \ge c\log q$. This can be done by following essentially the proof of Erdős, Silverman and Stein [48].

Almost nothing is known for (k,n)-arcs with constant $n > 2$. Taking $[n/2]$ disjoint conics, one can easily see that k can be at least $[n/2](q+1)$, but the upper bound is around $(n-1)q$. Even for $n = 3$ the best example known is to take a non-singular cubic curve giving a $(k,3)$-arc with $k \le q+2\sqrt{q}+1$. Aart Blokhuis has offered some money, unfortunately in Hungarian Forints, to decide whether there are $(k,3)$-arcs with $k \ge (1+\varepsilon)q$ and to disprove the existence of $(k,3)$-arcs with $k \ge (2-\varepsilon)q$.

In Section 6 (k,n)-arcs for a divisor n of q are considered, if k is sufficiently close to the upper bound by Barlotti, that is for $k \ge qn-q+n-\varepsilon$ and ε is small.

5 The generalized Menelaus' theorem and applications: large arcs

The method discussed in this section is due to B. Segre [88, 90], and is one of the most successful theories in finite geometry. It consists of three parts. The first part is a nice result in classical algebraic geometry, the generalization of Menelaus' theorem. The second part is crucial: it is a trick, sometimes called *Segre's "Lemma of tangents"* to associate algebraic curves to an arc in PG$(2, q)$. The third and final step is to use Weil's theorem (or other bounds for the number of GF(q)-rational points of algebraic curves) to obtain upper bounds for the size of a complete arc which is not an oval or hyperoval. Since this theory is discussed in detail already in the first edition of Hirschfeld's book [58], I mainly state the results, and only consider one particular case to illustrate how the method works. At the end of this section further applications of Segre's method are collected.

Let us fix a coordinate system in PG$(2, q)$ and let $A_1 : (1,0,0)$, $A_2 : (0,1,0)$, $A_3 : (0,0,1)$ be the fundamental triangle. If P is a point on one of the sides, then P can be written as $(0, c, 1)$, $(1, 0, c)$ or $(c, 1, 0)$. The element c $(\neq 0)$ is called the *coordinate of* P. The classical result of Ceva states that three points $(0, c, 1)$, $(1, 0, d)$ and $(e, 1, 0)$ are collinear if and only if $cde = -1$. This can be generalized to curves of degree n.

Theorem 5.1 (B. Segre, [88]) *Suppose that we are given a family*

$$G = \{(0, c_i, 1), (1, 0, d_j), (e_k, 1, 0) : i, j, k = 1, \ldots, n\}$$

of $3n$ *points and* $G \cap \{A_1, A_2, A_3\} = \emptyset$. *Then there is a curve of degree* n *intersecting the sides of the triangle* $A_1A_2A_3$ *in precisely the points of* G, *if and only if* $\prod_i c_i \prod_j d_j \prod_k e_k = (-1)^n$.

Note that G is a multiset, the same point can occur more than once. If a point has multiplicity r in G, then we require that the curve of degree n intersects the corresponding side of $A_1A_2A_3$ with multiplicity r. Actually, the proof is not difficult: the necessity comes from Vieta's formulas (for the product of the roots of an equation), and the sufficiency is proved by a dimension argument. In most of the applications, the dual version of this theorem is needed. For this, we have to define the coordinate of a line passing through a vertex of the fundamental triangle. Such a line has equation $X_2 = dX_3$, $X_3 = dX_1$ or $X_1 = dX_2$; the element d is the *coordinate of the line*. The dual of Ceva's theorem is this: three lines through the vertices of the fundamental triangle are concurrent if and only if the product of their coordinates is 1. This is called *Menelaus' theorem* in classical projective geometry. The dual of the above generalization is the following.

Theorem 5.2 *Suppose that we are given* n *lines (possibly with multiplicity), different from the sides of the fundamental triangle, through each vertex of the* *fundamental triangle. This multiset of $3n$ lines is contained in an algebraic envelope of class n if and only if the product of the $3n$ coordinates is* 1.

An algebraic envelope is just a polynomial, the name underlines that its zeroes are considered as coordinates of lines and not points. For traditional reasons we use class for the degree of the polynomial defining an envelope.

One can generalize further the above theorems, instead of 3 lines or points one can consider k lines or points. This generalization will be stated only in the dual form.

Theorem 5.3 (Segre's generalized Menelaus' theorem) *Suppose that we have a k-arc and at each point we are given a multiset of n lines, which are all tangents to the k-arc. Then these kn lines are contained in an algebraic envelope of class n if and only if for every 3 points of our k-arc the corresponding $3n$ lines are contained in an algebraic envelope of class n.*

So if there is an envelope locally for every triangle containing the $3n$ lines, then there is an envelope globally containing the kn lines.

The condition that the k points form a k-arc can be relaxed. Since only the original version of Segre's generalized Menelaus' theorem will be used later, we do not state this even more general version, just refer to the paper by Blokhuis, Cameron and Thas [21].

Note that the famous theorem of Segre, saying that the $(q+1)$-arcs in $\mathrm{PG}(2,q)$ (q odd) are conics, was proved using this method implicitly. Then Segre's idea was to embed arcs with size close to q in ovals or hyperovals. At the end of this section we shall make it clear what "close" means here. For q even it is a nice combinatorial exercise to prove that $(q+1)$-arcs are never complete; they are always contained in a $(q+2)$-arc (that is a hyperoval). Indeed, there is a unique tangent at each point, and the $q+1$ tangents cover all points; hence these line have to form a pencil. However, for the embedding of q-arcs there is no combinatorial proof; in fact, Menichetti [82] has constructed complete q-arcs in some non-desarguesian planes.

Let us illustrate Segre's method by proving the incompleteness of q-arcs in desarguesian planes of even order, a result first proved by Tallini [103].

Theorem 5.4 (Tallini) *In* $\mathrm{PG}(2,q)$, *q even, q-arcs are never complete.*

Proof Let Q be a q-arc. Through each point of Q there are exactly 2 tangents to the q-arc Q at that point, so in total there are $2q$ tangents. We wish to use Segre's generalized Menelaus' theorem for this set of $2q$ lines. According to the theorem, we only have to check the condition for triangles contained in Q. Let A_1, A_2, A_3 be three distinct points of Q and choose coordinates in such a way that $A_1=(1,0,0)$, $A_2=(0,1,0)$, $A_3=(0,0,1)$. Form a matrix of size $(q-3)\times 3$ whose rows are indexed by the points $P\in Q\setminus\{A_1,A_2,A_3\}$ and the columns by A_1, A_2 and A_3. Put the coordinate of the line A_iP in the position

(P, A_i). Then Menelaus' theorem implies that the product of the elements in one row is just 1. Thus the product of the elements in this matrix is 1^{q-3}, computing it row by row. Extend the matrix to a $(q-1) \times 3$ matrix by adding two rows. In these rows we write the coordinates of the two tangents at the points A_1, A_2 and A_3. In any column of this extended matrix each element of $\mathrm{GF}(q)$ occurs exactly once, hence their product is 1. Therefore, the product of all elements in the extended matrix is 1^3, computing it column by column. Hence the product of the coordinates of the tangents at A_1, A_2 and A_3 is 1, and Menelaus' theorem (for a triangle) implies that they are contained in an envelope of class 2. From the general theorem of Menelaus it also follows that this envelope of class two does not depend on the points A_1, A_2, A_3. The total number of tangents is $2q$, hence our curve contains at least $2q$ $\mathrm{GF}(q)$-rational points. An irreducible conic contains $q+1$ points, hence our curve is the union of two pencils, if $q \geq 4$. Geometrically this means that there are two points N, N' with the property that joining N or N' to any point of Q yields a tangent to Q. In other words, either of the two points can be added to Q. Hence Q cannot be complete. Note that this also implies, by the combinatorial argument showing the incompleteness of $(q+1)$-arcs, that Q can be embedded in a hyperoval by adding N and N' one after the other. In particular, this means that the line joining N and N' does not intersect Q. ∎

In general, one can apply Segre's Lemma of tangents to the set of all tangents to a k-arc and show that the tangents are contained in an envelope of class $q+2-k$ or $2(q+2-k)$ according as q is even or odd. This is the cornerstone of the proof of almost all embedding theorems for large arcs. For a detailed list of properties of these algebraic envelopes, see Hirschfeld [58, Theorems 10.3.1 and 10.4.1].

In the general case the envelope Γ_t or Γ_{2t} contains at least $kt = k(q+2-k)$ $\mathrm{GF}(q)$-rational points and a similar (but slightly more complicated) lower bound holds for the number of points of any component of it. Comparing this with the Weil bound gives a lower bound for the degree of non-linear components. The other possibility is to use the Weil bound for every component and adding these bounds up gives an upper bound for the number of $\mathrm{GF}(q)$-rational points of Γ_t or Γ_{2t}. The linear components, as in the above example, correspond to points that can be added to the arc, hence the method yields an upper bound for the size of a complete arc which is not an oval or hyperoval. This strategy is due to B. Segre [88, 90] and gives an upper bound for $m'(2,q)$.

In the final step of Segre's method one can use refinements and variants of Weil's theorem. Since most of them are technically difficult, we only mention an elementary upper bound for arbitrary q, and the Stöhr–Voloch bound for q prime. The first observation was used by Thas in [106].

Let C be a curve of degree d defined over $\mathrm{GF}(q)$, which does not have a linear component defined over $\mathrm{GF}(q)$. Then it has at most $N \leq qd - q + d$ points over $\mathrm{GF}(q)$.

The proof is easy: such a curve is a (N, d')-arc for some $d' \leq d$ and Barlotti's bound (see Section 4) immediately gives the upper bound.

The Stöhr–Voloch theorem is a bound which is in most cases stronger than Weil's bound. For an exposition, see Hirschfeld [61]. We state it in the simplest case, when there are no technical conditions.

Let C be an absolutely irreducible plane curve defined over GF(p), *p prime. If $3 \leq n \leq p/2$ and C has s double points, then*

$$N \leq \frac{2}{5}n(5(n-2)+p-10s), \tag{8}$$

where again N denotes the number of GF(p)*-rational points of C.*

The best results, obtained with Segre's strategy, and using the various upper bounds for the number of rational points, are the following.

Theorem 5.5 *For the size of the second largest complete arc, we have:*

(a) $m'(2,q) \leq q - \sqrt{q} + 1$, *if q is an even square (Segre [90], Thas [106]),*

(b) $m'(2,q) \leq q - \dfrac{\sqrt{q}}{4} + \dfrac{25}{16}$, *if q is an odd square (Thas [107]),*

(c) $m'(2,q) \leq q - \dfrac{\sqrt{pq}}{4} + \dfrac{29p}{16} + 1$, *if $q = p^n$, p prime, $p > 2$, $n \geq 3$, n odd (Voloch [114]),*

(d) $m'(2,q) \leq q - \sqrt{2q} + 2$, *if q is an even non-square (Voloch [114]),*

(e) $m'(2,q) \leq \dfrac{44q}{45} + 8/9$, *if $q = p$ is prime, $p > 2$ (Voloch [113]),*

(f) $m'(2,q) \leq q - \dfrac{\sqrt{q}}{2} + 5$, *if $q = p^h$, $p \geq 5$, (Hirschfeld, Korchmáros [62]),*

(g) $m'(2,q) \leq q - \dfrac{\sqrt{q}}{2} + 3$, *for $q = p^h$ with $p \geq 5$, if $q \geq 19^2$ and $q \neq 5^2$, (Hirschfeld, Korchmáros [63]).*

Let us see that the Stöhr–Voloch bound in the prime case indeed gives the bound in (e). The envelope containing the tangents has class $2(p+2-k)$. Take an irreducible component of class n. As in [58], $n \geq 3$ can be assumed. The Stöhr–Voloch bound (see (8)) gives the inequality

$$\frac{n}{2}(p+2-k) \leq \frac{2}{5}n(5(n-2)+p),$$

which implies that $2n - 4 \geq p/10 + 1 - k/2$. Using $n \leq 2k$ it gives $k \leq (44p + 40)/45$ indeed.

Similarly, the above mentioned elementary upper bound of Thas gives the bound in case (a).

Note that the arcs constructed by Kestenband [69], Ebert [46], Fisher, Hirschfeld and Thas [50] and Boros, Szőnyi [30] (see Section 4), show that for q square $q - \sqrt{q} + 1 \leq m'(2, q)$. In particular, part (a) is sharp. On the other hand, if q is not a square then the best lower bound known for $m'(2, q)$ is only $[q + 2\sqrt{q}]/2$, coming from Zirilli's construction (see Theorem 4.3).

Let us close this section with further applications of Segre's curve-making method. The first one is about sets having a large number of internal nuclei. Let K be a k-set in PG$(2, q)$. A point $P \in K$ is called an *internal nucleus* of K if each line through P meets K in at most two points including P. This notion was introduced for $k = q + 2$ by Bichara and Korchmáros [14] and generalized by Wettl [120]. At an internal nucleus P a line t is a tangent to K if it meets K in just P. From our point of view the main result of Wettl is the following: for q odd the tangents to K at the internal nuclei are contained in an algebraic envelope of class $2(q + 2 - k)$. Of course, the proof is based on Segre's Lemma of tangents. Wettl [120] used the theorem for characterizing $(q + 1)$-sets with at least 5 nuclei and collinear non-nuclei.

Therefore, the situation is quite analogous to the case of arcs, the only difference is that the curve does not contain too many points, since typically at most $(q + 1)/2$ points can be nuclei. This is the content of the next result, see Szőnyi [97].

Theorem 5.6 *Let K be a k-set with $k \geq q - \sqrt{q}/8 + c$ (q odd, $q > q_0(c)$). If K is not an arc, then it contains at most $(q + 1)/2$ internal nuclei.*

The next application is due to Blokhuis, Seress and Wilbrink: a set S is a *set without tangents* if every line either is disjoint from S or meets S in at least 2 points. The main result of [25] is the following.

Theorem 5.7 (Blokhuis, Seress, Wilbrink) *Let S be a set without tangents in* PG$(2, q)$, *q odd. Then $|S| > q + \sqrt{2q}/4 + 2$.*

The proof is a clever use of Segre's lemma of tangents. Using the generalized Menelaus' theorem (Theorem 5.3) and Bézout's theorem to glue together smaller curves, one obtains an envelope of class $2(|S| - q - 2)$, containing the lines that intersect S in more than two points. The inequality is obtained by using the upper bound for the number of singular points of a curve.

Let me finally mention the result, which led to the Blokhuis, Cameron, Thas [21] generalization of Segre's generalized Menelaus' theorem. This is due to Thas [108], and gives the following elegant characterization of Hermitian curves:

A Hermitian arc H is a Hermitian curve if and only if tangents of H at collinear points of H are concurrent.

6 The Rédei polynomial and applications: blocking sets and (k, n)-arcs

In this section a new approach of associating curves to blocking sets and (k, n)-arcs will be discussed. We shall follow the papers [23, 100, 101]. The first paper in which algebraic curves were used to prove results on blocking sets was the paper by Blokhuis, Pellikaan and Szőnyi [23]. Then in [100] a pair of curves was introduced and the bounds for the size of a minimal blocking set were obtained by using Bézout's theorem applied for this pair of curves.

Before sketching this new approach, let us summarize what was known for blocking sets in $\mathrm{PG}(2, q)$. For a more detailed account, together with historical remarks, see Blokhuis [19]. The first non-trivial result for blocking sets is due to Bruen and Pelikán;[†] Bruen [33, 34] proved that the size of a non-trivial blocking set is at least $q + \sqrt{q} + 1$. This is sharp for q square (see [33] and also Section 4). For q a non-square the bound was improved several times, the most recent one is Blokhuis's theorem.

Theorem 6.1 (Blokhuis [17, 19]) *If q is a prime, then the size of a non-trivial blocking set is at least $3(q + 1)/2$. If $q = p^h$ is neither a square nor a prime, then the size of a non-trivial blocking set is at least $q + \sqrt{pq} + 1$.*

Note that the bound in the prime case is sharp (see [77]) and it solved a thirty-year old conjecture of Jane di Paola [84]. The bound is also sharp in the case $q = p^3$; it solved a 25-year old conjecture of Bruen [34]. We shall return to the examples later.

Let B be a blocking set of $\mathrm{PG}(2, q)$. A point $P \in B$ is called *essential* if $B \setminus \{P\}$ is not a blocking set. The blocking set B is *minimal* (or *irreducible*) if and only if every $P \in B$ is essential. Geometrically this means that through each point of the blocking set B there is a line intersecting B in just one point. According to the standard terminology introduced in Section 4 such a line will be called a *tangent* and, more generally, a line intersecting B in r points will be called an *r-secant* (or a *line of length r*).

Let L be the line at infinity, and suppose that $\{(\infty)\} = L \cap B$, that is L is a tangent to B. Give affine coordinates to the points of $U := B \setminus L$; namely, let $U = \{(a_i, b_i) : i = 1, \ldots, q + k\}$. So $|B| = q + k + 1$.

Definition The *Rédei-polynomial* of U is defined as follows:

$$H(X, Y) := \prod_i (X + a_i Y - b_i) = X^{q+k} + h_1(Y) X^{q+k-1} + \cdots + h_{q+k}(Y). \quad (9)$$

Note that $\deg(h_j) \leq j$ for $j = 1, \ldots, q+k$. If $H(X, Y)$ is considered for a fixed $Y = y$ as a polynomial of X, then we write $H_y(X)$ (or just $H(X, y)$).

[†] Pelikán and Pellikaan are different people

Definition Let $\mathcal{C}$ be the affine curve of degree k defined by

$$f(X,Y) = X^k + h_1(Y)X^{k-1} + \cdots + h_k(Y). \tag{10}$$

By the remark for the degree of h_j in the previous definition, $f(X,Y)$ has degree k. The next proposition summarizes some important properties of the Rédei polynomial and of these curves.

Theorem 6.2 *(1) For a fixed $(y) \in L \setminus B$ the polynomial $(X^q - X)$ divides $H_y(X)$. Moreover, if $k < q-1$ then $H_y(X)/(X^q - X) = f(X,y)$ for every $(y) \in L \setminus B$; and $f(X,y)$ splits into linear factors over $\mathrm{GF}(q)$ for these fixed y's.*

(2) For a fixed $(y) \in L \setminus B$, the element x is an r-fold root of $H_y(X)$ if and only if the line with equation $Y = yX + x$ intersects U in exactly r points. If the line with equation $Y = y$ meets $f(X,Y)$ at (x,y) with multiplicity m, then the line with equation $Y = yX + x$ meets U in exactly $m+1$ points.

The first part of this theorem shows that f has a lot of $\mathrm{GF}(q)$-rational points, the second part helps us translate geometric properties of U into properties of f. The next lemma shows that the linear components of f correspond to points of B which are not essential.

Lemma 6.3 *If a point $P = (a,b) \in B$ is not essential, then $X + aY - b$ divides $f(X,Y)$ (as polynomials in two variables). Conversely, if $k < q$ and $X + aY - b$ divides $f(X,Y)$, then $(a,b) \in B$ and (a,b) is not essential.*

It will be convenient to suppose that not only the line at infinity but also the y-axis is a tangent to the blocking set B. Since the Rédei polynomial $H(X,Y)$ vanishes for all $(x,y) \in \mathrm{GF}(q) \times \mathrm{GF}(q)$ (see Theorem 6.2(1)), we can write it as

$$H(X,Y) = (X^q - X)f(X,Y) + (Y^q - Y)g(X,Y),$$

where $\deg(f), \deg(g) \le k$ as polynomials in two variables (see [2, 9]). Note that f here is the same as the one in (10). If one fixes $Y = y$ then $H(X,y)$ is divisible by $(X^q - X)$ and for an $(x,y) \in \mathrm{GF}(q) \times \mathrm{GF}(q)$ we have that $f(x,y) = 0$ if and only if the line with equation $Y = yX + x$ intersects U in at least two points (cf. Theorem 6.2(2)). The next theorem summarizes some important properties of the pair of curves (f,g).

Theorem 6.4 *The curves $f(x,y)$ and $g(x,y)$ have the same $\mathrm{GF}(q)$-rational affine points, but they do not have common components.*

The next result, already used in [23], gives a lower bound for the number of $\mathrm{GF}(q)$-rational points on a component of f.

Theorem 6.5 (Blokhuis, Pellikaan, Szőnyi) *Let h be a component of f with $h'_X \neq 0$. Let s denote the degree of h. Then the number of* GF(q)*-rational points of h is at least $qs - s(s-1)$.*

Using these properties of the curves f and g and the above lower bound for the number of rational points, one can show that for small blocking sets, that is when $k < (q+1)/2$, every component of f must have zero partial derivative with respect to X. Geometrically, this means the following: If B is a blocking set of size less than $3(q+1)/2$, then each line intersects it in 1 modulo p points. In particular, this gives a new proof of Blokhuis' famous lower bound $|B| \geq 3(p+1)/2$ (see Theorem 6.1) for blocking sets in PG$(2,p)$, p prime.

To say more in the case $q = p^n$, $n > 1$, components with zero partial derivative should be studied in detail and the following theorem can be proved.

Theorem 6.6 (Szőnyi [100]) *Let B be a minimal blocking set in* PG$(2,q)$, *$q = p^n$. Suppose that $|B| < 3(q+1)/2$. Then*

$$q+1+\frac{q}{p^e+2} \leq |B| \leq \frac{qp^e+1-\sqrt{(qp^e+1)^2-4q^2p^e}}{2}, \tag{11}$$

for some integer e, $1 \leq e \leq n/2$. If $p^e < 7$, then the upper bound has to be replaced by $3(q+1)/2$. This means that asymptotically

$$|B| \leq q+\frac{q}{p^e}+2\frac{q}{p^{2e}}+5\frac{q}{p^{3e}}+\cdots. \tag{12}$$

If $|B|$ lies in the interval belonging to e and $p^e \neq 4, 8$, then each line intersects B in 1 modulo p^e points.

To be more concrete, we have $|B| \leq q + 9q/(4p^e)$ for every p and e. A simple form for the upper bound is $|B| \leq q + q/(p^e - 3)$, which is still valid for every p^e. This is sharper than the previous one and has the advantage that it is similar to the form of the lower bound.

There are two new elements in this theorem. First of all, it is shown that the lines intersect the blocking set in 1 modulo p^e points, a result conjectured by Blokhuis [19]. The second new element, which is a consequence of the first, is that we obtained intervals for the possible sizes of minimal blocking sets; a phenomenon first discovered by Rédei [85] for a particular class of blocking sets. A blocking set B is *of Rédei type* if there is a line intersecting it in $|B| - q$ points. For Rédei type blocking sets the intervals are narrower; a combination of work by Blokhuis [19] and Rédei [85] gives

$$q+1+p^e\left\lceil\frac{q/p^e+1}{p^e+1}\right\rceil \leq |B| \leq q+(q-1)/(p^e-1).$$

In a very recent paper by Blokhuis et al. [20] the lower bound for Rédei type blocking sets is pushed to $q + 1 + q/p^e$ and it is also shown that only divisors of n can occur as e when $p^e > 3$. These blocking sets are actually characterized as GF(p^e)-subspaces of AG$(2, q)$ for $p^e > 3$. In particular, there exist minimal blocking sets of size $q + q/p^e + 1$ and $q + (q - 1)/(p^e - 1)$ as well as for some intermediate sizes if e divides n. This was first noticed by Brouwer and Wilbrink [32]. For more on Rédei type blocking sets, see [38] and [19]. It is important to note that all known blocking sets of size less than $3(q+1)/2$ are of Rédei type, hence the results of [20] give actually a complete characterization of known small blocking sets. Even for blocking sets of size $3(q + 1)/2$ there is only one sporadic example known for $q = 7$, which is not of Rédei type, see [19]. Rédei type blocking sets of size $3(q + 1)/2$ exist for every q odd, they were characterized by Lovász and Schrijver [77]. These examples are called projective triangles in Hirschfeld's book [58].

The consequence of Theorem 6.6 in the particular case $q = p^2$ is the following.

Theorem 6.7 ([100]) *Let $q = p^2$ and B be a minimal blocking set which is not a Baer subplane. Then $|B| \geq 3(q + 1)/2$.*

Actually, the same is true for blocking sets with $e = n/2$ and any square q. The theorem is sharp in the sense that projective triangles do not contain a subplane and they have size $3(q+1)/2$. At this point it is worthwhile to mention the following interpretation of this theorem: if B is a non-trivial blocking set of size $|B| < 3(q+1)/2$, then there is a Baer subplane contained in it. The bound is essentially better than the previously known ones. The first result in this direction was proved by Bruen–Thas [38] using combinatorial methods; they proved the non-minimality of B for $|B| = q + \sqrt{q} + 2$. Then Bruen–Silverman [37], Ball–Blokhuis [10] improved this result to $|B| < q + \sqrt{2q} + 1$ and to $|B| < q + 2\sqrt{q} + 1$, respectively. The best result obtained by the lacunary polynomial approach is due to Blokhuis, Storme and Szőnyi [26], who proved the non-minimality for $|B| < q + 1 + p\left\lceil \frac{1}{4} + \sqrt{(p+1)/2} \right\rceil$ if $q = p^2$, see later Theorem 6.11.

One can also relate Blokhuis' lacunary polynomial approach to the present one using algebraic curves, and can prove that the two exponents e are the same (see [100] and [24]). This immediately implies that the lower bound in Theorem 6.6 can be replaced by

$$q + 1 + p^e \left\lceil \frac{q/p^e + 1}{p^e + 1} \right\rceil \leq |B|.$$

This observation also permits us to use some recent results from the manuscript [26], which imply that for $p \geq 5$ it is not possible that $n/2 > e > n/3$. On the other hand, using the fact that each line intersects our blocking sets in 1 modulo p^e points, also the upper bound can slightly be improved. These observations are essentially due to Blokhuis and Polverino [24], who obtained the following theorem.

Theorem 6.8 (Blokhuis, Polverino) *With the notation of Theorem 6.6,*

$$q+1+p^e\left\lceil\frac{q/p^e+1}{p^e+1}\right\rceil \le |B| \le \frac{1+(p^e+1)(q+1)-\sqrt{\Delta}}{2},$$

where $\Delta=(1+(p^e+1)(q+1))^2-4(p^e+1)(q^2+q+1)$.

Note that this new upper bound is asymptotically $q+q/p^e+q/p^{2e}+2q/p^{3e}+\cdots$. This improvement can be used to determine the possible sizes of minimal blocking sets in $\mathrm{PG}(2,p^3)$ as the following corollary shows.

Corollary 6.9 (Blokhuis, Polverino) *Let B be a non-trivial minimal blocking set in* $\mathrm{PG}(2,p^3)$. *Then* $|B|=p^3+p^2+1$ *or* p^3+p^2+p+1 *or* $|B|\ge 3(p^3+1)/2$.

Again, the result extends immediately to blocking sets with $e=n/3$. Note that there are examples of minimal blocking sets (of Rédei type) for both cardinalities.

The same approach using algebraic curves can also be used to prove the affine blocking set theorem of Jamison [68] and Brouwer–Schrijver [31], and its generalization for multiple blocking sets, due to Bruen [36]. Concerning these problems, the reader is referred to Blokhuis' papers [18] and [19].

Using the previous results on blocking sets one can improve on the Lunelli–Sce bound for the cardinality of a complete arc, since the secants of a complete arc form a blocking set in the dual plane.

Theorem 6.10 (Blokhuis, Ball, Blokhuis–Polverino) *Let K be a complete k-arc in* $\mathrm{PG}(2,q)$, *and assume that* $q=p$, p^2 *or* p^3, *where p is a prime. Then* $k\ge\sqrt{3q}$.

Regarding multiple blocking sets many fewer results and examples are known. Taking the union of s disjoint Baer subplanes gives an s-fold blocking set, which will be characterized by the next theorem. Using the lacunary polynomial approach, Ball [9] (for the prime case), and Blokhuis, Storme and Szőnyi [26] proved the following lower bounds. Note that $|B|\ge s(q+1)$ is obvious.

Theorem 6.11 ([9, 26]) *Let B be an s-fold blocking set in* $\mathrm{PG}(2,q)$ *of size* $s(q+1)+c$. *Let* $c_2=c_3=2^{-1/3}$ *and* $c_p=1$ *for* $p>3$.

1. (Ball) *If* $q=p$ *prime and* $s\le(q-1)/2$, *then* $c\ge(q+1)/2$. *If* $s\ge(q+1)/2$ *then* $s+c\ge q$.
2. *If* $q=p^{2d+1}$ *and* $s<q/2-c_pq^{2/3}$, *then* $c\ge c_pq^{2/3}$.

3. *If $4 < q$ is a square, $s < q^{1/4}/2$ and $c < c_p q^{2/3}$, then $c \geq s\sqrt{q}$ and B contains the union of s disjoint Baer subplanes.*

4. *If $q = p^2$, $s < q^{1/4}/2$ and $c < p\left\lceil \frac{1}{4} + \sqrt{(p+1)/2} \right\rceil$, then $c \geq s\sqrt{q}$ and B contains the union of s disjoint Baer subplanes.*

The special cases $s = 2, 3$, that is the cases of double and triple blocking sets, were proved earlier by Ball and Blokhuis [10] and Ball [8]. Note that the case $s = 1$ yields considerable improvement on the bounds of Blokhuis mentioned in Theorem 6.1. It is worthwhile to mention that for $s = (q-1)/2$ the set of external points of a conic shows that Ball's bound is sharp. For $s = (q+1)/2$ the same set together with all points of the conic but one is an example. For small s the bounds do not seem to be sharp. In fact, Blokhuis conjectures that for $s = 2$ and $q = p$ prime $c \geq q - 2$. Still for $s = 2$ the bound does not seem to be exact, if q is not a square. Recently, L. Lovász and the author tried to extend the above method, that is the use of curves, to multiple blocking sets. To an s-fold blocking set is associated a set of $s+1$ curves with almost the same set of rational points. Since this research is still in progress let me only mention an improvement on the above bounds in the special case $s = 2$, $q = p^3$. This indicates that the use of curves can give more than the lacunary polynomial approach also in case of multiple blocking sets. For a general s, it is not yet clear whether the method using curves gives the extra term s in the bound or not.

Proposition 6.12 (Lovász, Szőnyi [78]) *A double blocking set of* $\mathrm{PG}(2, p^3)$ *has size at least* $2(p^3 + p^2 + 1 - 3p)$.

Now let us turn to (k, n)-arcs. Regarding the bounds for k we refer to Section 4. In the manuscript [101] the present author shows that for $n = p$, $q = p^h$, a (k, n)-arc K with $k \geq qn - q + n - \varepsilon$ points, $\varepsilon \leq c \cdot q^{1/4}$, can be embedded in a maximal arc. This result can probably be extended to any divisor n of q, if $n\varepsilon + \varepsilon^4$ is smaller than $c \cdot q$. The embeddability was only known before for $\varepsilon = 1$ (Thas [105]) and even for $\varepsilon = 2$ only partial results were known (Wilson [121]).

Theorem 6.13 ([101]) *In* $\mathrm{PG}(2, q)$, $q = p^h$ *let K be a (k, p)-arc with* $k \geq qp - q + p - cq^{1/4}$. *Then K can be embedded in a (k, p)-arc with* $k = qp - q + p$.

Since maximal arcs do not exist for $p > 2$ (see [12]), the above result simply means that for a (k, p)-arc in $\mathrm{PG}(2, q)$, $q = p^h$, $p > 2$, we have $k < qp - q + p - cq^{1/4}$.

Sketch Proof Let us sketch the proof. Consider a line ℓ which will be the line at infinity. Let $U = K \setminus \ell = \{(a_i, b_i)\}$ and write up the Rédei polynomial $H(X, Y)$. The key observation is that the fact that through $(y) \notin K$ there pass exactly s short lines can be interpreted in terms of $H(X,y)$. Namely, in this case the greatest common divisor of $H(X,y)$ and $H'_X(X,Y)$ has degree exactly $|U|-s$. The idea is that we can express the polynomial

$$r(X,y) = H(X,y)/\gcd(H(X,y), H'_X(X,y))$$

using the coefficients $h_1(y),\dots$ of $H(X,Y)$. It turns out that whenever we fix s this will be a polynomial of total degree at most s^2, and of degree s in X. So $r(X,Y)$ will be a curve of modest degree. Of course, this curve may depend on the line ℓ, but one can show that it is essentially the same for all lines. Using this observation it follows that the curve must be the union of some lines $X + a_jY - b_j$ if there are many points on ℓ having the same s. A combinatorial argument will show that there is indeed a typical index that can be associated to a line. Using the fact that the curve does not depend on the line ℓ it can be proved that the points (a_j, b_j), corresponding to the linear components, can be added to K.

Since this method is more technical than the one for blocking sets, let us illustrate the method by *proving the incompleteness of q-arcs in planes of even order* (Tallini, [103]). This was the example illustrating Segre's method based on the generalized Menelaus' theorem.

Let $U = \{(a_i, b_i)\}$ be a q-arc. Define the *index* of a point as the number of tangents through it. Suppose that the line at infinity is disjoint from our q-arc U. Then the Rédei polynomial $H(X,Y)$ will have the form $H(X,Y) = X^q + h_1(Y)X^{q-1} + h_2(Y)X^{q-2} + h_3(Y)X^{q-3} + \cdots$. The index of any point must obviously be even. Let us first see what does it mean that an infinite point (y) has index zero. It implies that through the point there are only 0-secants and 2-secants, and hence the polynomial $H(X,y)$ must be a square for such a (y). In particular, we see that if there are two such points, then the polynomial $h_1(Y)$ must be identically zero. If a point (y) has index 2, then $H(X,y)$ is the product of a square and two linear factors. It implies that $H'_X(X,y)$ must divide $H(X,y)$ for such a (y), since the derivative has degree $q-2$. If there are four infinite points of index 0, then $h_1(Y) = h_3(Y) = 0$ identically and therefore the index of every other point is either zero or at least four, since the derivative will automatically have degree at most $q-4$. In total there are $2q$ short lines (tangents), hence there are only two possibilities. Either there are at most 3 points of index zero, and most of the other points have index 2, or at least half of the points have index zero. In the first case we have found the typical index for the line at infinity; it is 2. In the second case the same argument shows that not only h_1, h_3 but also h_j $(j \le q/2-1)$ are identically zero, which means that the degree of $H'_X(X,y)$ is at most $q/2$ for any y. Thus the points having positive index must have index at least $q/2$. So there can be at most four points having positive index. But then there are at least $q-3$ points with index 0, and repeating the same argument we get that the points having positive index must have index at least $q-2$. Thus the only possibility is that there are two points having positive index; this index is exactly q, and the remaining points have index zero. Then the two points of index q can be added to the arc. Since the line at infinity was disjoint from our q-arc, the two points can be added simultaneously.

Let us consider now the case when there are at most three points of index zero. Then the typical index for the line at infinity is 2, and since the total number of short lines (tangents) is $2q$, there are at least $q-4$ points having index 2. As we remarked earlier, for these points (y) on the line at infinity $H'_X(X,y)$ divides $H(X,y)$ and the two tangents through such a point correspond to the roots of $H(X,y)/H'_X(X,y)$. This quotient can be computed simultaneously for these points. For these points $h_1(y) \neq 0$, since $\deg H'_X(X,y) = q-2$. Therefore,

$$h_1(y)H(X,y)/H'_X(X,y) = h_1(y)X^2 + h_1^2(y)X + h_1(y)h_2(y) + h_3(y), \qquad (13)$$

for at least $q-4$ infinite points (y). If the above equation is considered as the equation of a curve, then the curve has total degree 3, X-degree 2, and the number of its GF(q)-rational points is at least $2(q-4)$. The important property of this curve is that its rational points correspond to tangents passing through an infinite point of index 2. If q is large enough, then Weil's bound implies that the curve cannot be irreducible, but it does not follow automatically that it is the union of linear components, but there is at least one linear component containing X-terms. Let $P(a,b)$ be the point corresponding to such a linear component $X + aY - b$. Then the point P has index at least $q-4$, since the lines joining it to a point (y) of index two are tangents. Now change the line at infinity: suppose that P is an infinite point. Then for this new line at infinity there is one point of large index, whence the typical index for this new line at infinity can only be zero (see the first part of the proof). Then the first part of the proof actually shows that there is another point having large index and the two points can be added simultaneously. ∎

This proof indeed illustrates the difficulties in associating curves to large (k,n)-arcs. In general, the resulting curve has larger degree than the one obtained by Segre's method. In the present case a curve of degree 3 is obtained instead of a conic. Actually, Bézout's theorem shows that the two curves must be the same, so in our equation we can divide by $h_1(y)$. The next difficulty is that it is not enough in general to consider one line at infinity, we have to prove that the curve does not depend on the line at infinity. Finally, it may happen that the typical index depends on the line at infinity; then the arguments are somewhat similar to the first part of the above proof.

Acknowledgements

A large part of the research represented in this paper was done at the Department of Computer Science, Yale University, while the author was supported by a Hungarian State Eötvös Fellowship and the Discrete Mathematics program of Yale University.

Without the hospitality of the Department of Computer Science at Yale University and the several helpful discussions with Laci Lovász this work would have been impossible. I warmly thank Laci Lovász for his help and for the fruitful discussions.

This research was partially supported also by OTKA Grants 19367 and T-014302 and by the COST project with contract number ERBCIPACT930113.

I am grateful to the referee and Laci Lovász for valuable comments on the first draft of the paper.

References

[1] V. Abatangelo, A class of complete $((q+8)/3)$-arcs of PG$(2,q)$, with $q=2^h$ and h (>6) even, *Ars Combinatoria*, **16** (1983), 103–111.

[2] H. L. Abbott & A. Liu, Property of $B(s)$ and projective planes, *Ars Combinatoria*, **20** (1985), 217–220.

[3] N. Alon, Tools from higher algebra, Chapter 32, in *Handbook of Combinatorics* (eds. R. L. Graham, M. Grötschel & L. Lovász), North-Holland, Amsterdam (1995), pp. 1749–1783.

[4] L. Babai, A. Gál & A. Wigderson, Superpolynomial lower bounds for monotone span programs, submitted to *Combinatorica*.

[5] R. D. Baker & G. L. Ebert, On Buekenhout–Metz unitals of even order, *European Journal of Combinatorics*, **13** (1992), 109–117.

[6] R. D. Baker & G. L. Ebert, On Buekenhout–Metz unitals of odd order, *Journal of Combinatorial Theory, Series A*, **60** (1992), 67–84.

[7] R. D. Baker, G. L. Ebert, J. Hemmeter & A. Woldar, Maximal cliques in the Paley graph of square order, *Journal of Statistical Planning and Inference*, **56** (1996), 33–38.

[8] S. M. Ball, On the size of a triple blocking set in PG$(2,q)$, *European Journal of Combinatorics*, **17** (1996), 427–435.

[9] S. M. Ball, Multiple blocking sets and arcs in finite planes, *Journal of the London Mathematical Society (2)*, **54** (1996), 581–593.

[10] S. M. Ball & A. Blokhuis, On the size of a double blocking set in PG$(2,q)$, *Finite Fields and their Applications*, **2** (1996), 125–137.

[11] S. M. Ball & A. Blokhuis, An easier proof of the maximal arcs conjecture, manuscript, 1996.

[12] S. M. Ball, A. Blokhuis & F. Mazzocca, Maximal arcs in PG$(2,q)$, q odd do not exist, *Combinatorica*, in press.

[13] A. Barlotti, Su $\{k;n\}$-archi di un piano lineare finito, *Bollettino della Unione Matematica Italiana*, **11** (1956), 553–556.

[14] A. Bichara & G. Korchmáros, n^2-sets in a projective plane which determine exactly n^2+n lines, *Journal of Geometry*, **15** (1980), 175–181.

[15] A. Blokhuis, On subsets of GF(q^2) with square differences, *Indagationes Mathematicae*, **46** (1984), 369–372.

[16] A. Blokhuis, Polynomials in finite geometry and combinatorics, in *Surveys in Combinatorics, 1993* (ed. K. Walker), *London Mathematical Society Lecture Note Series*, 187, Cambridge University Press, Cambridge (1993), pp. 35–52.

[17] A. Blokhuis, On the size of a blocking set in PG$(2,p)$, *Combinatorica*, **14** (1994), 273–276.

[18] A. Blokhuis, On multiple nuclei and a conjecture of Lunelli and Sce, *Bulletin of the Belgian Mathematical Society, Simon Stevin*, **3** (1994), 349–353.

[19] A. Blokhuis, Blocking sets in Desarguesian planes, in *Combinatorics, Paul Erdős is Eighty, Volume 2* (eds. D. Miklós, V.T. Sós and T. Szőnyi), *Bolyai Society Mathematical Studies*, 2, Bolyai Society, Budapest (1996), pp. 133–155.

[20] A. Blokhuis, S. M. Ball, A. E. Brouwer, L. Storme & T. Szőnyi, On the number of slopes determined by a function on a finite field, manuscript, 1996.

[21] A. Blokhuis, P. J. Cameron & J. A. Thas, On a generalization of a theorem of B. Segre, *Geometriae Dedicata*, **43** (1992), 299–305.

[22] A. Blokhuis & K. Metsch, Large minimal blocking sets, strong representative systems and partial unitals, in *Finite Geometry and Combinatorics* (eds. F. De Clerck et al.), *London Mathematical Society Lecture Note Series*, 191, Cambridge University Press, Cambridge (1993), pp. 37–52.

[23] A. Blokhuis, R. Pellikaan & T. Szőnyi, Blocking sets of almost Rédei type, *Journal of Combinatorial Theory, Series A*, in press.

[24] A. Blokhuis & O. Polverino, private communication, 1996.

[25] A. Blokhuis, Á. Seress & H. A. Wilbrink, On sets of points without tangents, *Mitteilungen aus dem Mathematischen Seminar Giessen*, **201** (1991), 39–44.

[26] A. Blokhuis, L. Storme & T. Szőnyi, Multiple blocking sets and Baer-subplanes, manuscript, 1995.

[27] B. Bollobás, Geodesics in oriented graphs, *Annals of Discrete Mathematics*, **20** (1984), 67–73.

[28] B. Bollobás, *Random Graphs*, Academic Press, New York (1985).

[29] B. Bollobás & A. Thomason, Graphs which contain all small graphs, *European Journal of Combinatorics*, **2** (1981), 13–15.

[30] E. Boros & T. Szőnyi, On the sharpness of a theorem of B. Segre, *Combinatorica*, **6** (1986), 261–268.

[31] A. E. Brouwer & A. Schrijver, The blocking number of an affine space, *Journal of Combinatorial Theory, Series A*, **24** (1978), 251–253.

[32] A. E. Brouwer & H. A. Wilbrink, Blocking sets in translation planes, *Journal of Geometry*, **19** (1982), 200.

[33] A. A. Bruen, Baer subplanes and blocking sets, *Bulletin of the American Mathematical Society*, **76** (1970), 342–344.

[34] A. A. Bruen, Blocking sets in finite projective planes, *SIAM Journal on Applied Mathematics*, **21** (1971), 380–392.

[35] A. A. Bruen, Arcs and multiple blocking sets, in *Combinatorica*, *Symposia Mathematica*, 28, American Mathematical Society, Providence, Rhode Island (1986), pp. 15–29.

[36] A. A. Bruen, Polynomial multiplicities over finite fields and intersection sets, *Journal of Combinatorial Theory, Series A*, **60** (1992), 19–33.

[37] A. A. Bruen & R. Silverman, Arcs and blocking Sets II, *European Journal of Combinatorics*, **8** (1987), 351–356.

[38] A. A. Bruen & J. A. Thas, Blocking sets, *Geometriae Dedicata*, **6** (1977), 193–203.

[39] A. A. Bruen, J. A. Thas & A. Blokhuis, On M.D.S. codes, arcs in PG(n, q) with q even, and a solution of three fundamental problems of B. Segre, *Inventiones Mathematicae*, **92** (1988), 441–459.

[40] F. Buekenhout, Existence of unitals in finite translation planes of order q^2 with kernel of order q, *Geometriae Dedicata*, **5** (1976), 189–194.

[41] P. J. Cameron, Four lectures on projective geometry, in *Finite Geometry* (eds. C. A. Baker & L. M. Batten), *Lecture Notes in Pure and Applied Mathematics*, 103, Marcel Dekker, New York (1985), pp. 27–65.

[42] J. H. H. Chalk, The number of solutions of congruences in incomplete residue systems, *Canadian Journal of Mathematics*, **15** (1963), 291–296.

[43] F. R. K. Chung, R. L. Graham & R. M. Wilson, Quasi-random graphs, *Combinatorica*, **9** (1989), 345–362.

[44] A. Cossu, Su alcune proprietà dei $\{k,n\}$-archi di un piano proiettivo sopra un corpo finito, *Rendiconti di Matematica e delle sue Applicazioni*, **20** (1961), 271–277.

[45] R. H. F. Denniston, Some maximal arcs in finite projective planes, *Journal of Combinatorial Theory*, **6** (1969), 217–219.

[46] G. L. Ebert, Partitioning projective geometries into caps, *Canadian Journal of Mathematics*, **37** (1985), 1163–1175.

[47] P. Erdős, On a problem in graph theory, *Mathematical Gazette*, **47** (1963), 220–223.

[48] P. Erdős, R. Silverman & A. Stein, Intersection properties of families of sets of nearly the same size, *Ars Combinatoria*, **15** (1983), 247–259.

[49] P. Erdős & J. Spencer, *Probabilistic Methods in Combinatorics*, Akadémiai Kiadó and Academic Press, Budapest and New York (1974).

[50] J. C. Fisher, J. W. P. Hirschfeld & J. A. Thas, Complete arcs in planes of square order, *Annals of Discrete Mathematics*, **30** (1986), 243–250.

[51] A. Gács, On the number of directions determined by a pointset in $\mathrm{AG}(2,p)$, submitted to *Discrete Mathematics.*

[52] R. L. Graham & J. Spencer, A constructive solution to a tournament problem, *Canadian Mathematical Bulletin*, **14** (1971), 45–48.

[53] J. van der Geer & J. H. van Lint, *Introduction to Coding Theory and Algebraic Geometry*, Birkhäuser, Basel (1988).

[54] V. D. Goppa, *Geometry and Codes*, Kluwer, Dordrecht (1988).

[55] É. Hadnagy, Small complete arcs in Galois-planes of prime order, manuscript, 1996.

[56] R. Hill, Some problems concerning (k,n)-arcs in finite projective planes, *Rendiconti del Seminario Matematico di Brescia*, **7** (1984), 367–383.

[57] R. Hill & J. R. M. Mason, On (k,n)-arcs and the falsity of the Lunelli–Sce conjecture, in *Finite Geometries and Designs* (eds. P. J. Cameron, J. W. P. Hirschfeld & D. R. Hughes), *London Mathematical Society Lecture Note Series*, 49, Cambridge University Press, Cambridge (1980), pp. 153–168.

[58] J. W. P. Hirschfeld, *Projective Geometries over Finite Fields*, Clarendon Press, Oxford (1979). 2nd edition, in press.

[59] J. W. P. Hirschfeld, Maximum sets in finite projective spaces, in *Surveys in Combinatorics* (ed. E. K. Lloyd), *London Mathematical Society Lecture Note Series*, 82, Cambridge University Press, Cambridge (1983), pp. 55–76.

[60] J. W. P. Hirschfeld, The Weil conjectures in finite geometry, in *Combinatorial Mathematics X*, *Lecture Notes in Mathematics*, 1036, Springer, Berlin (1983), pp. 6–23.

[61] J. W. P. Hirschfeld, *Algebraic Curves, Arcs, and Caps over Finite Fields*, *Quaderni del Dipartimento di Matematica dell'Università di Lecce*, Q.5, (1986).

[62] J. W. P. Hirschfeld & G. Korchmáros, On the embedding of an arc into a conic in a finite plane, *Finite Fields and their Applications*, **2** (1996), 274–292.

[63] J. W. P. Hirschfeld & G. Korchmáros, The number of rational points on an algebraic curve over a finite field, manuscript, 1996.

[64] J. W. P. Hirschfeld & L. Storme, The packing problem in statistics, coding theory and finite geometry, *Journal of Statistical Planning and Inference*, in press.

[65] J. W. P. Hirschfeld & T. Szőnyi, Constructions of large arcs and blocking sets in finite planes, *European Journal of Combinatorics*, **12** (1991), 499–511.

[66] T. Illés, T. Szőnyi & F. Wettl, Blocking sets and maximal strong representative sytems in finite projective planes, *Mitteilungen aus dem Mathematischen Seminar Giessen*, **201** (1991), 97–107.

[67] S. Innamorati & A. Maturo, On irreducible blocking sets in projective planes, *Ratio Mathematica*, **2** (1991), 151–155.

[68] R. Jamison, Covering finite fields with cosets of subspaces, *Journal of Combinatorial Theory, Series A*, **22** (1977), 253–266.

[69] B. C. Kestenband, Unital intersections in finite projective planes, *Geometriae Dedicata*, **11** (1981), 107–117.

[70] B. C. Kestenband, A family of complete arcs in finite projective planes, *Colloquium Mathematicum*, **LVII** (1989), 59–67.

[71] J. H. Kim & V. H. Vu, private communication.

[72] J. Kollár, L. Rónyai & T. Szabó, Norm-graphs and bipartite Turán numbers, *Combinatorica*, **16** (1996), 399–406.

[73] G. Korchmáros, New examples of k-arcs in PG$(2,q)$, *European Journal of Combinatorics*, **4** (1983), 329–334.

[74] T. Kővári, V. T. Sós & P. Turán, On a problem of K. Zarankiewicz, *Colloquium Mathematicum*, **3** (1954), 50–57.

[75] L. Lombardo-Radice, Sul problema dei k-archi completi di $S_{2,q}$, *Bollettino dell'Unione Matematica Italiana*, **11** (1956), 178–181.

[76] R. Lidl & H. Niederreiter, *Finite Fields, Encyclopedia of Mathematics and its Applications*, 20, Addison–Wesley, Waltham (1983).

[77] L. Lovász & A. Schrijver, Remarks on a theorem of Rédei, *Studia Scientiarum Mathematicarum Hungarica*, **16** (1981), 449–454.

[78] L. Lovász & T. Szőnyi, Multiple blocking sets and algebraic curves, manuscript, 1996.

[79] L. Lunelli & M. Sce, Considerazioni aritmetiche e risultati sperimentali sui $\{K,n\}_q$-archi, *Rendiconti Istituto Lombardo, Accademia di Scienze e Lettere, A*, **98** (1964), 3–52.

[80] J. R. M. Mason, On the maximum sizes of certain (k,n)-arcs in finite projective geometries, *Mathematical Proceedings of the Cambridge Philosophical Society*, **91** (1982), 153–169.

[81] J. R. M. Mason, A class of $((p^n-p^m)(p^n-1),p^n-p^m)$-arcs in PG$(2,q)$, *Geometriae Dedicata*, **15** (1984), 355–361.

[82] G. Menichetti, q-archi completi nei piani di Hall di ordine $q=2^k$, *Atti dell'Accademia Nazionale dei Lincei, Classe di Scienze Fisiche Matematiche e Naturali, Rendiconti*, **56** (1974), 518–525.

[83] R. Metz, On a class of unitals, *Geometriae Dedicata*, **8** (1979), 125–126.

[84] J. di Paola, On minimum blocking coalitions in small projective plane games, *SIAM Journal on Applied Mathematics*, **17** (1969), 378–392.

[85] L. Rédei, *Lückenhafte Polynome über endlichen Körpern*, Akadémiai Kiadó and Birkhäuser Verlag, Budapest and Basel (1970).

[86] R. Schoof, Non-singular plane cubic curves over finite fields, *Journal of Combinatorial Theory, Series A*, **46** (1987), 183–211.

[87] B. Segre, Ovals in a finite projective plane, *Canadian Journal of Mathematics*, **7** (1955), 414–416.

[88] B. Segre, Le geometrie di Galois, *Annali di Matematica Pura ed Applicata*, **48** (1959), 1–97.

[89] B. Segre, Ovali e curve σ nei piani di Galois di caratteristica due, *Atti dell'Accademia Nazionale dei Lincei, Classe di Scienze Fisiche Matematiche e Naturali, Rendiconti*, **32** (1962), 785–790.

[90] B. Segre, Introduction to Galois geometries (edited by J. W. P. Hirschfeld), *Atti dell'Accademia Nazionale dei Lincei, Classe di Scienze Fisiche Matematiche e Naturali, Memorie*, **8** (1967), 133–236.

[91] R. A. Smith, The distribution of rational points on hypersurfaces defined over a finite field, *Mathematika*, **17** (1970), 328–332.

[92] K. O. Stöhr & J. F. Voloch, Weierstrass points and curves over finite fields, *Proceedings of the London Mathematical Society*, **52** (1986), 1–19.

[93] L. Storme, k-arcs in PG(n, q) and linear M.D.S. codes, *Academiae Analecta, Belgium*, **55** (1993), 88–126.

[94] T. Szőnyi, Small complete arcs in Galois planes, *Geometriae Dedicata*, **18** (1985), 161–172.

[95] T. Szőnyi, Note on the order of magnitude of k for complete k-arcs in PG$(2, q)$, *Discrete Mathematics*, **66** (1987), 263–266.

[96] T. Szőnyi, Arcs in cubic curves and 3-independent subsets of abelian groups, in *Combinatorics, Eger*, *Colloquia Mathematica Societatis János Bolyai*, 52, North-Holland, Amsterdam (1987), pp. 499–508.

[97] T. Szőnyi, k-sets in PG$(2, q)$ having a large set of internal nuclei, in *Combinatorics '88, Volume 2* (eds. A. Barlotti et al.), Mediterranean Press, Rende (1991), pp. 449–458.

[98] T. Szőnyi, Note on the existence of large minimal blocking sets in Galois planes, *Combinatorica*, **12** (1992), 227–235.

[99] T. Szőnyi, Arcs, caps, codes and 3-independent subsets, in *Giornate di Geometrie Combinatorie* (eds. G. Faina & G. Tallini), Università di Perugia, Perugia (1993), pp. 57–80.

[100] T. Szőnyi, Blocking sets in desarguesian affine and projective planes, *Finite Fields and their Applications*, in press.

[101] T. Szőnyi, On the embeddability of (k, p)-arcs, manuscript, 1996.

[102] E. Szekeres & G. Szekeres, On a problem of Schütte and Erdős, *Mathematical Gazette*, **49** (1965), 290–293.

[103] G. Tallini, Sui q-archi di un piano lineare finito di caratteristica $p = 2$, *Atti dell'Accademia Nazionale dei Lincei, Classe di Scienze Fisiche Matematiche e Naturali, Rendiconti*, **23** (1957), 242–245.

[104] M. Tallini Scafati, Archi completi in un $S_{2,q}$ con q pari, *Atti della Accademia Nazionale dei Lincei, Classe di Scienze Fisiche Matematiche e Naturali, Rendiconti*, **37** (1964), 48–51.

[105] J. A. Thas, Some results concerning $\{(q+1)(n-1), n\}$-arcs and $\{(q+1)(n-1)+1, n\}$-arcs in finite projective planes of order q, *Journal of Combinatorial Theory, Series A*, **19** (1975), 228–232.

[106] J. A. Thas, Elementary proofs of two fundamental theorems of B. Segre without using the Hasse-Weil theorem, *Journal of Combinatorial Theory, Series A*, **34** (1983), 381–384.

[107] J. A. Thas, Complete arcs and algebraic curves in $\mathrm{PG}(2, q)$, *Journal of Algebra*, **106** (1987), 451–464.

[108] J. A. Thas, A combinatorial characterization of Hermitian curves, *Journal of Algebraic Combinatorics*, **1** (1992), 97–102.

[109] J. A. Thas, Projective geometry over a finite field, Chapter 7, in *Handbook of Incidence Geometry* (ed. F. Buekenhout), North-Holland, Amsterdam (1995), pp. 295–347.

[110] A. Thomason, Random graphs, strongly regular graphs and pseudorandom graphs, in *Surveys in Combinatorics 1987* (ed. C. Whitehead), *London Mathematical Society Lecture Note Series*, 123, Cambridge University Press, Cambridge (1987), pp. 173–195.

[111] E. Ughi, On (k, n)-fold blocking sets which can be obtained as a union of conics, *Geometriae Dedicata*, **26** (1988), 241–246.

[112] V. H. Vu, A strongly regular N-full graph of small order, *Combinatorica*, **16** (1996), 295–299.

[113] J. F. Voloch, On the completeness of certain plane arcs, *European Journal of Combinatorics*, **8** (1987), 453–456.

[114] J. F. Voloch, Arcs in projective planes over prime fields, *Journal of Geometry*, **38** (1990), 198–200.

[115] J. F. Voloch, On the completeness of certain plane arcs II, *European Journal of Combinatorics*, **11** (1990), 491–496.

[116] J. F. Voloch, Complete arcs in Galois planes of non-square order, in *Advances in Finite Geometries and Designs* (eds. J. W. P. Hirschfeld, D. R. Hughes & J. A. Thas), Oxford University Press, Oxford (1991), pp. 401–406.

[117] W. Waterhouse, Abelian varieties over finite fields, *Annales Scientifiques de l'École Normale Supérieure*, **2** (1969), 521–560.

[118] A. Weil, *Sur les Courbes Algébriques et les Variétés qui s'en déduisent, Actualités Scientifiques et Industrielles*, 1041, Hermann & Cie, Paris (1948).

[119] A. Weil, On some exponential sums, *Proceedings of the National Academy of Science*, **34** (1948), 204–207.

[120] F. Wettl, On the nuclei of a finite projective plane, *Journal of Geometry*, **30** (1987), 157–163.

[121] B. J. Wilson, Incompleteness of $(nq + n - q - 2, n)$-arcs in finite projective planes of even order, *Mathematical Proceedings of the Cambridge Philosophical Society*, **91** (1982), 1–8.

[122] F. Zirilli, Su una classe di k-archi di un piano di Galois, *Atti della Accademia Nazionale dei Lincei, Classe di Scienze Fisiche Matematiche e Naturali, Rendiconti*, **54** (1973), 393–397.

Department of Computer Science, Eötvös Loránd University,
Múzeum krt. 6-8, H-1088 Budapest, Hungary
and
Department of Geometry, József Attila University,
Aradi vértanúk tere 1, H-6720 Szeged, Hungary
szonyi@cs.elte.hu

New Perspectives on Interval Orders and Interval Graphs

William T. Trotter

Summary Interval orders and interval graphs are particularly natural examples of two widely studied classes of discrete structures: partially ordered sets and undirected graphs. So it is not surprising that researchers in such diverse fields as mathematics, computer science, engineering and the social sciences have investigated structural, algorithmic, enumerative, combinatorial, extremal and even experimental problems associated with them. In this article, we survey recent work on interval orders and interval graphs, including research on on-line coloring, dimension estimates, fractional parameters, balancing pairs, hamiltonian paths, ramsey theory, extremal problems and tolerance orders. We provide an outline of the arguments for many of these results, especially those which seem to have a wide range of potential applications. Also, we provide short proofs of some of the more classical results on interval orders and interval graphs. Our goal is to provide fresh insights into the current status of research in this area while suggesting new perspectives and directions for the future.

1 Introduction

A complex process (manufacturing computer chips, for example) is often broken into a series of tasks, each with a specified starting and ending time. Task A *precedes* Task B if A ends before B begins. When A precedes B, the output of A can safely be used as input to B, and resources dedicated to the completion of A, such as machines or personnel, can now be applied to B. When A and B have overlapping time periods, they may be viewed as *conflicting* tasks, in the sense that they compete for limited resources.

This short paragraph is intended to motivate the formal definition of two of the most widely studied classes of discrete structures in all of combinatorial mathematics: interval orders and interval graphs. The main point to the discussion is that interval orders and interval graphs are important from an applications standpoint. This much is inescapable. They occur so naturally and with such frequency that they *must* be studied. Fortunately, the study of interval orders and interval graphs has yielded work of intrinsic interest and beauty, work that can be appreciated for its elegance independent of the fact that many find it useful and important.

The remainder of this section includes a brief summary of the notation and terminology necessary for the balance of the paper. For a more comprehensive treatment of background material, the reader is referred to Peter Fishburn's monograph *Interval Orders and Interval Graphs* [36]. Other recommended sources for background information are the author's survey articles [115], [116], [120], [121] and monograph [118] and the books by Golumbic [48] and

Roberts [98].

Throughout this paper, we consider a *partially ordered set* (or *poset*) $\mathbf{P} = (X, P)$ as a structure consisting of a set X and a reflexive, antisymmetric and transitive binary relation P on X. We call X the *ground set* of the poset $\mathbf{P}$, and we call P a *partial order* on X. The notations $x \leq y$ in P, $y \geq x$ in P and $(x, y) \in P$ are used interchangeably, and the reference to the partial order P is often dropped when its definition is fixed throughout the discussion. We write $x < y$ in P and $y > x$ in P when $x \leq y$ in P and $x \neq y$. When $x, y \in X$, $(x, y) \notin P$ and $(y, x) \notin P$, we say x and y are incomparable and write $x \| y$ in P. When $\mathbf{P} = (X, P)$ is a poset, we call the partial order $P^d = \{(y, x) : (x, y) \in P\}$ the *dual* of P and we let $\mathbf{P}^d = (X, P^d)$.

When P is a binary relation on X and $Y \subseteq X$, we denote the *restriction* of P to Y by $P(Y)$. When P is a partial order on X, $Q = P(Y)$ is a partial order on Y and $\mathbf{Q} = (Y, Q)$ is called a *subposet* of $\mathbf{P} = (X, P)$. Also, we call $\mathbf{Q}$ the subposet *determined* by Y. When $X_1, X_2, \ldots, X_r \subseteq X$, we will find it convenient to denote the subposets they determine by $\mathbf{X}_1, \mathbf{X}_2 \ldots, \mathbf{X}_r$, respectively. In this article, we tend not to distinguish between isomorphic posets, so we abuse language slightly and say that a poset $\mathbf{Q}$ is *contained* in another poset $\mathbf{P}$ when $\mathbf{Q}$ is isomorphic to a subposet of $\mathbf{P}$.

Although we are concerned primarily with *finite* posets, i.e., those posets with finite ground sets, we find it convenient to use the familiar notation $\mathbb{R}$, $\mathbb{Q}$, $\mathbb{Z}$ and $\mathbb{N}$ to denote respectively the reals, rationals, integers and positive integers equipped with the usual orders. Note that these four infinite posets are *total* orders; in each case, any two distinct points are comparable. Total orders are also called *linear* orders, or *chains*. When $X = X_1 \cup X_2 \cup \cdots \cup X_t$ is a partition and L_i is a linear order on X_i for each $i = 1, 2, \ldots, r$, we let $L = L_1 < L_2 < \cdots < L_r$ denote the linear order on X defined by $x < y$ in L if and only if $x \in X_i$, $y \in X_j$ and either $i < j$ or both $i = j$ and $x < y$ in L_i.

For a positive integer n, we let $\mathbf{n}$ denote the n-element chain $0 < 1 < \cdots < n-1$. Somewhat inconsistently, we let $[n]$ denote the n-element set $\{1, 2, \ldots, n\}$. Also, when X is a set, we let $\binom{X}{n}$ denote the set of all n-element subsets of X.

Let $\mathbf{P} = (X, P)$ be a poset, and let $\mathcal{F} = \{\mathbf{Q}_x = (Y_x, Q_x) : x \in X\}$ be a family of posets indexed by the elements of X. Define the *lexicographic sum* of $\mathcal{F}$ over $\mathbf{P}$, denoted $\sum_{x \in \mathbf{P}} \mathcal{F}$, as the poset $\mathbf{Q} = (Y, Q)$ where $Y = \{(x, y) : x \in X, y \in Y_x\}$ and $(x_1, y_1) < (x_2, y_2)$ in Q if and only if $x_1 < x_2$ in P, or if both $x_1 = x_2$ and $y_1 < y_2$ in Q_{x_1}. With this definition, a disjoint sum is just a lexicographic sum over a two-element antichain.

In the remainder of this article, we will assume some familiarity with the basic concepts for partially ordered sets. The author's survey article on partially ordered sets [120] provides a thorough overview of the combinatorial aspects. Other sources for background material on posets are Brightwell's survey article [17] and the author's other survey articles [115], [112], [117] and [121].

2 Interval orders and interval graphs

A poset $\mathbf{P} = (X, P)$ is called an *interval order* if there is a function I assigning to each element $x \in X$ a closed interval $I(x) = [a_x, b_x]$ of a linearly ordered set $\mathbf{L}$ (usually, we take $\mathbf{L}$ as the real line $\mathbb{R}$) so that for all x, $y \in X$, $x < y$ in P if and only if $b_x < a_y$ in L. We call I an *interval representation* of $\mathbf{P}$, or just a *representation* for short. For brevity, whenever we say that I is a representation of an interval order $\mathbf{P} = (X, P)$, we will use the alternate notation $[a_x, b_x]$ for the closed interval $I(x)$. Also, we let $|I(x)|$ denote the *length* of the interval, i.e., $|I(x)| = b_x - a_x$.

Note that end points of intervals used in a representation need not be distinct. In fact, distinct points x and y from X may satisfy $I(x) = I(y)$. We even allow degenerate intervals. On the other hand, a representation is said to be *distinguishing* if all intervals are non-degenerate and all end points are distinct. It is easy to see that every interval order has a distinguishing representation. In fact, since we are concerned only with finite posets, we could have just as well required that all intervals used in the representation be open.

Analogously, a graph $\mathbf{G} = (V, E)$ is an *interval graph* when there is a function I which assigns to each vertex $x \in V$ a closed interval $I(x) = [a_x, b_x]$ from a linearly ordered set $\mathbf{L}$ so that $\{x, y\} \in E$ if and only if $I(x) \cap I(y) \neq \emptyset$. As before, we call I an *interval representation* of $\mathbf{G}$ and note that, if desired, we may assume I is distinguishing.

Throughout this article, we will move back and forth between posets and graphs in discussions about a family of intervals. The interval graph determined by a family of intervals is just the incomparability graph of the interval order. Chains correspond to independent sets and antichains correspond to cliques.

3 Classical representation theorems

A good fraction of the early research on interval graphs and interval orders was focused on characterization issues. Recall that a graph is *triangulated* if it does not contain a cycle on four or more vertices as an induced subgraph. Also, a vertex x in a graph $\mathbf{G}$ is *simplicial* if its neighborhood is a complete subgraph of $\mathbf{G}$, so a graph is triangulated if and only if every induced subgraph has a simplicial vertex. Triangulated graphs are a well studied class of perfect graphs (see [58] and Chapter 4 of Golumbic's monograph [48], for example). Obviously, interval graphs are triangulated, but it is natural to ask whether all triangulated graphs are interval graphs. This is not true. In fact, not all trees are interval graphs, e.g. the subdivision of $\mathbf{K}(1, 3)$ is not an interval graph.

Three distinct vertices x, y and z in a graph $\mathbf{G}$ are said to form an *asteroidal triple* when for each two vertices in $\{x, y, z\}$, there is a path joining them, with no vertex on the path adjacent to the third. For example, the three leaves in a subdivision of $\mathbf{K}(1,3)$ form an asteroidal triple. In [83], Lekkerkerker and Boland proved that a triangulated graph is an interval graph if and only if it does not contain any asteroidal triples. They used this characterization theorem to provide a minimum list of forbidden subgraphs for interval graphs. This list includes the cycles on four or more vertices, three other infinite families and two isolated examples. One of these is the subdivision of $\mathbf{K}(1,3)$.

Other characterizations of interval graphs in terms of forbidden substructures have been provided by Gilmore and Hoffman [47] and by Ghouila-Houri [46]. Characterizations of interval graphs by forbidden subgraphs or forbidden substructures provide important structural information about the properties of interval graphs but do not necessarily yield a useful algorithm. Using a special kind of data structure called a *PQ–tree*, Booth and Lueker [15] produced an $O(n^2)$ algorithm for testing whether a graph $\mathbf{G}$ on n vertices is an interval graph and producing the representation when it is.

Characterization problems for interval graphs are closely related to characterization problems for comparability graphs. The classic paper of Gallai [45] provides a forbidden subgraph (again, in terms of induced subgraphs) characterization of comparability graphs with a minimum list including eight infinite families and 10 isolated examples. A comparability graph may have many different transitive orientations, but Gallai shows that if T_1 and T_2 are transitive orientations of the same comparability graph, then T_1 may be transformed into T_2 by a finite sequence of reversals applied to *autonomous* sets. Gallai's paper remains one of the deepest and most important contributions to this subject.

Next, we discuss three important representation theorems which are essential to understanding the material which follows. First, a finite poset $\mathbf{P} = (X, P)$ is called a *weak order* if there exists a function $f: X \to \mathbb{R}$ so that for all $x, y \in X$ with $x \neq y$,

1. $x < y$ in P if and only if $f(x) < f(y)$ in $\mathbb{R}$, and
2. $x \| y$ if and only if $f(x) = f(y)$.

The following elementary result is left as an exercise (see [36], e.g.).

Proposition 3.1 *Let $\mathbf{P} = (X, P)$ be a poset. Then the following are equivalent.*

1. *$\mathbf{P}$ is a weak order.*
2. *$\mathbf{P}$ does not contain $\mathbf{2} + \mathbf{1}$ as a subposet.*
3. *$\mathbf{P}$ is the lexicographic sum of a family of antichains over a chain.* ∎

Given a representation I of an interval order $\mathbf{P} = (X, P)$, there are two natural weak orders defined on X by the end points. The *ordering by left end points* L defined by $x < y$ in L if and only if $a_x < a_y$ in $\mathbb{R}$ and the *ordering by right end points* R defined by $x < y$ in R if and only if $b_x < b_y$ in $\mathbb{R}$. When the representation is distinguishing, these weak orders are linear orders.

Next, we have the following characterization theorem for interval orders due to Fishburn [35]. Our argument is motivated by proofs due to Bogart [5] and Greenough [51] and requires the following notation.

For a poset $\mathbf{P} = (X, P)$ and a subset $S \subset X$, let $D(S) = \{y \in X :$ there exists some $x \in S$ with $y < x$ in $P\}$. Also, let $D[S] = D(S) \cup S$. When $|S| = 1$, say $S = \{x\}$, we write $D(x)$ and $D[x]$ rather than $D(\{x\})$ and $D[\{x\}]$. Dually, for a subset $S \subseteq X$, we define $U(S) = \{y \in X :$ there exists some $x \in X$ with $y > x$ in $P\}$. As before, set $U[S] = U(S) \cup S$.

Theorem 3.2 *Let* $\mathbf{P} = (X, P)$ *be a poset. Then the following are equivalent.*

1. $\mathbf{P}$ *is an interval order.*
2. $\mathbf{P}$ *does not contain* $\mathbf{2} + \mathbf{2}$ *as a subposet.*
3. *Whenever* $x < y$ *and* $z < w$ *in* P*, then either* $x < w$ *or* $z < y$ *in* P.
4. *For every* $x, y \in X$*, either* $D(x) \subseteq D(y)$ *or* $D(y) \subseteq D(x)$.
5. *For every* $x, y \in X$*, either* $U(x) \subseteq U(y)$ *or* $U(y) \subseteq U(x)$.

Proof The equivalence of the last four statements is immediate. We now show that statement 1 is equivalent to 4. Suppose first that $\mathbf{P} = (X, P)$ is an interval order and that I is an interval representation of $\mathbf{P}$. Let $x, y \in X$; without loss of generality, we may assume $a_x \le a_y$ in $\mathbb{R}$. Then $D(x) \subseteq D(y)$.

Now suppose that statement 4 holds for a poset $\mathbf{P} = (X, P)$. We show that $\mathbf{P}$ is an interval order. Let $Y = \{D(x) : x \in X\}$, and let $m = |Y|$. Then define a linear order L on Y by $D < D'$ in L if $D \subsetneq D'$. Then label the sets in Y so that $D_1 < D_2 < \cdots < D_m$ in L. For each $x \in X$, let $F(x) = [i, j]$, where $D(x) = D_i$ and $j = m$ if x is maximal, and $D_{j+1} = \bigcap\{D(y) : x < y \text{ in } P\}$, otherwise. ∎

One advantage to the proof given here for Fishburn's representation theorem for interval orders is that the total number of end points used in the representation is minimal. Also, note that $|\{D(x) : x \in X\}| = |\{U(x) : x \in X\}|$, when $\mathbf{P} = (X, P)$ is an interval order, as pointed out by Greenough [51].

An interval order $\mathbf{P} = (X, P)$ is called a *semi-order* if there is a constant c for which it has an interval representation F such that the length of the interval $F(x)$ is exactly c, for every $x \in X$. From a modern perspective, it would perhaps be more natural if these objects were called *constant length* interval orders; however, after rescaling, it may be assumed that the constant length of intervals used in the representation of a semi-order is 1.

For semi-orders, we have the following representation theorem, the principal part of which is due to Scott and Suppes [104]. Simple proofs have been given by several authors, including Bogart [4], Fishburn [36] and Rabinovitch [92, 91].

Theorem 3.3 *Let* $\mathbf{P} = (X, P)$ *be an interval order. Then the following statements are equivalent.*

1. $\mathbf{P}$ *is a semi-order.*

2. $\mathbf{P}$ *does not contain* $\mathbf{3} + \mathbf{1}$ *as a subposet.*

3. *Whenever* $x < y < z$ *and* $w \| y$ *in* P, *then either* $x < w$ or $w < z$ *in* P.

4. *The binary relation* W *is a weak order on* X, *where*

$$\begin{aligned} W = &\{(x, y) \in X \times X : x = y\} \\ &\cup \{(x, y) \in X \times X : D(x) \subseteq D(y), U(y) \subsetneq U(x)\} \\ &\cup \{(x, y) \in X \times X : D(x) \subsetneq D(y), U(y) \subseteq U(x)\}. \end{aligned}$$

Proof The equivalence of statements 2, 3 and 4 is immediate. We show that statements 1 and 4 are equivalent. First let $\mathbf{P} = (X, P)$ be a semi-order and let I be an interval representation in which all intervals have length c. Let $I(x) = [a_x, a_x + c]$, for every $x \in X$. Then $(x, y) \in W$ if and only if $a_x \leq a_y$ in $\mathbb{R}$, so that W is a weak order on X.

Now suppose that statement 4 holds for a poset $\mathbf{P} = (X, P)$. We show that $\mathbf{P}$ is a semi-order. We actually prove something stronger. Let L be any linear order on X extending the weak order W. Proceeding by induction on $|X|$, we show that there exists a distinguishing interval representation I of $\mathbf{P}$ which assigns to each $x \in X$ a unit length interval $I(x) = [a_x, a_x + 1]$ such that for all $x, y \in X$, $a_x < a_y$ in $\mathbb{R}$ if and only if $x < y$ in L.

Noting that the claim holds trivially when $|X| = 1$, consider the inductive step. Suppose that L orders X as $x_1 < x_2 < \cdots < x_n$. Let $Y = X - \{x_n\}$, let $Q = P(Y)$ and $L' = L(Y)$. In the poset $\mathbf{Q} = (Y, Q)$, let W'' be the binary relation defined in statement 4 for the subposet $\mathbf{Q}$. Then let $W' = W(Y)$. Then $W'' \subseteq W' \subseteq L'$.

It follows that $\mathbf{Q}$ is a semi-order and that there exists a distinguishing representation I' of $\mathbf{Q}$ so that for all $y, z \in Y$, $a_y < a_z$ if and only if $y < z$ in L'. Also, $y < z$ in Q if and only if $a_y + 1 < a_z$. We now show that this representation can be extended by an appropriate choice of an interval $I(x_0) = [a_{x_n}, a_{x_n} + 1]$ for x_0. If $y < x_n$ for every $y \in Y$, let $a = \max\{a_y : y \in Y\}$ and set $a_{x_n} = 2 + a$.

So we may assume that $S = \{y \in Y : y \| x_n\} \neq \emptyset$. It follows that there is a positive integer i so that $S = \{x_i, x_{i+1}, \ldots, x_{n-1}\}$, and $[a, b] = \bigcap\{I(y) : y \in S\}$ is a nondegenerate interval. If $D(x_n) = \emptyset$, set $a' = a$; otherwise, set $a' = \max\{a_y + 1 : y < x_n\}$. In either case, note that $a' < b$ in $\mathbb{R}$. It follows that we may take a_n as any real number between a' and b distinct from any end point previously chosen. ∎

There is an important corollary to the Scott-Suppes theorem for semi-orders, a result which was first noted by Roberts [97]. Call an interval order

$\mathbf{P} = (X, P)$ *proper* if it admits an interval representation I so that if $x, y \in X$ and $x \neq y$, then $I(x) \not\subseteq I(y)$ and $I(y) \not\subseteq I(x)$. A semi-order is obviously a proper interval order, but as Roberts pointed out, a proper interval order is also a semi-order. We will revisit this theme in Section 19.

4 Dilworth's theorem for interval orders

The *width* of a finite poset is the maximum cardinality of an antichain, and the *height* is the maximum cardinality of a chain. Dilworth's theorem [26] asserts that a poset of width w can be partitioned into w chains. Dually [58], a poset of height h can be partitioned into h antichains. Short proofs of these results have been provided by many authors, e.g. see [120] or [118]. Here is one for the second result.

Theorem 4.1 *Let $\mathbf{P} = (X, P)$ be a poset of height h. Then there exists a partition*

$$X = A_1 \cup A_2 \cup \cdots \cup A_h, \tag{1}$$

with A_j an antichain, for each $j \in [h]$.

Proof For each $x \in X$, let $d(x)$ denote the height of the subposet determined by $D[x]$. Then for each $j \in [h]$, let $A_j = \{x \in X : d(x) = j\}$. ■

Here is a sketch of how a bipartite matching algorithm can be used to find the width w of a poset $\mathbf{P}$ and a partition into w chains. See Brightwell [17] for details. Given a poset $\mathbf{P} = (X, P)$, form a bipartite graph $G = (V, E)$ with $V = \{a_x : x \in X\} \cup \{b_x : x \in X\}$ and $E = \{\{a_x, b_y\} : x < y \text{ in } P\}$. Then let $\mathcal{M}$ be a maximum matching in $\mathbf{G}$. Define a chain partition of X by putting distinct $x, y \in X$ in the same chain when there exists a sequence $x = u_0, u_1, \ldots, u_s = y$ so that $\{a_{u_{i-1}}, b_{u_i}\} \in \mathcal{M}$, for every $i \in [s]$.

For interval orders, we can be more explicit by taking advantage of an important connection with graph coloring. Although this material is well known, we summarize it briefly as the concepts will be used later in this article. The argument comes from Hajnal and Suranyi's paper [58] on classes of perfect graphs.

Theorem 4.2 *Interval graphs are perfect, i.e., the chromatic number is equal to the maximum clique size. Furthermore, an optimal coloring always results from applying First Fit to the vertices in the order their right end points occur in a distinguishing representation.*

Proof Let I be a distinguishing representation of an interval graph $\mathbf{G} = (X, E)$, and let L be the linear order on X determined by the right end points, i.e., $x < y$ in L if and only if $b_x < b_y$ in $\mathbb{R}$. For each $x \in V$, let $N_L(x) = \{y \in X : \{x, y\} \in E\}$. Note that $N_L(x)$ is a complete subgraph of $\mathbf{G}$, for each

$x \in X$, as all the intervals corresponding to intervals in $N_L(x)$ contain the left end point a_x of $I(x)$.

When First Fit assigns color α to a vertex x, then x belongs to a complete subgraph of size α consisting of x and $\alpha - 1$ vertices from $N_L(x)$. It follows that if the maximum clique size of $\mathbf{G}$ is k, then First Fit will color $\mathbf{G}$ in exactly k colors. ∎

Equivalently, First Fit partitions the associated interval order $\mathbf{P} = (X, P)$ for which I is a distinguishing representation into w chains, where w is its width.

In material to follow, we will find it convenient to be even more explicit than the argument used for Theorem 4.1 in applications of the dual to Dilworth's theorem. Set $A_0 = \emptyset$ and $X_0 = X$. Now suppose that we have defined A_{j-1} and X_{j-1} for some $j \geq 1$. Set $X_j = X_{j-1} - A_{j-1}$. If $X_j \neq \emptyset$, let y_j be the unique element of X_j for which the right end point $r_j = b_{y_j}$ is minimum. Then let $A_j = \{x \in X_j : b_{y_j} \in I(x)\}$. When the algorithm halts, we have a partition $X = A_1 \cup A_2 \cup \cdots \cup A_h$ into h antichains, and we have a chain $C = \{y_1, y_2, \ldots, y_h\}$ of cardinality h. Furthermore, every interval in the representation intersects the right end point of at least one interval in C. We call C the *lexicographically least* maximum chain of $\mathbf{P}$, and we call the associated partition into antichains the *canonical minimum partition.*

5 Linear extensions and dimension

When P and Q are binary relations on a set X, we say Q is an *extension* of P when $P \subseteq Q$; a linear order L on X is called a *linear extension* of a partial order P on X when $P \subseteq L$. A set $\mathcal{R}$ of linear extensions of P is called a *realizer* of $\mathbf{P}$ when $P = \bigcap \mathcal{R}$, i.e., for all x, y in X, $x < y$ in P if and only if $x < y$ in L, for every $L \in \mathcal{R}$. The minimum cardinality of a realizer of $\mathbf{P}$ is called the *dimension* of $\mathbf{P}$ and is denoted $\dim(\mathbf{P})$. Note that if $\mathbf{P}$ contains $\mathbf{Q}$, then $\dim(\mathbf{Q}) \leq \dim(\mathbf{P})$.

It is natural to ask what causes a poset to have large dimension. Here is a partial answer. For integers $n \geq 2$ and $k \geq 0$, define the *crown* $\mathbf{S}_n^k$ as the poset of height two with $n+k$ minimal elements $a_1, a_2, \ldots, a_{n+k}$, $n+k$ maximal elements $b_1, b_2, \ldots, b_{n+k}$ and ordering $a_i < b_j$ if and only if $j \in \{i+k+1, i+k+2, \ldots, i-1\}$. In this definition, the subscripts are interpreted cyclically, so that $b_{n+k+1} = b_1$, etc. When $n \geq 3$, the dimension of the crown $\mathbf{S}_n^k$ is given by the following formula [110]:

$$\dim(\mathbf{S}_n^k) = \left\lceil \frac{2(n+k)}{k+2} \right\rceil \tag{2}$$

For each $k \geq 0$, the poset $\mathbf{S}_2^k$ is the disjoint sum of $k+2$ two-element chains, so these posets have dimension 2. When $n \geq 3$, the crown $\mathbf{S}_n^k$ always has dimension at least 3. Posets in the family $\mathcal{S} = \{\mathbf{S}_n^0 : n \geq 2\}$ are referred to as *standard examples*. Note that the dimension of $\mathbf{S}_n^0$ is exactly n. Furthermore, for each $n \geq 3$, $\mathbf{S}_n^0$ is n-irreducible, i.e., the removal of any point leaves a subposet having dimension $n-1$. Also note that when $n \geq 3$, the standard example $\mathbf{S}_n^0$ is isomorphic to the family of 1-element and $(n-1)$–element subsets of $\{1, 2, \ldots, n\}$ ordered by inclusion. The standard example $\mathbf{S}_2^0$ is somewhat of a special case. It has dimension two and is isomorphic to the disjoint sum of two 2-element chains, but it is not irreducible.

We summarize some basic facts about dimension in the following proposition, referring the reader to [118] for proofs and references.

Proposition 5.1 *Let* $\mathbf{P} = (X, P)$ *and* $\mathbf{Q} = (Y, Q)$ *be posets. Then:*

1. $\dim(\mathbf{P} + \mathbf{Q}) = \max\{2, \dim(\mathbf{P}), \dim(\mathbf{Q})\}$.
2. $\dim(\mathbf{P} \times \mathbf{Q}) \leq \dim(\mathbf{P}) + \dim(\mathbf{Q})$, *with equality holding if* $\mathbf{P}$ *and* $\mathbf{Q}$ *have greatest and least elements.*
3. *The removal of a point from* $\mathbf{P}$ *decreases* $\dim(\mathbf{P})$ *by at most one.*
4. *If* A *is a maximum antichain in* $\mathbf{P}$, *then* $\dim(\mathbf{P}) \leq |A|$ *and* $\dim(\mathbf{P}) \leq \max\{2, |X - A|\}$.
5. $\dim(\mathbf{P}) = \dim(\mathbf{P}^d)$. ∎

Note that the family of standard examples shows that inequalities 3 and 4 of Proposition 5.1 are best possible. We will also find it convenient to put inequality 1 in the preceding theorem in a more general setting. Here is the general formula for dimension and lexicographic sums (see [118]).

Proposition 5.2 *Let* $\mathbf{P} = (X, P)$ *be a poset, and let* $\mathcal{F} = \{\mathbf{Q}_x = (Y_x, P_x) : x \in X\}$ *be a family of posets. Then*

$$\dim\left(\sum_{x \in \mathbf{P}} \mathcal{F}\right) = \max\{\dim(\mathbf{P}), \max\{\dim(\mathbf{Q}_x) : x \in X\}\}. \quad \blacksquare \tag{3}$$

For additional background information on dimension, the reader is referred to the author's monograph [118], the survey article [63] on dimension by Kelly and Trotter and the survey articles [115] and [121]. The articles [112], [117] and [119] also discuss combinatorial problems for posets. Connections between dimension for posets and a wide range of combinatorial problems are discussed in [123], with greater detail provided in the monograph [118].

6 Linear extensions of interval orders

When $\mathbf{P} = (X, P)$ is a poset, $A, B \subset X$ and $A \cap B = \emptyset$, and L is a linear extension of P, we say B is *over* A in L when $b > a$ in L, whenever $a \in A$, $b \in B$ and $a \| b$ in P. In applying this definition, it is important to note that we do not require that $b > a$ in L, for *all* $a \in A$ and $b \in B$, only the incomparable pairs. The following elementary result was first discovered by Rabinovitch [94].

Theorem 6.1 *Let $\mathbf{P} = (X, P)$ be an interval poset, and let $A, B \subset X$ with $A \cap B = \emptyset$. Then there exists a linear extension L of P with B over A in L.*

Proof Let I be a distinguishing interval representation of $\mathbf{P}$. For each $x \in X$, let $p_x = a_x$ if $x \in A$ and $p_x = b_x$, otherwise. Then define a linear extension L by setting $x < y$ in L if and only if $p_x < p_y$ in $\mathbb{R}$. ■

More generally, the following proposition, first noted by Felsner in [28], is an easy exercise.

Proposition 6.2 *Let $\mathbf{P} = (X, P)$ be an interval order, and let I be any distinguishing interval representation of $\mathbf{P}$. If L is a linear extension of P, then it is possible to choose for each $x \in X$ a point $p_x \in I(x)$ so that $x < y$ in L if and only if $p_x < p_y$ in $\mathbb{R}$.* ■

7 Dimension of interval orders

It is natural to ask whether an interval order can have large dimension. If the answer is yes, it cannot be due to the presence of large standard examples, as no interval order contains any of them. Note that for each $n \geq 2$, the subposet of $\mathbf{S}_n^0$ determined by a_1, a_2, b_1 and b_2 is isomorphic to $\mathbf{2} + \mathbf{2}$.

Nevertheless, interval orders may have large dimension, and to explain how this may occur, we introduce a standard example of an interval order. For an integer $n \geq 2$, let $\mathbf{I}_n = (\binom{[n]}{2}, P_n)$ denote the interval order defined by the representation $I(\{i, j\}) = [i, j]$. To avoid confusion with the family of standard examples discussed previously, we call the interval orders in the family $\{\mathbf{I}_n : n \geq 2\}$ *canonical* interval orders.

The following result is due to Bogart, Rabinovitch and Trotter [10].

Theorem 7.1 *For every integer t, there exists an integer n_0 so that if $n \geq n_0$, then the dimension of the canonical interval order $\mathbf{I}_n$ is larger than t.*

Proof Evidently, $\dim(\mathbf{I}_n)$ is a non-decreasing function of n. We assume that $\dim(\mathbf{I}_n) \leq t$, for all $n \geq 2$ and obtain a contradiction when n is sufficiently large in terms of t. Let i, j, and k be distinct integers with $1 \leq i < j < k \leq n$. Then $\{i, j\} \| \{j, k\}$ in P_n, so if $\mathcal{R} = \{L_1, L_2, \ldots, L_t\}$ is a realizer of P_n, then we may choose $\alpha \in \{1, 2, \ldots, t\}$ so that $\{i, j\} > \{j, k\}$ in L_α. This is a coloring of the 3-element subsets of $\{1, 2, \ldots, n\}$ with t colors. If n is sufficiently large, there exists a 4-element subset $S = \{i < j < k < l\}$ and an integer $\alpha \in \{1, 2, \ldots, t\}$ so that all 3-element subsets of S are mapped to α. This implies that $\{i, j\} > \{j, k\} > \{k, l\} > \{i, j\}$ in L_α, which is a contradiction. ■

Now that we know that interval orders can have large dimension, we pause to discuss some of the special properties interval orders exhibit.

Let $\mathbf{P} = (X, P)$ be a poset and let $X = X_1 \cup X_2$ be a partition of X into disjoint non-empty subsets. It is natural to ask whether one can say anything about the dimension of dim(**P**) given information about dim($\mathbf{X}_1$) and dim($\mathbf{X}_2$). For posets in general, the answer is no. For example, for each $n \geq 2$, consider the partition of the point set of the standard example $\mathbf{S}_n^0$ into minimal elements and maximal elements. The dimension of $\mathbf{S}_n^0$ is n but the two antichains have dimension 2.

For interval orders, things are different. The next result follows easily from Proposition 6.2.

Lemma 7.2 *Let $\mathbf{P} = (X, P)$ be an interval order, and let $X = X_1 \cup X_2$ be a partition of X into disjoint non-empty subsets. If L_1 and L_2 are linear extensions of the subposets $\mathbf{P}_1$ and $\mathbf{P}_2$ induced by X_1 and X_2 respectively, then there exists a linear extension L of P so that $L_1 = L(X_1)$ and $L_2 = L(X_2)$.* ∎

Theorem 7.3 *Let $\mathbf{P} = (X, P)$ be an interval order, and let $X = X_1 \cup X_2$ be a partition of X into disjoint non-empty subsets. Then*

$$\dim(\mathbf{P}) \leq 2 + \max\{\dim(\mathbf{X}_1), \dim(\mathbf{X}_2)\}. \tag{4}$$

Proof Let $t = \max\{\dim(\mathbf{X}_1), \dim(\mathbf{X}_2)\}$. From Lemma 7.2, we know that there exists a family $\mathcal{F} = \{L_1, L_2, \ldots, L_t\}$ of linear extensions of P so that $\mathcal{F}_i = \{L_1(X_i), L_2(X_i), \ldots, L_t(X_i)\}$ is a realizer of $\mathbf{X}_i$, for $i = 1, 2$. Then let M_1 and M_2 be linear extensions of P so that X_1 is over X_2 in M_1 and X_2 is over X_1 in M_2. It follows that $\{M_1, M_2\} \cup \mathcal{F}$ is a realizer of P. ∎

When one of the two sets in the partition is the set of maximal elements, we can do a little better. We leave the proof as an exercise.

Theorem 7.4 *Let $\mathbf{P} = (X, P)$ be an interval order which is not an antichain. If X_1 is the set of all maximal elements, and $X_2 = X - X_1$, then*

$$\dim(\mathbf{P}) \leq 1 + \dim(\mathbf{X}_2). \quad ∎ \tag{5}$$

Now it is natural to ask whether we can say anything about what must be contained in an interval order of large dimension. Here we present a partial answer. In Section 10, we will give a more complete answer. For now, we are content to show that an interval order of large dimension must contain a long chain.

Theorem 7.5 *An interval order of height h has dimension at most $h + 1$.*

Proof Let $\mathbf{P} = (X, P)$ be an interval order of height h, let I be a distinguishing representation of $\mathbf{P}$ and let $X = A_1 \cup A_2 \cup \cdots \cup A_h$ be the canonical partition into antichains. Note that if $x < y$ in P, $x \in A_i$ and $y \in A_j$, then $i < j$.

For each $i \in [h]$, let L_i be a linear extension of P with A_i over $X - A_i$ in L_i. Then let $L_{h+1} = L_1^d(A_1) < L_1^d(A_2) < \cdots < L_1^d(A_h)$. ∎

Although interval orders can have large dimension, this is not true for semi-orders. The following result is due to Rabinovitch [93].

Theorem 7.6 *If* $\mathbf{P} = (X, P)$ *is a semi-order, then* $\dim(\mathbf{P}) \leq 3$.

Proof Let $\mathbf{P} = (X, P)$ be a semi-order, let I be a distinguishing representation of $\mathbf{P}$ and let let $X = A_1 \cup A_2 \cup \cdots \cup A_h$ be the canonical partition into antichains.

Let $\mathcal{O} = \bigcup\{A_j : 1 \leq j \leq h,\ j \text{ odd}\}$ and $\mathcal{E} = \bigcup\{A_j : 1 \leq j \leq h,\ j \text{ even}\}$. Let L_1 and L_2 be linear extensions of P with $\mathcal{E}$ over $\mathcal{O}$ in L_1 and $\mathcal{O}$ over $\mathcal{E}$ in L_2. Then let $L_3 = L_1^d(A_1) < L_1^d(A_2) < \cdots < L_1^d(A_h)$. ∎

A semi-order has bounded dimension, not just because it has a representation in which all the intervals have the same length, but rather because there is no element incomparable with all the points in a long chain. We leave the following lemma as an exercise.

Lemma 7.7 *For every* $k \geq 1$, *there exists an integer* s_k *so that if* $\mathbf{P} = (X, P)$ *is an interval order in which the maximum size of a chain* C *for which there exists a point* x *incomparable to all points of* C *is at most* k, *then* $\dim(\mathbf{P}) \leq s_k$. ∎

8 Critical pairs and alternating cycles

In arguments to follow, we will find it convenient to take advantage of a technical detail in the proof of Theorem 7.6. Let L be an arbitrary linear order on X. Define linear extensions L_d and L_u of P as follows. Set $x < y$ in L_d if and only if one of the following conditions holds:

1. $D(x) \subsetneq D(y)$.
2. $D(x) = D(y)$ and $U(y) \subsetneq U(x)$.
3. $D(x) = D(y)$, $U(y) = U(x)$, and $x < y$ in L.

Dually, set $x < y$ in L_u if and only if one of the following conditions holds:

1. $U(y) \subsetneq U(x)$.
2. $U(y) = U(x)$ and $D(x) \subsetneq D(y)$.
3. $U(y) = U(x)$, $D(x) = U(x)$, and $x > y$ in L.

Now let $\mathcal{F}$ be a family of linear extensions of P. Then $\{L_d, L_u\} \cup \mathcal{F}$ is a realizer of P if and only if for every $x, y \in X$ with $x \| y$ in P, $D(x) \subsetneq D(y)$, and $U(y) \subsetneq U(x)$, there exists $L \in \mathcal{F}$ with $x > y$ in L.

This last observation is a special case of a somewhat more general situation. For an arbitrary poset $\mathbf{P} = (X, P)$, let $\text{inc}(\mathbf{P}) = \{(x, y) \in X \times X : x \| y \text{ in } P\}$. Then a family $\mathcal{R}$ of linear extensions of P is a realizer of P if and only if for every $(x, y) \in \text{inc}(\mathbf{P})$, there exists $L \in \mathcal{R}$ so that $x > y$ in L. Call a pair

$(x, y) \in \text{inc}(\mathbf{P})$ a *critical pair* if $u < x$ in P implies $u < y$ in P and $v > y$ in P implies $v > x$ in P, for all $u, v \in X$. Then let $\text{crit}(\mathbf{P})$ denote the set of all critical pairs. It follows that $\mathcal{R}$ is a realizer of P if and only if for every $(x, y) \in \text{crit}(\mathbf{P})$, there exists some $L \in \mathcal{R}$ so that $x > y$ in L.

We say L *reverses* the incomparable pair (x, y) when $x > y$ in L. Let $S \subset \text{inc}(\mathbf{P})$. We say that L *reverses* S when $x > y$ in L, for every $(x, y) \in S$. For an integer $k \geq 2$, a subset $S = \{(x_i, y_i) : 1 \leq i \leq k\} \subset \text{inc}(\mathbf{P})$ is called an *alternating cycle* when $x_i \leq y_{i+1}$ in P, for all $i = 1, 2, \ldots, k$. In this last definition, the subscripts are interpreted cyclically, i.e., $y_{k+1} = y_1$. An alternating cycle $S = \{(x_i, y_i) : 1 \leq i \leq k\}$ is *strict* if $x_i \leq y_j$ in P if and only if $j = i + 1$, for all $i, j = 1, 2, \ldots, k$. When an alternating cycle is strict, the following three statements hold:

1. The elements in $\{x_1, x_2, \ldots, x_k\}$ form a k–element antichain.

2. The elements in $\{y_1, y_2, \ldots, y_k\}$ form a k–element antichain.

3. If $i, j \in [k]$ and $x_i \geq y_j$, then $j = i + 1$ and $x_i = y_j$.

The following elementary result is due to Trotter and Moore [124]. See [118] for a short proof and a number of applications.

Theorem 8.1 *Let* $\mathbf{P} = (X, P)$ *be a poset and let* $S \subseteq \text{inc}(\mathbf{P})$. *Then the following statements are equivalent.*

1. *There exists a linear extension* L *of* P *which reverses* S.

2. S *does not contain an alternating cycle.*

3. S *does not contain a strict alternating cycle.* ■

9 Interval orders and shift graphs

Although it has been known for many years that interval orders of large height must contain long chains, it has only been in the last few years that relatively tight bounds have been found. The relationship between height and dimension in interval orders is best explained via a connection with a graph coloring problem. Fix integers n and k with $1 \leq k < n$. We call an ordered pair (A, B) of k-element sets a (k, n)-*shift pair* if there exists a $(k+1)$-element subset $C = \{i_1 < i_2 < \cdots < i_{k+1}\} \subseteq \{1, 2, \ldots, n\}$ so that $A = \{i_1, i_2, \ldots, i_k\}$ and $B = \{i_2, i_3, \ldots, i_{k+1}\}$. We then define the (k, n)-*shift graph* $\mathbf{S}(k, n)$ as the graph whose vertex set consists of all k-element subsets of $\{1, 2, \ldots, n\}$ with a k-element set A adjacent to a k-element set B exactly when (A, B) is a (k, n)-shift pair. The shift graph $\mathbf{S}(1, n)$ is just a complete graph on n vertices. It is customary to call a $(2, n)$-shift graph just a *shift* graph; similarly, a $(3, n)$-shift graph is called a *double shift graph.*

One of the folklore results of graph theory is the following formula for the chromatic number of the shift graph (throughout this paper, we use the notation $\lg n$ to denote the base 2 logarithm of n).

Theorem 9.1 *The chromatic number of the shift graph* $\mathbf{S}(2,n)$ *is* $\lceil \lg n \rceil$. ∎

The proof of Theorem 7.1 establishes the following lower bound.

Proposition 9.2 *The dimension of the canonical interval order* $\mathbf{I}_n$ *is at least as large as the chromatic number of the double shift graph* $\mathbf{S}(3,n)$. ∎

In turn, the next result relates the determination of the dimension of the family of canonical interval orders to the classical enumeration problem known as Dedekind's problem: estimate the number of antichains in the poset $\mathbf{2^t}$, the cartesian product of t two-element chains. This poset is just the subset lattice, the family of all subsets of $[t]$ partially ordered by inclusion. The next four results are due to Füredi, Hajnal, Rödl and Trotter [44].

Theorem 9.3 *The chromatic number of the double shift graph* $\mathbf{S}(3,n)$ *is the least positive integer* t *for which there are at least* n *antichains in the subset lattice* $\mathbf{2^t}$.

Proof Suppose that there are n antichains in the subset lattice $\mathbf{2^t}$. We show that the chromatic number of $\mathbf{S}(3,n) \leq t$. Let Q be the partial order defined on the antichains of $\mathbf{2^t}$ by setting $\mathcal{A} \leq \mathcal{B}$ in Q if and only if for every $S \in \mathcal{A}$, there exists $B \in \mathcal{B}$ so that $A \subseteq B$. Then let L be any linear extension of Q, and suppose that $\mathcal{A}_1 < \mathcal{A}_2 < \cdots < \mathcal{A}_n$ in L. For each i, j with $1 \leq i < j \leq n$, let $B(i,j) \in \mathcal{A}_j$ be a set so that there is no set $A \in \mathcal{A}_i$ with $A \subseteq B(i,j)$. Then for each i, j, k with $1 \leq i < j < k \leq n$, choose an element $\alpha \in B(j,k) - B(i,j)$, and set $\phi(\{i,j,k\}) = \alpha$. Then ϕ is a coloring of $\mathbf{S}(3,n)$.

Conversely, if the chromatic number of $\mathbf{S}(3,n)$ is at most t and ϕ: $\binom{[n]}{3} \to [t]$ is a coloring, we define for each i, j with $1 \leq i < j \leq n$, the set $B(i,j) = \{\phi(\{i,j,k\}) : j < k \leq n\}$. Then for each $i \in [n]$, set $\mathcal{B}_i = \{B(i,j) : i < j \leq n\}$. Partial order each $\mathbf{B}_i$ by inclusion and let $\mathcal{A}_i$ be the maximal elements. Then each $\mathcal{A}_i$ is an antichain in $\mathbf{2^t}$ and $\mathcal{A}_{i_1} \neq \mathcal{A}_{i_2}$ when $i_1 \neq i_2$. ∎

Although no closed form solution to Dedekind's problem has been found, relatively tight estimates are known (see [77], e.g.). For our purposes, we may use the estimate which results as follows. There are $\binom{t}{\lceil t/2 \rceil}$ subsets of size $\lceil t/2 \rceil$. Any subset of these sets forms an antichain in $\mathbf{2^t}$.

Theorem 9.4 *The chromatic number of the double shift graph satisfies:*

$$\chi(\mathbf{S}(3,n)) = \lg\lg n + (1/2 + o(1)) \lg\lg\lg n. \quad ∎ \tag{6}$$

Hopefully, the reader has noticed the following subtlety to the dimension problem for canonical interval orders. The lower bound depends heavily on the use of repeated end points, a phenomenon which we can eliminate by modifying the representation. After the modification, it is conceivable that the dimension problem is much harder. However, this turns out not to be the case.

Theorem 9.5 *Let n and t be positive integers with $n \geq 2$. If*

$$n \leq 2^{\binom{t}{\lceil t/2 \rceil}}, \tag{7}$$

then the dimension of the canonical interval order $\mathbf{I}_n$ is at most $t+3$.

Proof We let M_1 and M_2 be the following two linear extensions of $\mathbf{I}_n$. Let $[i_1, j_1]$ and $[i_2, j_2]$ be distinct elements of $\mathbf{I}_n$. Set $[i_1, j_1] < [i_2, j_2]$ in M_1 if $i_1 < i_2$, or if both $i_1 = i_2$ and $j_1 > j_2$. Dually, set $[i_1, j_1] < [i_2, j_2]$ in M_2 if $j_1 < j_2$, or if both $j_1 = j_2$ and $i_1 > i_2$. It remains to find $t+1$ additional linear extensions $L_1, L_2, \ldots, L_{t+1}$ so that when $i - 1 < i_2 \leq j_1 < j_2$, there is at least one L_α so that $[i_1, j_1] > [i_2, j_2]$ in L_α.

Let $s = \binom{t}{\lceil t/2 \rceil}$ and let $S_1, S_2, \ldots, S_s$ be a listing of all the $\lceil t/2 \rceil$-element subsets of $[t]$. Note that $A = \{S_1, S_2, \ldots, S_s\}$ is an antichain in the subset lattice $\mathbf{2^t}$. Also let X denote the set of all 0-1 sequences of length s, i.e., the elements of X are functions from $[s]$ to $\{0, 1\}$. We let L be the lexicographic order on X. By this, we mean that if $f, g \in X$ and j is the least integer for which $f(j) \neq g(j)$, then $f < g$ in L if and only if $f(j) < g(j)$. This implies $f(j) = 0$ and $g(j) = 1$. Now let $L' = [f_1, f_2, \ldots, f_n]$ be the restriction of L to n distinct elements of X.

Now let $[i_1, j_1]$ and $[i_2, j_2]$ be elements of $\mathbf{I}_n$ with $i_1 < i_2 \leq j_1 < j_2$. Note that we allow the possibility that $i_2 = j_1$. Set $E = \{i_1, j_2, j_1, j_2\}$ so that E has either 3 or 4 elements. Choose the least integer k_1 so that $|\{f_i(k_1) : i \in E\}| > 1$. Note that $f_{i_1}(k_1) = 0$ and $f_{j_2}(k_1) = 1$. Furthermore, exactly one of the following statements holds:

1. For every $i \in E$ with $i \neq i_1$, $f_i(j_1) = 1$.

2. For every $i \in E$ with $i \neq j_2$, $f_i(j_1) = 0$.

3. $f_{i_2}(k_1) = 0$ and $f_{j_1}(k_1) = 1$.

When the third of these statements holds, we must have $|E| = 4$, but the first two may occur with either $|E| = 3$ or $|E| = 4$. Also, when the third statement holds, we will require that $[i_1, j_1] > [i_2, j_2]$ in L_{t+1}.

When the first statement holds, set $E' = E - \{i_1\}$, and when the second statement holds, set $E' = E - \{j_2\}$. In either case, let k_2 be the least element where $|\{f_i(k_1) : i \in E'\}| > 1$. If the first statement holds, choose $\alpha \in S_{k_1} - S_{k_2}$ and require $[i_1, j_1] > [i_2, j_2]$ in L_α. If the second statement holds, choose $\alpha \in S_{k_2} - S_{k_1}$ and require $[i_1, j_1] > [i_2, j_2]$ in L_α.

We leave it as an exercise that such linear extensions exist. ■

The preceding theorem shows that the lower bound provided in inequality (6) is also an upper bound. With a little more work, the same kind of estimate works for arbitrary interval orders (see [44] for details).

Theorem 9.6 *The maximum dimension $d(h)$ of an interval order of height h satisfies:*

$$d(h) = \lg\lg h + (1/2 + o(1)) \lg\lg\lg h. \quad \blacksquare \tag{8}$$

Before closing this section, we comment that the dimension problem for interval orders is closely related to the problem of determining the dimension of the poset consisting of all 1-element and 2-element subsets of $\{1, 2, \ldots, n\}$, partially ordered by inclusion. Spencer [107] was the first to establish the connection between this problem and the classic result of Erdős and Szekeres concerning monotonic subsequences of a sequence of integers. In recent years, there has been rapid progress in estimating the dimension of posets consisting of layers of the subset lattice. A summary of this work together with additional references is provided by Trotter in [121].

10 Interval orders and overlap graphs

A graph $\mathbf{G} = (V, E)$ is called an *overlap graph* when there exists a function I assigning to each vertex $x \in V$ a closed interval $I(x) = [a_x, b_x]$ of $\mathbb{R}$ so that for all $x, y \in V$, $\{x, y\} \in E$ if and only if $I(x) \cap I(y) \neq \emptyset$, $I(x) \not\subseteq I(y)$ and $I(y) \not\subseteq I(x)$, i.e., the intervals intersect, but neither is contained in the other. Again, we call the function I a *representation* of the overlap graph $\mathbf{G}$. If required, we may assume that a representation of an overlap graph is *distinguishing*.

In general, overlap graphs need not be perfect, e.g. a cycle on 5 vertices is an overlap graph. However, when all the intervals used in the representation intersect, then the graph is perfect, and it is easy to color such graphs.

Proposition 10.1 *Let I be a distinguishing representation for an overlap graph $\mathbf{G} = (V, E)$. If $I(x) \cap I(y) \neq \emptyset$, for all $x, y \in V$, then $\mathbf{G}$ is the comparability graph of a poset $\mathbf{P} = (V, P)$ with $\dim(\mathbf{P}) \leq 2$, so $\mathbf{G}$ is perfect. Furthermore, the First Fit algorithm will provide an optimal coloring of $\mathbf{G}$ if the vertices are colored in the order determined by the left end points.*

Proof Let L_1 and L_2 be the linear orders on V determined by left and right end points, respectively. Then let $P = L_1 \cap L_2$. Clearly, $\{x, y\}$ is an edge of $\mathbf{G}$ if and only if x and y are comparable in the the poset $\mathbf{P} = (V, P)$. From its definition, we know that $\dim(\mathbf{P}) \leq 2$. The First Fit algorithm applied to the vertices in the order of their left end points is the minimum antichain partition described in the proof of Theorem 4.1. ∎

When the intervals used in the representation do not share a common point, it is not immediately clear that there is any bound on the chromatic number of an overlap graph in terms of the maximum clique size. The first proof of this fact is due to Gyárfás [55] and the best bounds to date are due to Kostochka and Kratochvíl [78]. Recall that the notation $f(k) = \Omega(g(k))$ means that there exists a positive constant c and an integer k_0 so that $f(k) \geq cg(k)$, for all $k > k_0$.

Theorem 10.2 *Let* $m(k) = \max\{\chi(\mathbf{G}) : \mathbf{G}$ *is an overlap graph with maximum clique size* $k\}$. *Then*

1. $m(k) = \Omega(k \log k)$.

2. $m(k) = O(2^k)$. ∎

Quite recently, the concepts used in the proof of Theorem 10.2 have been applied by Kierstead and Trotter [75] to solve a long standing problem in dimension theory for interval orders. We outline this work, but we do not aim for the best possible constants.

Theorem 10.3 *For every interval order* $\mathbf{Q} = (Y, Q)$, *there exists an integer* t_0 *so that if* $\mathbf{P} = (X, P)$ *is any interval order with* $\dim(\mathbf{P}) > t_0$, *then* $\mathbf{P}$ *contains a subposet isomorphic to* $\mathbf{Q}$. ∎

Since every interval order $\mathbf{Q}$ is isomorphic to a subposet of the canonical interval order $\mathbf{I}_n$, provided n is sufficiently large, Theorem 10.3 is equivalent to showing that for every integer $n \geq 2$, there exists an integer t_n so that if $\mathbf{P}$ is an interval order with $\dim(\mathbf{P}) > t_n$, then $\mathbf{P}$ contains a subposet isomorphic to the canonical interval order $\mathbf{I}_n$.

Let $\mathbf{P} = (X, P)$ be an interval order. For an integer $m \geq 2$, we call a subposet $\mathbf{T}$ of $\mathbf{P}$ an *m–tower* $\mathbf{T}$ when

1. $\mathbf{T}$ contains an m–element chain $Z = \{z_1 < z_2 < \cdots < z_m\}$, and

2. For every pair i, j with $1 \leq i < j \leq m$, $\mathbf{T}$ contains an element $w(i, j)$ which is incomparable with $z_i, z_{i+1}, \ldots, z_j$ and comparable with all other elements of Z.

It is an easy exercise to show that if $\mathbf{P}$ contains a $3n$–tower, it contains a subposet isomorphic to $\mathbf{I}_n$. So, Theorem 10.3 is also equivalent to the following somewhat more technical result.

Theorem 10.4 *For every integer* $m \geq 2$, *there exists an integer* t_m *so that if* $\mathbf{P}$ *is an interval order with* $\dim(\mathbf{P}) \geq t_m$, *then* $\mathbf{P}$ *contains a subposet isomorphic to an* m*–tower.*

Proof We proceed by induction on m. An interval order which does not contain a 2–tower is a weak order and has dimension at most 2. So it suffices to take $t_2 = 3$. Now consider a value of $m \geq 3$ and assume that there exists an integer t_{m-1} so that any interval order whose dimension is at least t_{m-1}

contains an $(m-1)$–tower. Now let $\mathbf{P} = (X, P)$ be any interval order whose dimension t is at least $t_{m-1} + 9$. We show that $\mathbf{P}$ contains an m–tower.

The key idea in the remainder of the proof is the notion of distance in the overlap graph. Let I be a distinguishing representation of $\mathbf{P} = (X, P)$. We proceed to build a realizer of P, starting with the two linear extensions M_1 and M_2, the orderings determined by the left and right end points respectively in the representation I. The important thing to notice is that it only remains to reverse critical pairs of the form (x, y), where $a_x < a_y < b_x < b_y$. In particular, x and y are adjacent in the overlap graph.

Let $\mathbf{G}$ denote the overlap graph determined by I, and let $\mathbf{G}_1, \mathbf{G}_2, \ldots, \mathbf{G}_s$ denote the components of $\mathbf{G}$. Then let $\mathbf{X}_i = (X_i, P_i)$ be the subposet determined by the vertex set of $\mathbf{G}_i$. For each $i \in [s]$, define the *root* of $\mathbf{G}_i$ to be the unique vertex in $\mathbf{G}_i$ whose left end point is minimal. We denote the root of $\mathbf{G}_i$ by $\mathbf{r}_i$. For every vertex $x \in \mathbf{G}_i$, let $d(x, \mathbf{r}_i)$ denote the distance from x to $\mathbf{r}_i$ in $\mathbf{G}_i$. The following key lemma is due to Gyárfás [55]. We leave the proof as an exercise.

Lemma 10.5 *Let $i \in [s]$ and let $j \geq 0$. Then let x, y and z each be at distance j from the root $\mathbf{r}_i$ of a component $\mathbf{G}_i$ of $\mathbf{G}$. If $I(z) \subset I(x) \cap I(y)$, and $\mathbf{r}_i = u_0, u_1, \ldots, u_i = z$ is a shortest path from $\mathbf{r}_i$ to z in $\mathbf{H}$, then $I(u_{i-1}) \not\subseteq I(x) \cup I(y)$.* ∎

Next we classify all vertices of $\mathbf{G}$ as either *left* or *right*, and we denote the set of all left vertices by $\mathcal{L}$ and the set of all right vertices by $\mathcal{R}$. A vertex x belongs to $\mathcal{L}$ if and only if there exists a shortest path $\mathbf{r}_i = u_0, u_1, \ldots, u_i = x$ from the root of the component to which x belongs to x so that the left end point of $I(u_{i-1})$ is less than the left end point of $I(x)$. Then set $\mathcal{R} = X - \mathcal{L}$.

Similary, we classify all vertices as either *even* or *odd* and denote these two sets by $\mathcal{E}$ and $\mathcal{O}$, respectively. A vertex x belongs to $\mathcal{E}$ if and only if the distance from x to the root of the component to which it belongs is even. Then set $\mathcal{O} = X - \mathcal{E}$.

Then let M_3, M_4, M_5 and M_6 be linear extensions of P with

1. $\mathcal{L}$ over $\mathcal{R}$ in M_3;
2. $\mathcal{R}$ over $\mathcal{L}$ in M_4;
3. $\mathcal{E}$ over $\mathcal{O}$ in M_5; and
4. $\mathcal{O}$ over $\mathcal{E}$ in M_5.

It follows that we may assume that there is a pair i, j with $i \in [s]$ and $j \geq 2$ so that the subposet $\mathbf{Q} = (Y, Q)$ determined by all left vertices at distance j from the root $\mathbf{r}_i$ of component $\mathbf{G}_i$ has dimension at least $t_{m-1} + 3$.

Consider the following recursive definition. Set $Y_0 = Y$. If Y_k has already been defined for some $k \geq 0$ and the dimension of the subposet $\mathbf{Y}_k$ is less than $t_{m-1}+1$, set $Z_{k+1} = Y_k$ and halt. If on the other hand, the dimension of $\mathbf{Y}_k$ is at least $t_{m-1}+1$, let y_{k+1} be the unique element of Y_k with $\dim(W(y_{k+1}, Y_i)) \geq t_{m-1}+1$ whose left end point is as small as possible. Then set $B_{k+1} = \{y \in Y_{k+1} : b_{y_{k+1}} \in I(y)\}$, $Z_{k+1} = W(y_{k+1}, Y_k) - B_{k+1}$ and $Y_{i+1} = Y_i - W(y_{k+1}, Y_k)$. It follows that $\dim(\mathbf{Y}_{i+1}) = t_{m-1}+1$ and $\dim(\mathbf{Z}_{i+1}) = t_{m-1}$.

Suppose this recursive definition halts in a partition $Y = Z \cup B$, where $Z = Z_1 \cup Z_2 \cup Z_s$ and $B = B_1 \cup B_2 \cup B_{s-1}$. Then $\dim(\mathbf{Z}) = t_{m-1}$, so that $\dim(\mathbf{B}) \geq t_{m-1}+1$. Also, note that for each $i = 1, 2, \ldots, s-1$, the inductive hypothesis implies that $\mathbf{Z}_i$ contains an $(m-1)$–tower.

Since the dimension of $\mathbf{B}$ is large, it follows (being very generous) that there are integers k_1 and k_2 with $1 < k_1 < k_2 - 3$ and elements $b_{k_1} \in B_{k_1}, b_{k_2} \in B_{k_2}$ so that $b_{k_1} \| b_{k_2}$ in P. It follows that the interval for b_{k_1} properly contains intervals from two disjoint $(m-1)$–towers, one from Z_{k_1+1} and the other from Z_{k_1+2}. Choose a vertex x from Z_{k_1+1}, and consider the $(m-1)$–tower T from Z_{k_2+2}. For each vertex, $y \in T$, the interval corresponding to the vertex just before y on the shortest path from $\mathbf{r}_i$ to y properly overlaps the interval for y. By the lemma, this interval also has a left end point which precedes the left end point of b_{k_1}. Thus this interval also intersects x. It follows that P contains an m–tower. ∎

11 Semi-orders and balancing pairs

Let $\mathbf{P} = (X, P)$ be a poset and let $\mathcal{F} = \{M_1, \ldots, M_t\}$ be a multiset of linear extensions of P. Consider the linear extensions of $\mathcal{F}$ as outcomes in a uniform sample space. For a distinct pair $x, y \in X$, the *probability* that x is over y in $\mathcal{F}$, denoted $\mathrm{Prob}_{\mathcal{F}}[x > y]$, is defined by

$$\mathrm{Prob}_{\mathcal{F}}[x > y] = \frac{1}{t}|\{i : 1 \leq i \leq t,\ x > y \text{ in } M_i\}|. \tag{9}$$

In this section, we are concerned with the family $\Lambda(P)$ consisting of all linear extensions of P. We let $\lambda(P) = |\Lambda(P)|$. For this family, we drop the subscript and just write $\mathrm{Prob}[x > y]$. Note that $\mathrm{Prob}[x > y] = 0$, when $x < y$ in P; $\mathrm{Prob}[x > y] = 1$, when $x > y$ in P and $0 < \mathrm{Prob}[x > y] < 1$, when $x \| y$ in P. In 1969, S. S. Kislitsyn [76] made the following conjecture, which remains one of the most intriguing problems in the combinatorial theory of posets.

Conjecture 11.1 *If* $\mathbf{P} = (X, P)$ *is a finite poset which is not a chain, then there exists an incomparable pair* $x, y \in X$ *so that*

$$1/3 \leq \mathrm{Prob}(x > y) \leq 2/3. \quad ∎ \tag{10}$$

This conjecture was made independently by both M. Fredman and N. Linial, and many papers on this subject attribute the conjecture to them. It is now known as the 1/3–2/3 conjecture. If true, the conjecture would be best possible, as shown by $\mathbf{2}+\mathbf{1}$.

The first major breakthrough in this area came in 1984, when Kahn and Saks [62] used the Alexandrov/Fenchel inequalities for mixed volumes to prove the following result.

Theorem 11.2 *If* $\mathbf{P} = (X, P)$ *is a finite poset which is not a chain, then there exists an incomparable pair* $x, y \in X$ *so that*

$$\frac{3}{11} < \mathrm{Prob}[x > y] < \frac{8}{11}. \quad \blacksquare \tag{11}$$

Recently, there has been a slight improvement in this result using a special case of a conjecture called the Cross-product conjecture. The result is due to Brightwell, Felsner and Trotter [18].

Theorem 11.3 *If* $\mathbf{P} = (X, P)$ *is a finite poset which is not a chain, then there exists an incomparable pair* $x, y \in X$ *so that*

$$\frac{5-\sqrt{5}}{10} < \mathrm{Prob}[x > y] < \frac{5+\sqrt{5}}{10}. \quad \blacksquare \tag{12}$$

As pointed out in [18], there is an infinite semi-order for which the inequality in Theorem 11.3 is best possible, so that the 1/3–2/3 conjecture is *false* if one attempts to extend it to infinite posets. However, for finite semi-orders, we can do even better. For a poset $\mathbf{P} = (X, P)$, we say x *covers* y and write $x :> y$ in P when $x > y$ in P and if $x \geq z \geq y$ in P, then either $x = z$ or $y = z$. The next result is due to Brightwell [16].

Theorem 11.4 *If* $\mathbf{P} = (X, P)$ *is a finite semi-order which is not a chain, then there exists an incomparable pair* $x, y \in X$ *so that*

$$\frac{1}{3} \leq \mathrm{Prob}[x > y] \leq \frac{2}{3}. \tag{13}$$

Proof Suppose that the theorem is false. Choose a counterexample $\mathbf{P} = (X, P)$ with $|X| = n$ minimum. Then let I be a distinguishing representation. Label the points of X as $x_1, x_2, \ldots, x_n$ in the order determined by left end points. Define a linear order L on P by setting $x < y$ in L if and only if $\mathrm{Prob}[x > y] < 1/3$. Clearly, L is a linear extension of P. Furthermore L orders X as $x_1 < x_2 < \cdots < x_n$.

We claim that $x_i \| x_{i+1}$, for all $i = 1, 2, \ldots, n-1$. To the contrary, suppose $x_i < x_{i+1}$ in P. Then $\mathbf{P}$ is the lexicographic sum over a two-element chain of the subposets determined by $\{x_1, x_2, \ldots, x_i\}$ and $\{x_{i+1}, x_{i+2}, \ldots, x_n\}$. One of these posets is not a chain, and we immediately contradict our choice of $\mathbf{P} = (X, P)$ as a minimum counterexample.

We say that a point x_j *separates* x_i *and* x_{i+1} *from above* if $x_j :> x_i$ and $x_j \| x_{i+1}$ in P. Dually, we say x_j *separates* x_i *and* x_{i+1} *from below* if $x_{i+1} :> x_j$

and $x_i \| x_j$ in P. Finally, we say x_j *separates* x_i and x_{i+1} if it separates them from above or from below. Note that if x_j separates x_i and x_{i+1} from above, then $x_k < x_j$ in P, for all $k = 1, 2, \ldots, i$. Dually, if x_k separates x_i and x_{i+1} from below, then $x_j < x_k$ in P, for all $k = i+1, i+2, \ldots, n$. So each x_j separates at most two pairs, one from above and one from below. Furthermore, x_1 and x_2 do not separate pairs from below, while x_{n-1} and x_n do not separate pairs from above. It follows that there at most $2(n-4)+4 = 2n-4$ pairs (i,j) so that x_j separates x_i and x_{i+1}. From this we conclude that there is an integer i (in fact, there are at least two such values) for which there is at most one integer j so that x_j separates x_i and x_{i+1}. We show that $1/3 \le \text{Prob}[x_i > x_{i+1}] \le 2/3$.

Let $\Lambda(P)$ be the set of all linear extensions of P, and let $|\Lambda(P)| = t$. Set $\Lambda_1 = \{L \in \lambda : x_i < x_{i+1}$ in L, but there is no element of X which separates x_i and x_{i+1} between them in $L\}$; $\Lambda_2 = \{L \in \Lambda - \Lambda_1 : x_i < x_{i+1}$ in $L\}$; and $\Lambda_3 = \Lambda - (\Lambda_1 \cup \Lambda_2)$. Then $|\Lambda_3|/t = \text{Prob}[x_i > x_{i+1}] < 1/3$. Consider the map $h\colon \Lambda_1 \to \Lambda_3$ defined as follows. For a linear extension $L \in \Lambda_1$, form $h(L)$ by exchanging x_i and x_{i+1}. Clearly, the map h is an injection. It follows that $|\Lambda_1| \le |\Lambda_3|$. Furthermore, $|\Lambda_3|/t = \text{Prob}[x_i > x_{i+1}] < 1/3$, so $|\Lambda_2| > t/3$. In particular, there exists a unique element x_j which separates x_i and x_{i+1}. If x_j separates from above, $1/3 < \text{Prob}[x_{i+1} > x_j] < 2/3$. If x_j separates from below, then $1/3 < \text{Prob}[x_j > x_i] < 2/3$. ∎

There are some other special classes of posets for which the 1/3–2/3 conjecture is known to be true. For example, Fishburn, Gehrlein and Trotter [39] showed that it is valid for all posets of height 2.

In a poset $\mathbf{P} = (X, P)$, a sequence $(x_1, x_2, \ldots, x_n)$ of length $n \ge 3$ is called a *linear extension majority cycle*, or just an LEM cycle for short, when $\text{Prob}[x_i > x_{i+1}] > \frac{1}{2}$, for all $i \in [n]$. It is an easy exercise to show that semi-orders do not contain LEM cycles, but Brightwell, Fishburn and Winkler [19] show that LEM cycles can exist in interval orders—in fact, even in interval orders having dimension at most two.

12 Interval orders and extremal problems

Here are two interesting extremal problems involving semi-orders. The first problem is investigated by Fishburn and Trotter in [41]. For integers n and k with $0 \le k \le \binom{n}{2}$, let $Q(n,k)$ denote the family of all posets with n points and k comparable pairs. Then set $e(n,k) = \max\{|\Lambda(P)| : \mathbf{P} = (X,P) \in Q(n,k)\}$.

Theorem 12.1 *Every poset $\mathbf{P} = (X,P) \in Q(n,k)$ with $|\Lambda(P)| = e(n,k)$ is a semi-order.*

Proof Suppose that $\mathbf{P} = (X,P) \in Q(n,k)$, $|\Lambda(P)| = e(n,k)$, but that $\mathbf{P}$ is not a semi-order. Suppose further that $\mathbf{P}$ is not an interval order. Then $\mathbf{P}$ contains a subposet isomorphic to $\mathbf{2}+\mathbf{2}$. Label the 4 points in the copy of $\mathbf{2}+\mathbf{2}$ as $\{x, y, u, v\}$, so that $u \in D(x) - D(y)$ and $v \in D(y) - D(x)$. Of all copies of

$\mathbf{2}+\mathbf{2}$ in $\mathbf{P}$, we may assume that we have chosen one so that $|U(x)|+|U(y)|$ is minimum. It follows that one of $U(x)$ and $U(y)$ is a subset of the other. Without loss of generality, we assume that $U(x) \subseteq U(y)$. Let $\mathbf{P}'=(X,P')$ be the poset obtained from $\mathbf{P}$ by replacing the relations $z<y$ by $z<x$ for all $z \in D(y)-D(x)$. Then $\mathbf{P}' \in Q(n,k)$.

Interchanging the points x and y transforms a linear extension from $\Lambda(P)-\Lambda(P')$ into a linear extension from $\Lambda(P')-\Lambda(P)$. Furthermore, this map is an injection. It is not a surjection, because any linear extension with $y<u<v<x$ is not in the image of the map. The contradiction shows that $\mathbf{P}$ is an interval order.

Now assume that $\mathbf{P}$ contains a subposet isomorphic to $\mathbf{3}+\mathbf{1}$, and label the elements in the 3-element chain so that $x<y<z$ in P. Label the element incomparable to these three points as w. Now form a poset $\mathbf{P}''=(X,P'') \in Q(n,k)$ by replacing relations $t<y$ by $t<w$ for all $t \in D(y)-D(w)$. Then $\mathbf{P}' \in Q(n,k)$. As before, $\mathbf{P}''$ has more linear extensions than $\mathbf{P}$. ∎

The second problem sounds similar. It was posed to me by Peter Winkler. Define the *flexibility* of a poset $\mathbf{P}=(X,P)$, denoted flex($\mathbf{P}$), by

$$\text{flex}(\mathbf{P}) = \sum_{x\in X} |U(x)+D(x)|^2. \tag{14}$$

Then the same kind of argument used to prove Theorem 12.1 can be used to show that among all posets with n points and k comparable pairs, those with maximum flexibility are semi-orders. Despite our knowledge about the structure of the extremal posets, little progress has been made in solving either of these problems in full generality.

Now here is an interesting extremal problem for posets on which significant results have been obtained for interval orders. When L is a linear extension of $\mathbf{P}=(X,P)$, let $j(L,P)$ count the number of consecutive pairs of elements in L which are incomparable in P. The *jump number* of $\mathbf{P}$ is then the minimum value of $j(L,P)$ taken over all linear extensions of $\mathbf{P}$. In [89], Mitas shows that determining the jump number of an interval order is NP-complete. However, Mitas [89], Felsner [29] and Syslo [108] have (independently) given a polynomial algorithm for approximating the jump number within a ratio of $3/2$.

Bogart and Stellpflug [11] define the *representation length* of a semi-order as the least positive integer k for which it has a a representation using intervals of length k with integer endpoints. For each $k \geq 1$, they provide a forbidden subposet characterization of semi-orders with representation length k.

For interval orders, we have the following natural extremal problem posed by Peter Fishburn in [36]. Given an interval order $\mathbf{P}$, find the least positive integer k for which $\mathbf{P}$ has a representation using intervals having k distinct lengths. This parameter is called the *interval count* of $\mathbf{P}$. Two interesting questions are immediate. First, what is the maximum value of the interval count of an interval order on n points? Second, can the removal of a single point drop the interval count by an arbitrarily large amount?

13 Interval orders and hamiltonian paths

Considered as a graph, the diagram of a poset of height h cannot have chromatic number exceeding h. However, the "partite" construction of Nešetřil and Rődl [90] shows that for every integer h, there exists a poset $\mathbf{P}$ of height h so that the chromatic number of the diagram of $\mathbf{P}$ is exactly h.

For interval orders, the situation is completely different, and the chromatic number of the diagram of an interval order of height h is much less than h. The open intervals with integer end points in $\{1, 2, \dots, h+1\}$ form an interval order of height h. Furthermore, the diagram of this interval order is just the shift graph $S(2, h+1)$, a graph whose chromatic number is exactly $\lceil \lg(h+1) \rceil$. Surprisingly, this is not far from best possible.

Let t be a positive integer, and let $\mathcal{S} = (S_0, S_1, \dots, S_h)$ be a sequence of sets. Felsner and Trotter [34] called $\mathcal{F}$ an *α-sequence* if $S_1 \not\subseteq S_0$ and $S_j - (S_i \cup S_{i-1}) \neq \emptyset$, for all i, j with $1 \leq i < j \leq h$. Define $\alpha(t)$ to be the maximum h for which there exists an α–sequence $(S_0, S_1, \dots, S_h)$, with each S_i a subset of $[t]$. For example, $\alpha(3) = 5$ as evidenced by the α-sequence $\mathcal{A} = (\emptyset, \{1\}, \{2\}, \{3\}, \{1,3\}, \{1,2,3\})$. Note that any subsequence of consecutive terms from an α–sequence is again an α–sequence.

Let $D(h)$ denote the maximum chromatic number of an interval order of height h. Clearly, $D(1) = 1$ and $D(2) = 2$.

Theorem 13.1 *For each $h \geq 2$, $D(h)$ is the least t for which $\alpha(t) \geq h$.*

Proof We first show that if $\alpha(t) > h$, then $D(h) \leq t$. Let $\mathcal{S} = (S_0, S_1, \dots, S_h)$ be an α–sequence of subsets of $[t]$, and let $\mathbf{P}$ be an interval order of height h. Then let I be a distinguishing representation of $\mathbf{P}$. Let be the lexicographically least maximum chain be $C = \{y_1 < y_2 < \cdots < y_h\}$, and let the canonical partition into antichains be $X = A_1 \cup A_2 \cup \cdots \cup A_h$. For each $i \in [h]$, let $r_i = b_{y_i}$. Then let r_0 be any real number with $r_0 < a_x$ in $\mathbb{R}$, for every $x \in X$.

For each $x \in X$, set $i(x) = \max\{i : 0 \leq i \leq h,\ r_i < a_x\}$ and $j_x = \max\{j : 1 \leq j \leq h,\ r_j \in I(x)\}$. Note that $i_x < j_x$, for every $x \in X$. We then define a coloring $\phi\colon X \to [t]$ as follows. If $i_x = 0$, choose $\phi(x) \in S_{j_x} - S_0$. If $i_x > 1$, choose $\phi(x) \in S_{j_x} - (S_{i_x} \cup S_{i_x - 1})$.

We claim that ϕ is a proper coloring of the diagram of $\mathbf{P}$. Suppose that $x <: y$ in $\mathbf{P}$. Then either $i_y = j_x$ or $i_y = 1 + j_x$. In either case, note that $\phi(x) \neq \phi(y)$.

We now sketch the proof that if $D(h) \leq t$, then $\alpha(t) \geq h$. For integers $h, m \geq 2$, let $\mathbf{P}(h, m)$ denote the interval order determined by the family of all closed intervals with length at least $m - 1$ having integer end points from $\{1, 2, \dots, m(h+1) - 1\}$. Note that the height of $\mathbf{P}(h, m)$ is h. We now show that for each $h \geq 2$, there exists an integer m_0 so that if $m > m_0$ and the chromatic number of the diagram of $\mathbf{P}(h, m)$ is t, then $\alpha(t) \geq h$. In fact, we show that the choice $m_0 = 2^{h^2}$ works.

Fix $h \geq 2$ and then let m be any integer with $m > m_0$. Suppose that the chromatic number of the diagram of $\mathbf{P}(h,m)$ is t. Note $t \leq h$. Now suppose that ϕ is a coloring of the diagram of $\mathbf{P}(m,h)$ using colors from $[t]$. For each $j = 1, 2, \ldots, m(h+1) - 1$, let $A_j = \{\phi([i,j]) : 1 \leq i \leq j - m + 1\}$. Then for each $i = 0, 1, \ldots, m-1$, let $V_i = (A_{m+i}, A_{2m+i}, \ldots, A_{hm+i})$. Each V_i is a vector of length h with each entry a subset of $[t]$. Since there are at most 2^{h^2} such vectors, it follows that there exist integers i_1, i_2 with $0 \leq i_1 < i_2 \leq m-1$ for which $V_{i_1} = V_{i_2}$.

Set $S_0 = \emptyset$ and $S_k = A_{mk+i_1}$, for $k = 1, 3, \ldots, h$. We claim that the sequence $(S_0, S_1, \ldots, S_h)$ is an α–sequence. Clearly, $S_1 \neq \emptyset$ as S_1 contains the color ϕ assigns to the interval $[1+i_1, m+i_1]$. Thus $S_1 \not\subseteq S_0$. Now suppose that $1 < i < j \leq h+1$ and $S_j \subseteq S_{i-1} \cup S_i$. Suppose that ϕ assigns color $\beta \in [t]$ to the interval $x = [1+mi+i_1, mj+i_2]$. Then there is an interval $y \in S_{i-1} \cup S_i$ with $\phi(y) = \beta$. Since $S_i = A_{mi+i_1} = A_{mi+i_2}$, and $S_{i-1} = A_{m(i-1)+i_1} = A_{m(i-1)+i_2}$, there is an interval y with $b_y \in \{mi+i_1, m(i-1)+i_2\}$ so that $\phi(y) = \phi(x) = \beta$. This is a contradiction, since $x :> y$. ∎

Felsner and Trotter [34] conjecture that

$$\alpha(t) = 2^{t-1} + \left\lfloor \frac{t-1}{2} \right\rfloor . \tag{15}$$

If this conjecture is true, then an α–sequence $\mathcal{S}$ of subsets of $[t]$ of maximum size has the following property. If we form a new sequence $\mathcal{H}$ from $\mathcal{S}$ by inserting between two consecutive sets in $\mathcal{S}$ their union, when the first set is not a subset of the second, then we get a listing of all 2^t subsets of $[t]$. For example, from the 6 term α–sequence of subsets of [3] given above, this listing is $(\emptyset, \{1\}, \{1,2\}, \{2\}, \{2,3\}, \{3\}, \{1,3\}, \{1,2,3\})$. This listing is a special kind of hamiltonian path in the t–cube. Whenever a set appears in the list, all of its subsets, with at most a single exception, appear previously. If there is an exception, it is listed next.

We call such a path an *order-preserving hamiltonian path* in the t–cube. This is a slight abuse of the concept of order-preserving, but it is the strongest notion that makes sense. It is known that there are order-preserving hamiltonian paths in the t–cube for $1 \leq t \leq 8$, but the general question is open.

We should point out that attempts to settle whether equation (15) is always valid have produced the best known partial result on the well known "middle two levels" problem. The origins of the problem are a bit unclear, but it was first told to me by Ivan Havel during a visit to Prague.

Problem 13.2 *Is the diagram of the poset consisting of all k–element and $(k+1)$–element subsets of a $(2k+1)$–element set, partially ordered by inclusion, a hamiltonian graph?* ∎

We refer the reader to [34] and [99] for details.

14 On-line and un-cooperative coloring

An on-line optimization problem, such as on-line graph coloring, can be considered as a two-person game involving a *Builder* and a *Colorer*. The game is played in a series of rounds with the players alternating turns. Each instance of on-line coloring involves two parameters: an integer t and a graph $\mathbf{G}$. If $\mathbf{G}$ has n vertices, the game lasts at most n rounds. In Round i, where $1 \leq i \leq n$, Builder presents the vertex v_i of $\mathbf{G}$ and describes all edges joining v_i with vertices in $\{v_j : 1 \leq j < i\}$. This information is complete and correct. In particular, if the game lasts all n rounds, then Builder must have correctly specified the entire graph.

After receiving the information for the new vertex v_i, Colorer must then assign to v_i a color from the set $\{1, 2, \ldots, t\}$ so that this color is distinct from those previously assigned to neighbors of v_i. These assignments are permanent.

The $(t, \mathbf{G})$ game ends at Round i and Builder is the winner if Colorer has no legitimate choice of a color for the new vertex v_i. If on the other hand, Colorer is able to respond with a legitimate color for each of the n vertices of $\mathbf{G}$, then Colorer is the winner. The *on-line chromatic number* of a graph $\mathbf{G}$ is then the least t for which Colorer has a winning strategy for the $(t, \mathbf{G})$ game—regardless of the strategy employed by Builder.

In [71], Kierstead and Trotter prove the following foundational result.

Theorem 14.1 *The on-line chromatic number of an interval graph of maximum clique size k is at most $3k - 2$.*

Proof Here's the winning strategy for Colorer. Given a new vertex x by Builder, Colorer assigns x to a set S_i where i is the least positive integer for which there is no complete subgraph of size $i + 1$ containing x and i other vertices previously assigned to $S_1 \cup S_2 \cup \cdots \cup S_i$. Note that S_1 is just an independent set, so it can be colored with a single color. For each $i \geq 2$, we show that First Fit will color S_i with the 3 colors from the set $\{3i - 4, 3i - 3, 3i - 2\}$. We accomplish this by showing that for $i \geq 2$, S_i is the disjoint sum of paths.

Fix $i \geq 2$. When a vertex u is presented by Builder and assigned by Colorer to S_i, as opposed to S_{i-1}, there is a clique K_u consisting of u and $i-1$ vertices from $S_1 \cup S_2 \cup \cdots \cup S_{i-1}$. Then the intersection of the intervals from K_u is a nonempty interval I_u, which is contained in the interval corresponding to u. Now let u and v be adjacent vertices from S_i. Then it is easy to see that I_u does not intersect the interval corresponding to v and I_v does not intersect the interval corresponding to u. From this, it follows easily that S_i is triangle free and that each vertex from S_i has at most 2 neighbors in S_i. ∎

The algorithm presented in the preceding theorem has one feature in common with the First Fit algorithm discussed in Theorem 4.2: it is not necessary to know the maximum clique size in advance. If First Fit is used to color an interval graph, and the vertices are not processed in the order of left end points, then it is not clear how many colors will be used. In [65], Kierstead showed that First Fit will use at most $40k$ colors on an interval graph with maximum clique size k, regardless of the order in which the vertices are processed. Subsequently, Kierstead and Qin [70] improved this upper bound to $26k$. From below, Chrobak and Slusarek [23] showed that no on-line algorithm can color all interval graphs with maximum clique size k with fewer than $4.4k$ colors.

Kierstead's analysis of the performance of First Fit in coloring interval graphs provided a solution to an important long standing problem in computer science called the *Dynamic Storage Allocation* problem. The standard two dimensional bin packing problem is to pack a family of rectangles in $\mathbb{R}^2$, with sides parallel to the coordinate axes, into a region of minimum area. The Dynamic Storage Allocation problem is to pack the rectangles into a region of minimum height—when the projections of the rectangles onto the horizontal coordinate axis form a fixed interval graph. Of course, by "pack," we mean that the rectangles are to be placed so that their interiors are disjoint. So if the maximum sum of the heights of rectangles whose projections have a common point is t, then t is a lower bound on the height required for a packing, and it was conjectured that a height of $O(t)$ would suffice.

One proposed approach to finding a reasonably good packing was to assume all rectangles had height a power of 2. This assumption would at most double the optimal height required for a packing. These rectangles would then be partitioned into subrectangles of height one. Finally, First Fit would be used to color the rectangles (intervals) with all intervals formed from the same rectangle colored consecutively. The number of colors used by First Fit would then be an upper bound on the minimum height required for a packing. Accordingly, Kierstead and Qin's bound implies that the rectangles can in fact be packed into a region of height $52t$.

We refer the reader to [66] for a full discussion. As an added bonus, this paper provides an alternative approach to the dynamic storage allocation problem which stands as the best solution to date. This approach uses the same partition of rectangles, but colors them with a modified version of the on-line algorithm used in Theorem 14.1 rather than with First Fit. The end result is to show that the rectangles can be packed into a height of $6t - 4$.

Ironically, the research which led to the proof of Theorem 14.1 was motivated, not by the Dynamic Storage Allocation problem, but by the on-line version of Dilworth's theorem. In [64], Kierstead proved that there is an on-line algorithm which will partition a poset built one point at a time into $(5^w - 1)/4$ chains, where w denotes the width of the poset. When the poset is known to be an interval order, then the preceding theorem asserts that $3w - 2$ chains suffice.

Kierstead's on-line chain partitioning algorithm requires knowledge of the order. Just knowing whether points are comparable is not enough. However, for interval graphs, our algorithm only makes use of the comparability graph. For many years, it remained an open problem to determine whether a comparability graph of independence number k can be partitioned on-line into a bounded number of complete subgraphs. An affirmative answer was provided by Kierstead, Penrice and Trotter in [69].

In [68], Kierstead, McNulty and Trotter investigate *on-line dimension.* Here the game is between a Realizer and a Builder, with Realizer building a family $\mathcal{R}$ of linear extensions of a poset $\mathbf{P}$ which Builder is constructing one point at a time. They show that the on-line dimension of a class of width 4 posets is infinite. However, the posets in this class all contain the 3-dimensional crown $\mathbf{S}_3^0$. They then proceed to show that the on-line dimension of posets of bounded width is well defined, provided that the posets are *crown-free*, i.e., do not contain any 3-dimensional crown $\mathbf{S}_3^k$.

Theorem 14.2 *The on-line dimension of a crown-free poset of width k is at most $t!$, where $t = (5^{k+1} - 1)/4$.* ∎

On the surface, this result has nothing to do with interval orders, but the proof makes use of an auxiliary order at a critical point in the argument. This structure turns out to be an interval order, and the order-theoretic properties this structure gains from being an interval order are key elements of the proof.

Other sources of information about on-line coloring include [72] and the more recent survey by Kierstead [67]. In particular, this last paper contains a concise treatment of the recent breakthrough where Kierstead succeeded in showing that for all $k \geq 3$, there exists an $\epsilon > 0$ so that the on-line chromatic number of any k-colorable graph on n vertices is at most $n^{1-\epsilon}$. Kierstead's argument shows that $\epsilon = O(1/k!)$. Probably, this can be improved to $O(1/k)$. Another good source of problems (some of which are on-line) concerning interval graphs and other classes of perfect graphs in Gyárfás' survey paper [54].

In [27], Faigle, Kern, Kierstead and Trotter consider the following game theoretic problem for graphs. Two players, Alice and Bob, color a graph $\mathbf{G}$ using elements of the set $[t]$ as colors. They alternate turns with Alice having the first move. Alice wins if the graph is eventually colored and Bob (an uncooperative partner) wins if at some step before the graph is colored, there is no legitimate move. The *game chromatic number* of $\mathbf{G}$ is the least t for which Alice has a winning strategy. For example, it is shown in [27] that the game chromatic number of a tree is at most 4; furthermore, this result is best possible. In [74], Kierstead and Trotter show that a planar graph has game chromatic number at most 33; they also show that there exists a planar graph with game chromatic number at least 8.

For interval graphs, the following result, given in [27], provides the best known bound on the game chromatic number of an interval graph.

Theorem 14.3 *The game chromatic number of an interval graph $\mathbf{G} = (V, E)$ with maximum clique size k is at most $3k - 2$.*

Proof Let I be a distinguishing representation of an interval graph $\mathbf{G}$ with maximum clique size k. When it is her turn to color, Alice prefers to color a vertex x adjacent to the vertex just colored by Bob. Among such, she prefers those whose intervals intersect the interval corresponding to the vertex just colored by Bob. Finally among such vertices, Alice prefers the one whose interval has right end point as large as possible. She then colors this vertex by First Fit.

We claim that Alice and Bob can never reach an impasse if the number of colors is $3k-2$. It suffices to show that the strategy given for Alice can be used by either player. Let x be the vertex to be colored. It suffices to show that x has at most $3k-3$ colored neighbors. Split the colored neighbors into three sets $N_1 \cup N_2 \cup N_3$, where

1. N_1 is the set of colored neigbors of x whose intervals contain the right end point of $I(x)$;
2. N_2 is the set of colored neighbors of x whose intervals are properly contained in $I(x)$; and
3. N_3 is the set of colored neighbors of x whose intervals contain the left end point of $I(x)$ but not the right.

Clearly, $|N_1| \leq k-1$ and $|N_3| \leq k-1$, so our claim follows if we can show that $|N_2| \leq k-1$. Now our strategy for Alice insures that she will not have colored any of the vertices in N_2, since she will always prefer to color x. So all vertices in N_2 are colored by Bob, and at every turn—except possibly the last one—Alice has selected a vertex other than x to color. Such a vertex must have an interval containing the interval corresponding to the vertex in N_2 just colored by Bob, and its right end point is greater than the right end point of x. Therefore Alice's response was to color a vertex from N_1. It follows that $|N_2| \leq k$. Now suppose that $|N_2| = k$. Among the vertices in N_2, let y be the unique vertex whose right end point is as large as possible. Then y, x and the $k-1$ vertices in N_2 form a clique of size $k+1$. ∎

The reader should note that it is just a coincidence that the expression $3k-2$ appears in both the preceding two theorems. In the first case, we know that it is best possible, but in the second, we believe it is not. We leave it as an exercise to show that, for each $k \geq 2$, there exists an interval graph **G** whose maximum clique size is k and whose game chromatic number is at least $2k$.

15 Fractional dimension and ramsey theory for probability spaces

It is often useful to consider a fractional version of an integer valued combinatorial parameter as, in many cases, the resulting LP relaxation sheds light on the original problem. In [20], Brightwell and Scheinerman proposed to investigate fractional dimension for posets.

Let $\mathbf{P} = (X, P)$ be a poset, and let $\mathcal{F} = \{M_1, \ldots, M_t\}$ be a multiset of linear extensions of P. Brightwell and Scheinerman [20] call $\mathcal{F}$ a *k–fold realizer* of P if, for each incomparable pair (x, y), there are at least k linear extensions in $\mathcal{F}$ which reverse the pair (x, y), i.e., $|\{i : 1 \leq i \leq t,\ x > y \text{ in } M_i\}| \geq k$. The *fractional dimension* of $\mathbf{P}$, denoted by $\text{fdim}(\mathbf{P})$, is then defined as the least real number $q \geq 1$ for which there exists a k–fold realizer $\mathcal{F} = \{M_1, \ldots, M_t\}$ of P so that $k/t \geq 1/q$ (it is easily verified that the least upper bound of such real numbers q is indeed attained and is therefore a rational number). Using this terminology, the *dimension* of $\mathbf{P}$ is just the least t for which there exists a 1–fold realizer of P. It follows immediately that $\text{fdim}(\mathbf{P}) \leq \dim(\mathbf{P})$, for every poset $\mathbf{P}$.

The dimension or fractional dimension of a class of posets is defined to be the least upper bound of $\dim(\mathbf{P})$ (respectively $\text{fdim}(\mathbf{P})$) over all posets $\mathbf{P}$ in the class. We have seen that $\dim(\mathcal{I}) = \infty$ for the class $\mathcal{I}$ of interval orders, but Brightwell and Scheinerman showed that $\text{fdim}(\mathcal{I}) \leq 4$. To see this, observe that if $\mathbf{P} = (X, P)$ is an interval order and $A \subset X$, there is a linear extension L of P with $x > y$ in L for any incomparable pair (x, y) with $x \in A$ and $y \notin A$. Building a realizer from one such L for each subset A of X of size $\lfloor |X|/2 \rfloor$ gives $\text{fdim}(\mathbf{P}) < 4$.

Brightwell and Scheinerman conjectured in [20] that $\text{fdim}(\mathcal{I}) = 4$, even though no example of an interval order of fractional dimension even as high as 3 was then known. In the remainder of this section, we sketch the approach taken by Trotter and Winkler in [126] to settle this conjecture in the affirmative.

First, the following preliminary result is required. Intuitively, this theorem asserts that in a sufficiently long sequence of events, one cannot do substantially better than toss a fair coin in trying to balance between events being true and events being false. See [126] for the proof.

Theorem 15.1 *For every $\epsilon > 0$, there exists an integer m_0 so that if $m \geq m_0$ and $\{U_i : 1 \leq i \leq m\}$ is any sequence of events in a probability space, then there exist integers i and j with $1 \leq i < j \leq m$ so that* $\text{Prob}[U_i\overline{U_j}] < \frac{1}{4} + \epsilon$. ∎

With this result in hand, we can now sketch the proof of the solution. The argument makes extensive use of ramsey theory to make certain statements about sets hold in a uniform manner. To be precise, these statements involve small errors, and the argument takes some care to show that the errors can be kept under control. In this sketch, we ignore these errors.

Theorem 15.2 *For every $\epsilon > 0$, there exists an integer n_0 so that if $n > n_0$, the fractional dimension of the canonical interval order $\mathbf{I}_n$ is at least $4 - \epsilon$.*

Proof Let $\epsilon > 0$, and suppose that $\text{fdim}(\mathbf{I}_n) < 4 - \epsilon$, regardless of the size of n. We argue to a contradiction, provided n is sufficiently large.

Let $S = \{s_1, s_2, \ldots, s_{2m}\}$ be a $2m$–element subset of $[n]$, with $s_1 < s_2 < \cdots < s_{2m}$. Then let $U(S)$ denote the event that for some i with $1 < i \leq m$,

$[s_1, s_{m+1}] > [s_i, s_{m+i}]$. Using ramsey theory, it is relatively easy to see that for fixed m, if n is sufficiently large, we may assume that the probability of $U(S)$ is constant, for all $2m$–element subsets of $[n]$. But Trotter and Winkler show more. They show that one may also assume that the event $U(S)$ depends only on s_1 and s_{m+1}. We denote this event by $U(x, y)$, where $x = s_1$ and $y = s_{m+1}$.

Dually, let $D(S)$ denote the event that for some j with $1 \leq j < m$, $[s_j, s_{m+j}] > [s_m, s_{2m}]$. This time, the event $D(S)$ depends only on s_m and s_{2m}. So we can just write $D(x, y)$, where $x = s_m$ and $y = s_{2m}$.

It follows that one can find a large homogeneous subset H so that $U(x, y) \cap D(x, y) = \emptyset$, for every $x, y \in H$ with $x < y$ in $\mathbb{R}$. Furthermore, if $x < y < z < w$ in H, then $U(x, z) \cap U(y, w) = \emptyset$. If the homogenous set H has more than $2m_0$ elements, the result follows from Theorem 15.1. ∎

The dimension problem for interval orders is closely related to the graph coloring problem for shift graphs, a subject of independent interest. Similarly, the research on the fractional dimension of interval orders has led to many new and interesting concepts. We give hints to one of these in the sketch of the proof of Theorem 15.2, namely the development of a general Ramsey theory for probability spaces. However, there are several concrete combinatorial problems which are also quite attractive.

Fix integers n and k with $1 \leq k < n$. Suppose we have a probability space containing an event E_S for every k–element subset S of $[n]$. We abuse notation and just refer to this event as S. Now consider the minimum probability $\text{Prob}(A\overline{B})$ taken over all (k, n)–shift pairs. In turn, take the maximum value of this probability over all probability spaces and let n go to infinity. The resulting value is called $f(k)$. For example, from Theorem 15.1, it follows that $f(1) = \frac{1}{4}$. In [126], Trotter and Winkler prove that $f(2) = \frac{1}{3}$, $f(3) \geq \frac{3}{8}$ and $f(4) \geq \frac{2}{5}$. In general, they prove that $f(k)$ is strictly increasing and converges to $\frac{1}{2}$.

The relaxation of dimension to fractional dimension is an appealing concept. In [33], Felsner and Trotter show that the fractional dimension of a poset in which each point is comparable with at most k others is at most $k + 1$. They also prove several other inequalities linking fractional dimension with width and cardinality. Nevertheless, there are many challenging open questions in this area.

16 Higher-dimensional analogues for graphs

In recent years, there has been a steady stream of results providing higher dimension analogues of interval graphs. Perhaps the first of these is due to Roberts [96] who defined the *boxicity* of a graph $\mathbf{G} = (V, E)$ as the least t for which there exists a function B assigning to each vertex $x \in V$ a sequence $(I_x(1), I_x(2), \ldots, I_x(t))$ of closed intervals of $\mathbb{R}$ so that $\{x, y\} \in E$ if and only if $I_x(i) \cap I_y(i) \neq \emptyset$, for all $i \in [t]$. Equivalently, the boxicity of a graph is just the least t so that the graph is the intersection of boxes in $\mathbb{R}^t$. So interval graphs are graphs with boxicity one. Roberts showed that the boxicity of a graph on n vertices is at most $\lfloor n/2 \rfloor$, when $n \geq 2$. For example the graph $\mathbf{H}_n$, obtained by taking the complement of a matching on n edges has $2n$ vertices and boxicity n. In [127], Wittenshausen showed that for all $n \geq 1$, $\mathbf{H}_n$ is the only graph with $2n$ vertices and boxicity n.

However, when a graph has $2n+1$ vertices and boxicity n, the situation is modestly more complicated. For example, the cycle $\mathbf{C}_5$ on 5 vertices has boxicity 2. Also, the graph $\mathbf{W}_3$ with vertex set $\{1, 2, \ldots, 7\}$ and edges joining i to $i+2$, $i+3$, $i+4$ and $i+5$ (cyclically) has boxicity 3. In [113], Trotter showed that a graph $\mathbf{G}$ on $2n+1$ vertices has boxicity n if and only if one of the following conditions holds:

1. $\mathbf{G}$ contains $\mathbf{H}_n$.

2. $\mathbf{G}_n$ contains the join of $\mathbf{C}_5$ and $\mathbf{H}_{n-2}$.

3. $\mathbf{G}_n$ contains the join of $\mathbf{W}_3$ and $\mathbf{H}_{n-3}$.

In [109], Thomassen showed that the boxicity of a planar graph is at most 3; in fact, the boxes corresponding to adjacent vertices may be required to intersect on a face.

Many of the basic concepts for interval graphs have natural interpretation for digraphs. In [3], Beineke and Zamfirescu introduced the notion (with different terminology) of an *interval digraph*. By this we mean that for each vertex x in a digraph $\mathbf{D}$, there are two intervals of the real line R_x and S_x, so that $\mathbf{D}$ contains a directed arc from x to y if and only if $R_x \cap S_y \neq \emptyset$. Structural questions for interval digraphs are studied in [84], [85], [105] and [106].

Define the *interval number* of a graph $\mathbf{G} = (V, E)$ as the least t for which $\mathbf{G}$ is the intersection graph of a family of sets, with each set being the union of t pairwise disjoint closed intervals of $\mathbb{R}$. In [53], Griggs and West show that if the maximum degree of $\mathbf{G}$ is d, then the interval number of $\mathbf{G}$ is at most $\lceil (d+1)/n \rceil$. This inequality is tight if $\mathbf{G}$ is triangle-free. Griggs and West also showed that there exists an absolute constant $c > 0$ so that the interval number of a graph with q edges is at most $c\sqrt{q}$. In [103], Scheinerman and West showed that the interval number of a planar graph is at most 3, and Scheinerman [100] showed that there exists an absolute constant $c' > 0$ so that the interval number of a graph of genus γ is at most $c'\sqrt{\gamma}$.

In [122], Trotter and Harary show that the interval number of a complete bipartite graph $\mathbf{K}(m, n)$ is $\lceil (mn+1)/(m+n) \rceil$. The fact that the interval number of $\mathbf{K}(m, n)$ is at least this large follows from the following elementary observation. The interval number of a triangle-free graph $\mathbf{G}$ with n vertices and q edges is at least $\lceil (q+1)/n \rceil$. This inequality follows from the fact that if the interval number of the graph is t, then the nt intervals used in a representation form the intersection graph of a forest on nt vertices and at least q edges. This requires $q \leq nt - 1$.

Somewhat surprisingly, the determination of the interval numbers for complete multipartite graphs proved to be more challenging. The interval number of the complete multipartite graph $\mathbf{K}(n_1, n_2, \ldots, n_s)$, with $n_1 \geq n_2 \geq \cdots \geq n_t \geq 2$, is at least as large as the interval number of $\mathbf{K}(n_1, n_2)$. Call this quantity t_0. In [59], Hopkins, Trotter and West show the interval number of $\mathbf{K}(n_1, n_2, \ldots, n_s)$ is at most $t_0 + 1$. Furthermore, they show that it is equal to t_0, except possibly for the two cases $(n_1, n_2) = (7, 5)$ and $n_1 = n_2^2 - n_2 - 1$. In both these exceptional cases, the interval number of $\mathbf{K}(n_1, n_2, \ldots, n_t)$ may equal $t_0 + 1$, provided there are enough other parts of appropriate size.

Motivated by the formula for complete bipartite graphs, Trotter and Harary [122] conjectured that the maximum interval number of a graph on n vertices is $\lceil (n+1)/4 \rceil$. This conjecture was proved by Griggs in [52].

In [22], Chang and West introduced the concept of *interval number* for digraphs. For a digraph $\mathbf{D}$, the interval number of $\mathbf{D}$ is just the least positive integer i for which there exists a function F assigning to each vertex x two subsets R_x, S_x of the real numbers so that

1. For each node x in $\mathbf{D}$, R_x and S_x are each the union of at most t pairwise disjoint intervals of $\mathbb{R}$, and

2. $\mathbf{D}$ contains an arc from x to y if and only if $R_x \cap S_y \neq \emptyset$.

Chang and West showed that the maximum interval number of a digraph on n nodes is $\Theta(n/\log n)$. They also defined the concept of *boxicity* for digraphs and showed that the maximum boxicity of a digraph on n nodes is $\lceil n/2 \rceil$.

Aigner and Andreae [1] introduced an interesting variation of interval number. For an graph $\mathbf{G} = (V, E)$, they defined the *total interval number* of $\mathbf{G}$ as the least positive integer t for which there exists a function F assigning to each vertex x of $\mathbf{G}$ a set $F(x)$ which is the union of t_x pairwise disjoint closed intervals of $\mathbb{R}$ so that:

1. For every $x, y \in V$, $\{x, y\} \in E$ if and only if $F(x) \cap F(y) \neq \emptyset$, and

2. $\sum_{x \in X} t_x = t$.

Aigner and Andreae [1] produced upper bounds on total interval number for several classes of graphs. For example, they showed that the maximum total interval number of a tree on n nodes is $\lfloor (5n-3)/4 \rfloor$. In [80], Kratzke and West showed that the maximum total interval number of an outerplanar graph on n nodes is $\lfloor 3n/2 - 1 \rfloor$ while the maximum total interval number of a general graph on n nodes is $\lceil (n^2+1)/4 \rceil$. These results settled conjectures made by Aigner and Andreae in [1]. Other results on total interval number are given by Kostochka and West in [79]; in particular, they bound the total interval number in terms of the maximum degree, and characterize graphs for which the bound is sharp. The components of these graphs are balanced complete bipartite graphs.

In [81], Kratzke and West provide a linear time algorithm for computing the total interval number of a tree, and they show that it is NP-complete to test whether the total interval number of a graph is exactly one more than the number of edges, even for the class of triangle-free, 3–regular planar graphs.

Given a poset $\mathbf{P} = (X, P)$ and points $x, y \in X$, with $x \leq y$ in P, the *interval* $[x, y]$ is just the set $\{u \in X : x \leq u \leq y \text{ in } P\}$. The *poset boxicity* of a graph $\mathbf{G}$ is the least t for which there exists a t-dimensional poset $\mathbf{P}$ for which $\mathbf{G}$ is the intersection graph of intervals in $\mathbf{P}$. In [125], Trotter and West show that there exists an absolute constant $c > 0$ so that the poset boxicity of a graph on n vertices is at most $c \log \log n$. They also show that there exist graphs with arbitrarily large poset boxicity.

In [56], Gyárfás and West consider the *multitrack interval number* of a graph as the least t for which the graph is the union of t interval orders. We will discuss analogous concepts for posets in Sections 17 and 19.

17 Higher dimensional analogues for orders

The investigation of higher dimensional analogues of interval orders has also produced a steady stream of results. First, let $\mathcal{P}$ be any hereditary class of orders which contains the linear orders. Then we can define the $\mathcal{P}$–dimension of a poset $\mathbf{P} = (X, P)$ as the least t for which P is the intersection of t orders from $\mathcal{P}$. The hereditary property serves to ensure that the $\mathcal{P}$–dimension of $\mathbf{P}$ is at most the $\mathcal{P}$–dimension of $\mathbf{Q}$ when $\mathbf{P}$ is contained in $\mathbf{Q}$. Of course, the $\mathcal{P}$–dimension of $\mathbf{P}$ is at most $\dim(\mathbf{P})$, and to emphasize the distinction between the original definition of dimension and variants discussed in the remainder of this paper, the dimension is also called the *ordinary* dimension.

In [14], Bogart and Trotter defined the *interval dimension* of a poset $\mathbf{P} = (X, P)$ as the least t for which P is the intersection of t interval orders on X. So a poset has interval dimension 1 if and only if it is an interval order. Posets with interval dimension at most 2 have also been studied extensively. In [114], Trotter gave a forbidden subposet characterization of height two posets having interval dimension at most 2. This characterization results in a complete listing of all minimal posets of height 2 having interval dimension 3. Polynomial time recognition algorithms for posets having interval dimension at most 2 have been provided by several authors, but the best to date is due to Ma and Spinrad [86].

One of the most appealing aspects of interval dimension is the positive solution of the *removable pair* conjecture. For ordinary dimension, Trotter conjectured (see [118], for example) that if $\mathbf{P}$ is a poset having three or more points, then there is always a pair of points whose removal decreases the dimension by at most 1. In fact, he conjectured that the removal of a critical pair always decreases the dimension by at most 1. Although the removable pair conjecture remains open, this second conjecture was disproved by Reuter [95], and an infinite family of counterexamples was then constructed by Kierstead and Trotter [73].

However, for interval dimension, we have the following elementary result.

Theorem 17.1 *Let* $\mathbf{P} = (X, P)$ *be a poset and let* $(x, y) \in \text{crit}(\mathbf{P})$. *If* $\mathbf{Q} = (Y, Q)$ *is the subposet determined by* $Y = X - \{x, y\}$, *then the interval dimension of* $\mathbf{P}$ *is at most one more than the interval dimension of* $\mathbf{Q}$.

Proof Let $Q_1, Q_2, \ldots, Q_t$ be interval orders on Y whose intersection is Q. For each $i \in [t]$, let P_i be an interval order on X so that $P_i(Y) = Q_i$. Then let L be any linear extension of Y with $D(x) < Y - D(x)$ and $Y - U(y) < U(y)$ in L . Define a partial order P_{t+1} on X by setting $P_{t+1} = P \cup L$. It is easy to see that P_{t+1} is an interval order and that $P = P_1 \cap P_2 \cap \cdots \cap P_{t+1}$. ■

Another appealing aspect of the concept of interval dimension is that there is a relatively simple characterization of posets having maximal dimension for a given number of points (see Bogart and Trotter [13]), while the corresponding problem for ordinary dimension is considerably more difficult. Several other inequalities relating interval dimension to other combinatorial parameters are simpler than the corresponding results for ordinary dimension, e.g., compare the forbidden subposet characterization of the inequality $\dim(P, X) \leq \max\{2, |X - A|\}$, when A is an antichain, for ordinary dimension [111] with the result for interval dimension in [13].

Other aspects of the interplay between dimension and interval dimension are discussed in [30]. In [57], Habib, Kelly and Möhring show that the property of a poset having interval dimension at most 2 is a comparability invariant, i.e., it depends only on the underlying comparability graph and not on the specific order.

Bogart and Trotter also defined the semi-order dimension of a poset and noted that if the semi-order dimension of $\mathbf{P}$ is t, then the ordinary dimension of $\mathbf{P}$ is at most $3t$. This result is tight when $t = 1$, but it is not known whether it is best possible when $t \geq 2$. In [31], Felsner and Möhring show that the property of a poset having semi-order dimension at most 2 is a comparability invariant.

In a somewhat different direction, more closely connected to the concepts discussed in the preceding section, Madej and West [87] define the *interval inclusion number* of a poset $\mathbf{P} = (X, P)$ as the least integer t for which there exist a function F assigning to each $x \in X$ a set $F(x) \subset \mathbb{R}$ so that:

1. For each $x \in X$, $F(x)$ is the union of at most t pairwise disjoint closed intervals of $\mathbb{R}$, and

2. For each $x, y \in X$, $x \leq y$ in P if and only if $F(x) \subseteq F(y)$.

In [88], Madej and West show that "almost all" posets on n points have interval number $o(n)$, but it is still not known whether there exists a positive real number c so that for all n, there exists a poset on n points with interval inclusion number exceeding cn. It is easy to see that the interval inclusion number of an n-dimensional poset is at most $\lceil n/2 \rceil$. Furthermore, as Madej and West note in [88], the set of all subsets of an n–element set, ordered by inclusion, shows that this last inequality is tight. However, as they point out, the n-dimensional standard example $\mathbf{S}_n$ has interval inclusion number 2, for all $n \geq 2$.

18 Intervals, angles and spheres

Over the past 10 years, there has been a flurry of work on geometric problems which arise when posets are represented by a family of sets (usually some geometrically defined objects) ordered by inclusion. These structures are called *inclusion orders*, and they are the natural order theoretic analogue of intersection graphs. For example, as is well known, a poset has dimension at most 2 if and only if it isomorphic to the inclusion order determined by a family of intervals of the real line. Space limitations do not allow us to discuss the full range of research on inclusion orders, but we will attempt to highlight those which related directly to interval orders.

Fishburn and Trotter [40] define a poset $\mathbf{P} = (X, P)$ to be an *angle order* when $\mathbf{P}$ is the inclusion order of a family of subsets of the euclidean plane, with each set being an angular region determined by two rays emanating from a common point. They show that every interval order is an angle order and that every poset with dimension at most 4 is an angle order. They also showed that there exists a 7-dimensional poset which is not an angle order. Subsequently, several authors showed that there exists a 5-dimensional poset which is not an angle order, but the most elegant proof of this fact results from the theory of "degrees of freedom" developed by Alon and Scheinerman in [2].

A *d–sphere* with center $\mathbf{x}$ and radius r is the set of all points in $\mathbb{R}^d$ whose distance to $\mathbf{x}$ is at most r. A 1–sphere is just a closed interval of $\mathbb{R}$. Call a poset $\mathbf{P}$ a *sphere order* if there is some d so that $\mathbf{P}$ is isormorphic to the inclusion order determined by a family of d–spheres in $\mathbb{R}^d$. When $\mathbf{P}$ is a sphere order, we may define the *sphere dimension* of a poset $\mathbf{P} = (X, P)$ as the least d for which $\mathbf{P}$ is the inclusion order of a family of d–spheres. So a poset has sphere dimension 1 if and only if it has ordinary dimension at most 2.

The problem of determining whether every finite poset is a sphere order is posed by Brightwell and Winkler in [21], and it is widely believed that the answer is negative.

When $d = 2$, there are some interesting results and one especially vexing problem. For historical reasons, posets with sphere dimension at most 2 are called *circle orders*, although it might have been more accurate to call them *disk orders*. The recent article [101] contains a number of interesting perspectives on the problem of representing order by circles in the plane. The range and extent of the connections with other combinatorial problems is most surprising.

In [37], Fishburn shows that every interval order is a circle order. Trivially, every poset with ordinary dimension at most 2 is a circle order—in fact, we can require that the circles all have centers on a fixed line in the plane. By the Alon/Scheinerman theory, there exist 4-dimensional posets which are not circle orders. However, it is not known whether every finite 3-dimensional poset is a circle order. Scheinerman and Weirman [102] showed that the countably infinite 3-dimensional poset $\mathbb{N}^3$ is not a circle order. Subsequently, a somewhat shorter proof of this result was given by Hurlbert [60]. The sharpest result to date is due to Fon-der-Flaass [43] who showed that $\mathbf{2} \times \mathbf{3} \times \mathbb{N}$ is not a sphere order, but that $\mathbf{2} \times \mathbf{2} \times \mathbb{N}$ is a circle order.

On the other hand, it is an easy exercise to show that if $\mathbf{P}$ is a finite poset with ordinary dimension at most 3 and $n \geq 3$, then $\mathbf{P}$ is the inclusion order of a family of regular n–gons in the euclidean plane, and it is easy to suspect that when n is quite large relative to $|X|$, these polygons are extremely close to being circles. However, I would conjecture that there is a finite 3-dimensional poset which is not a sphere order.

In this discussion, the metric used to determine distance plays a critical role. Of course, if $\mathbf{x} = (x_1, x_2, \ldots, x_d)$ and $\mathbf{y} = (y_1, y_2, \ldots, y_d)$, then the ordinary distance from $\mathbf{x}$ to $\mathbf{y}$ is $\sqrt{\sum_{i=1}^{d}(x_i - y_i)^2}$. But if we change this to $\max\{|x_i - y_i| : 1 \leq i \leq d\}$, then a d–sphere is just a *cube.* Furthermore, it is an easy exercise to show that every poset with dimension at most $d+1$ is the inclusion order of a family of cubes in $\mathbb{R}^d$. Again, by the Alon/Scheinerman theory, this is best possible, meaning that there are $(d+2)$-dimensional posets which cannot be represented by cubes in $\mathbb{R}^d$ ordered by inclusion.

19 Tolerances, thresholds and gaps

In the preceding two sections, we discussed higher dimensional analogues for interval graphs and interval orders. In this section, we discuss generalizations which arise when just one interval is assigned but more complex rules are used to determine edges and comparabilities. Here is the basic motivation. If we have an indexed family $\mathcal{F} = \{I(x) : x \in X\}$ of closed intervals with distinct end points, then an interval graph results when we define an edge set E by $\{x, y\} \in X$ if and only if $|I(x) \cap I(y)| > 0$. From an applications standpoint, the problem with this definition is that we take two vertices to be adjacent when their intervals intersect *regardless* of how small this intersection might be. Similarly, an interval order assigns x to be less than y only when $F(x)$ lies entirely to the left on $F(y)$. But there are many scheduling problems where we want to consider one job as preceding another even when there is some overlap in time.

We begin with generalizations of interval graphs. Golumbic and Monma [49] proposed the following definition. Given an indexed family $\mathcal{F} = \{I(x) : x \in X\}$ of closed intervals of $\mathbb{R}$ and a subset $T = \{t_x : x \in X\}$ of the non-negative real numbers $\mathbb{R}_0$, define the *tolerance graph* $\mathbf{G} = \mathbf{G}(\mathcal{F}, T) = (X, E)$ by setting

$E = \{\{x, y\} : x, y \in X,\ x \neq y \text{ and } |I(x) \cap I(y)| \geq \min\{t_x, t_y\}\}$.

It is easy to see that an interval graph is a tolerance graph. Just take a distinguishing representation and give each vertex a tolerance smaller than the distance between any two end points used in the representation. Also, the complement of an interval graph is a tolerance graph. In this case, for each $x \in X$, set $t_x = |I(x)|$. A tolerance graph is *bounded* if $0 \leq t_x \leq |I(x)|$, for all $x \in X$. Golumbic and Monma [49] showed that a bounded tolerance graph is the complement of a comparability graph and is therefore perfect. This argument does not work for tolerance graphs which are not bounded, but in [50], Golumbic, Monma and Trotter showed that all tolerance graphs are perfect. The proof in the general case follows by showing that the complement of a tolerance graph is perfectly orderable. Note that Chvatál's concept of a graph being perfectly orderable [24] is a weakening of the key property used to show that interval graphs are perfect.

Here is an interesting way in which tolerance graphs differ from interval graphs. Recall that an interval graph is proper if and only if it has a representation using only intervals of length 1. This is not true for tolerance graphs. In [6], Bogart, Fishburn, Isaak and Langley show that the class of proper tolerance graphs is strictly larger than the class of tolerance graphs in which all intervals have unit length.

In the last several years, a number of new concepts for generalizing tolerance graphs have been introduced. Perhaps the most general is due to Jacobson, McMorris and Mulder [61] who proposed to study graphs defined by a family of intervals $\{I_x : x \in X\}$, a subset $T = \{t_x : x \in X\}$ of tolerances drawn from the set $\mathbb{R}_0$ of non-negative reals, and a function $\phi\colon \mathbb{R}_0 \times \mathbb{R}_0 \to \mathbb{R}_0$ by setting the edge set E to consist of all 2-element sets $\{x, y\}$ for which $|I(x) \cap I(y)| > \phi(t_x, t_y)$. The original definition of a tolerance graph is just the function $\phi(t_x, t_y) = \min\{t_x, t_y\}$.

Now here are some of the new ideas for posets. McMorris and Jacobsen (see [8]) propose to study a generalization of interval orders in which extra conditions are imposed on the gaps between intervals corresponding to comparable pairs of points. The definition requires an indexed family $\{I(x) = [a_x, b_x] : x \in X\}$ of closed intervals of $\mathbb{R}$, a subset $T = \{t_x : x \in X\}$ of the non-negative reals $\mathbb{R}_0$, and a function $\phi\colon \mathbb{R}_0 \times \mathbb{R}_0 \to \mathbb{R}_0$. We then define a relation P on X by setting $(x, y) \in P$ if and only if (1) $x = y$ or (2) $b_y - a_x > \phi(t_x, t_y)$. We call these posets *$\phi$–gap* orders to reflect that the relation P is defined in terms of the gap between the two intervals.

In certain cases, P will be a partial order on X. For example, this is always the case if ϕ satisfies the triangle inequality: $\phi(t_x, t_y) + \phi(t_y, t_z) \geq \phi(t_x, t_z)$, for all $t_x, t_y, t_z \in \mathbb{R}_0$. In particular, P is always a partial order if $\phi(t_x, t_y) = \max\{t_x, t_y\}$. On the other hand, we may fail to get a partial order if $\phi(t_x, t_y) = \min\{t_x, t_y\}$. The special case where $\phi(t_x, t_y) = \max\{t_x, t_y\}$ is called a *max-gap* order.

In another direction, Bogart and Trenk [12] call a poset $\mathbf{P} = (X, P)$ a *bi-tolerance order* when there exists a triple (I, F, G) where:

1. I assigns to each $x \in X$ a closed interval $I(x) = [a_x, b_x]$ of $\mathbb{R}$;
2. $F = \{f_x : x \in X\} \subset \mathbb{R}$, $G = \{g_x : x \in X\} \subset \mathbb{R}$;
3. $a_x \le f_x, g_x \le b_x$ in $\mathbb{R}$, for every $x \in X$;
4. $x < y$ in P if and only if $b_x < f_y$ and $g_x < a_y$ in $\mathbb{R}$.

For a vertex x, the value $f_x - a_x$ is called the *left tolerance* of x, and the value $b_x - g_x$ is called the *right tolerance* of x. When $f_x - a_x = b_x - g_x$, for all $x \in X$, we call the poset a *tolerance order*. It is an easy exercise to show that the tolerance orders are just the posets which arise from ordering the complement of a bounded tolerance graph. For this reason, bi-tolerance orders were originally called *bounded* bi-tolerance orders, but with time the adjective "bounded" seems to have been discarded.

Every interval order is a bi-tolerance order with $f_x = a_x$ and $g_x = b_x$, for every vertex x, but it is easy to see that there are bi-tolerance orders which are not interval orders. It is an easy exercise to show that if $\mathbf{P} = (X, P)$ is a max-gap order, as evidenced by the intervals $\{[a_x, b_x] : x \in x\}$ and the subset $T = \{t_x : x \in X\} \subset \mathbb{R}_0$, then $\mathbf{P}$ is also a bi-tolerance order, as evidenced by the intervals $\{[a_x - t_x, b_x + t_x] : x \in X\}$ and the families $F = \{a_x : x \in X\}$ and $G = \{b_x : x \in X\}$.

In the definition of a bi-tolerance order, no restriction is placed on the order of f_x and g_x. However, if desired, we can always assume that $g_x \le f_x$. This follows from the observation that if M is a positive number, then we may modify the representation by setting:

1. $a'_x = f_x - M$, $g'_x = g_x - M$, $b'_x = b_x + M$ and $f'_x = f_x + M$.

When M is sufficiently large, we always have $g'_x \le f'_x$. We say the representation (I, F, G) of a bi-tolerance order is *separated* if $g_x < f_y$, for every $x, y \in X$. Note that if M is very large, then the new representation is separated. In fact, if desired, one can assume that $\{a_x : x \in X\} = \{1, 2, \ldots, n\}$ and $\{b_x : x \in X\} = \{n+1, n+2, \ldots, 2n\}$, where $n = |X|$. This last observation makes use of the fact that it is only the order on the various values that matters. On the other hand, the question as to which bi-tolerance orders have totally bounded representations is not as well understood, at least not for posets of arbitrary height.

The notion of separation makes it clear that for every $x \in X$, we have two intervals $I_1(x) = [a_x, g_x]$ and $I_2(x) = [f_x, b_x]$. Furthermore, the intervals $\{I_1(x) : x \in X\}$ form an interval order $\mathbf{P}_1$, and the intervals in $\{I_2(x) : x \in X\}$ form another interval order $\mathbf{P}_2$. Since x, y in P if and only if $I_1(x) < I_1(y)$ and $I_2(x) < I_2(y)$, it follows that the bi-tolerance orders are just the posets with interval dimension at most two.

Another interesting translation of the concept of bi-tolerance orders involves the geometric insight gained when the intervals $I_1(x)$ and $I_2(x)$ are selected from two parallel lines in the plane, an interpretation first proposed by Dagan, Golumbic and Pinter [25]. The convex hull in $\mathbb{R}^2$ determined by the two intervals is a trapezoid, and the posets determined by a family of trapezoids is called a *trapezoid order*. So the trapezoid orders are again just the posets with interval dimension at most two. These insights are critical to the argument given by Habib, Kelly and Möhring [57] to show that interval dimension 2 is a comparability invariant. Also, in [9], Bogart, Möhring, and Ryan show that the family of unit area trapezoid orders is strictly contained in the family of proper trapezoid orders. In [32], Felsner, Müller and Wernisch provide fast algorithms for finding optimal partitions of trapezoid orders into chains and antichains. Their results even allow for weights to be assigned to the elements of the poset, and they are able to extend them to posets of large (but fixed) interval dimension.

Returning to Bogart and Trenk's definition of a bi-tolerance order, we should comment that their formulation has prompted the study of a number of interesting families of posets which can be described in terms of restrictions on the triple (I, F, G). As before, we may use the adjectives *proper* and *unit* to describe bi-tolerance orders (also, tolerance orders) in which the intervals are incomparable under inclusion and have unit length, respectively. Following Fishburn [38], we say a bi-tolerance order is *split* when $f_x = g_x$, for all $x \in X$, and we say a split bi-tolerance order is a *50% tolerance order* when $f_x = g_x = (a_x + b_x)/2$, for all $x \in X$. Finally, we say the bi-tolerance order is *totally bounded* if $f_x \leq g_x$, for all $x \in X$.

The main theorem of [12] asserts that almost all these definitions coincide on posets of height two. Here are some of the equivalencies proved in [12].

Theorem 19.1 *Let* **P** *be a poset of height* 2. *Then the following statements are equivalent.*

1. **P** *is a proper bi-tolerance order.*
2. **P** *is a unit bi-tolerance order.*
3. **P** *is a tolerance order.*
4. **P** *is a unit tolerance order.*
5. **P** *is a 50%–tolerance order.*
6. **P** *is a totally bounded bi-tolerance order.* ∎

For posets of arbitrary height, the various classes begin to separate, and equivalencies become more surprising. Recall that there is a distinction between unit and proper interval graphs. This distinction also holds between unit and proper tolerance orders. However, for bi-tolerance orders, we have the following result of Bogart and Isaak [7].

Theorem 19.2 *Let* **P** *be a poset. Then the following statements are equivalent.*

1. **P** *is a unit bi-tolerance order.*

2. **P** *is a proper bi-tolerance order.*

Proof A distinguishing representation of a unit bi-tolerance order shows that it is also a proper bi-tolerance order. Now let (I, F, G) be a distinguishing representation which evidences that a poset $\mathbf{P} = (X, P)$ is a proper bi-tolerance order. Without loss of generality, we may assume this representation is separated; we may also assume that $\{a_x : x \in X\} = \{1, 2, \ldots, n\}$ and $\{b_x : x \in X\} = \{n+1, n+2, \ldots, 2n\}$, where $|X| = n$. This is now a representation in which each interval has length n. ∎

The next equivalency, due to Langley [82], is somewhat more surprising.

Theorem 19.3 *Let* **P** *be a poset. Then the following statements are equivalent.*

1. **P** *is a unit bi-tolerance order.*

2. **P** *is a split interval order.*

Proof Let (I, F, G) be a distinguishing unit representation of a poset $\mathbf{P} = (X, P)$. Modify the representation as follows.

$$a'_x = f_x;\ f'_x = g'_x = b_x \text{ and } b'_x = g_x + 1. \tag{16}$$

It follows easily that

$$b_x < f_y \text{ and } g_x < a_y \text{ if and only if } b'_x < f'_y \text{ and } g'_x < a'_y. \tag{17}$$

This transformation is easily seen to be reversible. ∎

Note that an interval order is also a split interval order. To see this, just consider a distinguishing representation of an interval order and set $f_x = g_x = a_x$. On the other hand, there are split interval orders which are not interval orders, e.g. $\mathbf{2}+\mathbf{2}$.

Bogart, Fishburn, Isaak and Langley prove the following equivalency in [6], which is now an immediate corollary to Theorem 19.3.

Corollary 19.4 *Let* **P** *be a poset. Then the following statements are equivalent.*

1. **P** *is a unit tolerance order.*

2. **P** *is a 50% tolerance order.* ∎

Although the family of tolerance orders includes the interval orders and therefore contains posets of arbitrarily large dimension, this is not the case for many of the families involving restrictions on lengths and tolerances. For example, here is a very recent result of Fishburn and Trotter [42].

Theorem 19.5 *If* $\mathbf{P}$ *is a split semi-order, then* $\dim(\mathbf{P}) \leq 6$. ∎

Bi-tolerance orders must have large height in order to have large dimension, and it would be interesting to determine (or estimate) the maximum value of a bi-tolerance order of height h.

Acknowledgements

In the preparation of this article, the author has benefited from many conversations and communications with valued colleagues, especially Ken Bogart, Peter Fishburn and Hal Kierstead, and he would like to express his deep appreciation for their assistance.

This research is supported in part by the Office of Naval Research.

References

[1] M. Aigner & T. Andreae, The total interval number of a graph, *Journal of Combinatorial Theory, Series A*, **46** (1989), 7–21.

[2] N. Alon & E. R. Scheinerman, Degrees of freedom versus dimension for containment orders, *Order*, **5** (1988), 11–16.

[3] L. Beineke & C. M. Zamfirescu, Connection digraphs and second order line graphs, *Discrete Mathematics*, **39** (1982), 237–254.

[4] K. P. Bogart, A discrete proof of the Scott-Suppes representation theorem for semiorders, Technical Report PMA-TR91-173, Dartmouth College, 1991.

[5] K. P. Bogart, An obvious proof of Fishburn's interval order theorem, *Discrete Mathematics*, **118** (1993), 239–242.

[6] K. P. Bogart, P. C. Fishburn, G. Isaak & L. Langley, Proper and unit tolerance graphs, *Discrete Applied Mathematics*, (1995), 99–117.

[7] K. P. Bogart & G. Isaak, Proper and unit bitolerance orders and graphs, Technical Report PMA-TR96-187, Dartmouth College, 1996.

[8] K. P. Bogart, M. Jacobson, L. Langley & F. R. McMorris, Tolerance orders and bipartite unit tolerance graphs, Technical Report PMA-TR93-107, Dartmouth College, 1993.

[9] K. P. Bogart, R. Möhring & S. P. Ryan, Proper and unit trapezoid orders and graphs, in press.

[10] K. P. Bogart, I. Rabinovitch & W. T. Trotter, A bound on the dimension of interval orders, *Journal of Combinatorial Theory, Series A*, **21** (1976), 219–238.

[11] K. P. Bogart & K. Stellpflug, Discrete representation theory for semiorders, Technical Report PMA-TR93-104, Dartmouth College, 1993.

[12] K. P. Bogart & A. N. Trenk, Bipartite tolerance orders, *Discrete Mathematics*, (1994), 11–22.

[13] K. P. Bogart & W. T. Trotter, Maximal dimensional partially ordered sets III: A characterization of Hiraguchi's inequality for interval dimension, *Discrete Mathematics*, **15** (1976), 389–400.

[14] K. P. Bogart & W. T. Trotter, On the complexity of posets, *Discrete Mathematics*, **16** (1976), 71–82.

[15] K. S. Booth & G. S. Lueker, Testing for consecutive ones property, interval graphs, and graph planarity using PQ–tree algorithms, *Journal of Computer and System Sciences*, **13** (1976), 335–379.

[16] G. R. Brightwell, Semiorders and the 1/3–2/3 conjecture, *Order*, **5** (1989), 369–380.

[17] G. R. Brightwell, Partial orders, in *Graph Connections* (eds. L. W. Beineke & R. J. Wilson), Oxford University Press, Oxford (1997), pp. 52–69.

[18] G. R. Brightwell, S. Felsner & W. T. Trotter, Balancing pairs and the cross product conjecture, *Order*, **12** (1995), 327–349.

[19] G. R. Brightwell, P. C. Fishburn & P. W. Winkler, Interval orders and linear extension cycles, *Ars Combinatoria*, **36** (1993), 283–288.

[20] G. R. Brightwell & E. R. Scheinerman, Fractional dimension of partial orders, *Order*, **9** (1992), 139–158.

[21] G. R. Brightwell & P. W. Winkler, Sphere orders, *Order*, **6** (1989), 235–240.

[22] Y.-W. Chang & D. B. West, Interval number and boxicity of digraphs, in *Proceedings of the 8th International Graph Theory Conference*, in press.

[23] M. Chrobak and M. Slusarek, On some packing problems relating to dynamical storage allocation, in *RAIRO Informatique Theoretique*, in press.

[24] V. Chvatál, Perfectly orderable graphs, in *Topics on Perfect Graphs* (eds. C. Berge & V. Chvatál), North-Holland, Amsterdam (1984), 63–65.

[25] I. Dagan, M. C. Golumbic & R. Y. Pinter, Trapezoid graphs and their coloring, *Discrete Applied Mathematics*, **21** (1988), 35–46.

[26] R. P. Dilworth, A decomposition theorem for partially ordered sets, *Annals of Mathematics, Series 2*, **51** (1950), 161–166.

[27] U. Faigle, W. Kern, H. A. Kierstead & W. T. Trotter, The game chromatic number of some classes of graphs, *Ars Combinatoria*, **35** (1993), 143–150.

[28] S. Felsner, Tolerance graphs, interval orders: Combinatorial structure and algorithms, Ph. D. Thesis, Technische Universität Berlin, 1992.

[29] S. Felsner, A 3/2–approximation algorithm for the jump number of interval orders, *Order*, **6** (1990), 325–334.

[30] S. Felsner, M. Habib & R. Möhring, On the interplay between interval dimension and dimension, *SIAM Journal on Discrete Mathematics*, **7** (1994), 11–22.

[31] S. Felsner & R. Möhring, Note: Semi-order dimension two is a comparability invariant, in press.

[32] S. Felsner, R. Müller & L. Wernisch, Trapezoid graphs and generalizations, geometry and algorithms, Technical Report B 94-02, Freie Universität Berlin, 1994.

[33] S. Felsner & W. T. Trotter, On the fractional dimension of partially ordered sets, *Discrete Mathematics*, **136** (1994), 101–117.

[34] S. Felsner & W. T. Trotter, Colorings of diagrams of interval orders and α–sequences of sets, *Discrete Mathematics*, **144** (1995), 23–31.

[35] P. C. Fishburn, Intransitive indifference with unequal indifference intervals, *Journal of Mathematical Psychology*, **7** (1970), 144–149.

[36] P. C. Fishburn, *Interval Orders and Interval Graphs*, John Wiley, New York (1985).

[37] P. C. Fishburn, Interval orders and circle orders, *Order*, **5** (1988), 225–234.

[38] P. C. Fishburn, Generalizations of semiorders: A review note, in press.

[39] P. C. Fishburn, W. Gehrlein & W. T. Trotter, Balance theorems for height-2 posets, *Order*, **9** (1992), 43–53.

[40] P. C. Fishburn & W. T. Trotter, Angle orders, *Order*, **1** (1985), 333–343.

[41] P. C. Fishburn & W. T. Trotter, Linear extensions of semiorders: A maximization problem, *Discrete Mathematics*, **103** (1992), 25–40.

[42] P. C. Fishburn & W. T. Trotter, The dimension of split semi-orders, in press.

[43] D. G. Fon-der-Flaass, A note on sphere containment orders, *Order*, **10** (1993), 143–146.

[44] Z. Füredi, P. Hajnal, V. Rödl & W. T. Trotter, Interval orders and shift graphs, in *Sets, Graphs and Numbers* (eds. A. Hajnal & V. T. Sós), *Colloquia Mathematica Societatis János Bolyai*, 60, (1991), 297–313.

[45] T. Gallai, Transitiv orientierbare Graphen, *Acta Mathematica Academiae Scientiarum Hungaricae*, **18** (1967), 25–66.

[46] A. Ghouila-Houri, Caractérisation des graphes non orientés dont on peut orienter les arêtes de manière à obtenir le graphe d'une relation d'order, in *Comptes Rendus de l'Académie des Sciences de Paris* **254** (1962), 1370–1371.

[47] P. C. Gilmore & A. J. Hoffman, A characterization of comparability graphs and of interval graphs, *Canadian Journal of Mathematics*, **16** (1964), 539–548.

[48] M. C. Golumbic, *Algorithmic Graph Theory and Perfect Graphs*, Academic Press, New York (1980).

[49] M. C. Golumbic & C. Monma, A generalization of interval graphs with tolerances, *Congressus Numerantium*, **35** (1982), 321–331.

[50] M. C. Golumbic, C. Monma and W. T. Trotter, Tolerance graphs, *Discrete Applied Mathematics*, **9** (1984), 157–170.

[51] T. L. Greenough, The representation and enumeration of interval orders, Ph. D. Thesis, Dartmouth College, 1974.

[52] J. Griggs, Extremal values of the interval number of a graph II, *Discrete Mathematics*, **28** (1979), 37–47.

[53] J. R. Griggs & D. B. West, Extremal values of the interval number of a graph, *SIAM Journal on Algebraic and Discrete Methods*, **1** (1980), 1–7.

[54] A. Gyárfás, Problems from the world surrounding perfect graphs, *Zastowania Matematyki Applicationes Mathematicae*, **29** (1985), 413–441.

[55] A. Gyárfás, On the chromatic number of multiple interval graphs and overlap graphs, *Discrete Mathematics*, **55** (1985), 161–166.

[56] A. Gyárfás & D. B. West, Multitrack interval number, *Congressus Numerantium*, **109** (1995), 109–116.

[57] M. Habib, D. Kelly & R. Möhring, Interval dimension two is a comparability invariant, *Discrete Mathematics*, **88** (1991), 211–229.

[58] A. Hajnal & J. Suŕanyi, Über die Auflosung von Graphen in vollständige Teilgraphen, *Annales Universitatis Scientiarum Budapestinensis de Rolando Eötvös Nominatae. Sectio Mathematica*, **1** (1958), 113–121.

[59] L. Hopkins, D. West & W. T. Trotter, The interval number of a complete multi-partite graph, *Discrete Applied Mathematics*, **18** (1984), 163–189.

[60] G. Hurlbert, A short proof that $\mathbb{N}^3$ is not a circle order, *Order*, **5** (1988), 235–237.

[61] M. Jacobson, F. R. McMorris & H. Mulder, An introduction to tolerance intersection graphs, in *Graph Theory, Combinatorics and Applications* (eds. Y. Alavi, et al.), John Wiley, New York (1991), 705–723.

[62] J. Kahn & M. Saks, Balancing poset extensions, *Order*, **1** (1984), 113–126.

[63] D. Kelly & W. T. Trotter, Dimension theory for ordered sets, in *Proceedings of the Symposium on Ordered Sets*, (ed. I. Rival, et al.), Reidel, Dordrecht (1982), 171–212.

[64] H. A. Kierstead, An effective version of Dilworth's theorem, *Transactions of the American Mathematical Society*, **268** (1981), 63–77.

[65] H. A. Kierstead, The linearity of first-fit coloring of interval graphs, *SIAM Journal on Discrete Mathematics*, **1** (1988), 526–530.

[66] H. A. Kierstead, A polynomial time approximation algorithm for dynamic storage allocation, *Discrete Mathematics*, **88** (1991), 231–237.

[67] H. A. Kierstead, Coloring graphs on-line, in press.

[68] H. A. Kierstead, G. McNulty & W. T. Trotter, A theory of recursive dimension for ordered sets, *Order*, **1** (1984), 67–82.

[69] H. A. Kierstead, S. G. Penrice & W. T. Trotter, On-line coloring and recursive graph theory, *SIAM Journal on Discrete Mathematics*, **7** (1994), 72–89.

[70] H. A. Kierstead & J. Qin, Coloring interval graphs with first-fit, *Discrete Mathematics*, **144** (1995), 47–57.

[71] H. A. Kierstead & W. T. Trotter, An extremal problem in recursive combinatorics, *Congressus Numerantium*, **33** (1981), 143–153.

[72] H. A. Kierstead & W. T. Trotter, On-line graph coloring, in *On-Line Algorithms* (eds. L. McGeoch & D. Sleator), *DIMACS Series in Discrete Mathematics and Theoretical Computer Science*, (1992), 85–92.

[73] H. A. Kierstead & W. T. Trotter, A note on removable pairs, in *Graph Theory, Combinatorics and Applications, Vol. 2* (eds. Y. Alavi, et al.), John Wiley, New York (1991), 739–742.

[74] H. A. Kierstead & W. T. Trotter, Planar graph coloring with an uncooperative partner, *Journal of Graph Theory*, **18** (1994), 569–584.

[75] H. A. Kierstead & W. T. Trotter, The ramsey complexity of dimension for interval orders, in press.

[76] S. S. Kislitsyn, Finite partially ordered sets and their associated sets of permutations, *Matematicheskiye Zametki*, **4** (1968), 511–518.

[77] D. J. Kleitman & G. Markovsky, On Dedekind's problem: The number of isotone boolean functions, II, *Transactions of the American Mathematical Society*, **213** (1975), 373–390.

[78] A. Kostochka & J. Kratochvíl, Covering and coloring polygon circle graphs, in press.

[79] A. Kostochka & D. B. West, Total interval number for graphs of bounded degree, *Journal of Graph Theory*, in press.

[80] T. M. Kratzke & D. B. West, The total interval number of a graph I: Fundamental classes, *Discrete Mathematics*, **118** (1993), 145–156.

[81] T. M. Kratzke & D. B. West, The total interval number of a graph II: Trees and complexity, *SIAM Journal on Discrete Mathematics*, **9** (1996), 339–348.

[82] L. Langley, Interval tolerance orders and dimension, Ph. D. Thesis, Dartmouth College, 1993.

[83] C. G. Lekkerkerker & J. Ch. Boland, Representation of a finite graph by a set of intervals on the real line, *Fundamenta Mathematicae*, **51** (1962), 45–64.

[84] I.-J. Lin & D. B. West, Interval digraphs that are indifference digraphs, in *Graph theory, Combinatorics, and Algorithms* (eds. Y. Alavi and A. Schwenk), Wiley, New York (1995), 751–765.

[85] I.-J. Lin, M. K. Sen & D. B. West, Classes of interval digraphs and $0,1$ matrices, in press.

[86] T.-H. Ma & J. P. Spinrad, An $O(n^2)$ recognition algorithm for the 2–chain cover problem and related problems, in *Proceedings of the Second Annual ACM-SIAM Symposium on Discrete Algorithms*, ACM Press, New York (1991), pp. 363–372.

[87] T. Madej & D. B. West, The interval inclusion number of a partially ordered set, *Discrete Mathematics*, **88** (1991), 259–277.

[88] T. Madej & D. B. West, The interval number of special posets and random posets, *Discrete Mathematics*, **144** (1995), 67–74.

[89] J. Mitas, Tackling the jump number of interval orders, *Order*, **8** (1991), 115–132.

[90] J. Nešetřil & V. Rödl, A short proof of the existence of highly chromatic graphs without short cycles, *Journal of Combinatorial Theory, Series B*, **27** (1979), 225–227.

[91] I. Rabinovitch, The dimension theory of semiorders and interval orders, Ph. D. thesis, Dartmouth College, 1973.

[92] I. Rabinovitch, The Scott-Suppes theorem on semiorders, *Journal of Mathematical Psychology*, **15** (1977), 209–212.

[93] I. Rabinovitch, The dimension of semiorders, *Journal of Combinatorial Theory, Series A*, **25** (1978), 50–61.

[94] I. Rabinovitch, An upper bound on the dimension of interval orders, *Journal of Combinatorial Theory, Series A*, **25** (1978), 68–71.

[95] K. Reuter, Removing critical pairs, *Order*, **6** (1989), 107–118.

[96] F. S. Roberts, On the boxicity and cubicity of a graph, in *Recent Progress in Combinatorics* (ed. W. T. Tutte), Academic Press, New York (1969), 301–310.

[97] F. S. Roberts, Indifference graphs, in *Proof Techniques in Graph Theory* (ed. F. Harary), Academic Press, New York (1969), 139–146.

[98] F. S. Roberts, *Discrete Mathematical Models, with Applications to Social, Biological and Environmental Problems*, Prentice-Hall, Englewood Cliffs (1976).

[99] C. D. Savage & P. W. Winkler, Monotone Gray codes and the middle levels problem, *Journal of Combinatorial Theory, Series A*, in press.

[100] E. R. Scheinerman, The maximum interval number of graphs of given genus, *Journal of Graph Theory*, **11** (1987), 441–446.

[101] E. R. Scheinerman, The many faces of circle orders, *Order*, **9** (1992), 343–348.

[102] E. R. Scheinerman & J. C. Weirman, On circle containment orders, *Order*, **4** (1988), 315–318.

[103] E. R. Scheinerman & D. B. West, The interval number of a planar graph—three intervals suffice, *Journal of Combinatorial Theory, Series B*, **35** (1983), 224–239.

[104] D. Scott & P. Suppes, Foundational aspects of theories of measurement, *Journal of Symbolic Logic*, (1958), 113–128.

[105] M. Sen, S. Das, A. B. Roy & D. B. West, Indifference digraphs: An analogue of interval graphs, *Journal of Graph Theory*, **13** (1989), 189–202.

[106] M. Sen & B. K. Sanyal, Indifference digraphs: A generalization of indifference graphs, *SIAM Journal on Discrete Mathematics*, **7** (1994), 157–165.

[107] J. Spencer, Minimal scrambling sets of simple orders, *Acta Mathematica Academiae Scientiarum Hungaricae*, **22** (1972), 349–353.

[108] M. Syslo, The jump number problem on interval orders: A 3/2 approximation algorthm, *Discrete Mathematics*, **144** (1995), 119–130.

[109] C. Thomassen, Interval representations of planar graphs, *Journal of Combinatorial Theory, Series B*, **40** (1986), 9–20.

[110] W. T. Trotter, Dimension of the crown $\mathbf{S}_n^k$, *Discrete Mathematics*, **8** (1974), 85–103.

[111] W. T. Trotter, A forbidden subposet characterization of an order dimension inequality, *Mathematical Systems Theory*, **10** (1976), 91–96.

[112] W. T. Trotter, Combinatorial problems in dimension theory for partially ordered sets, in *Problèmes Combinatoires et Theorie des Graphes*, *Colloques Internationeaux C.N.R.S.*, 260, (1978), 403–406.

[113] W. T. Trotter, A forbidden subgraph characterization of Roberts' inequality for boxicity, *Discrete Mathematics*, **28** (1979), 303–313.

[114] W. T. Trotter, Stacks and splits of partially ordered sets, *Discrete Mathematics*, **35** (1981), 229-256.

[115] W. T. Trotter, Graphs and partially ordered sets, in *Selected Topics in Graph Theory 2* (eds. L. W. Beineke & R. J. Wilson), Academic Press, New York (1983), 237–268.

[116] W. T. Trotter, Interval graphs, interval orders, and their generalizations, in *Applications of Discrete Mathematics* (eds. R. Ringeisen & F. Roberts), SIAM, Philadelphia (1988), 45–58.

[117] W. T. Trotter, Problems and conjectures in the combinatorial theory of ordered sets, *Annals of Discrete Mathematics*, **41** (1989), 401–416.

[118] W. T. Trotter, *Combinatorics and Partially Ordered Sets: Dimension Theory*, The Johns Hopkins University Press, Baltimore (1992).

[119] W. T. Trotter, Progress and new directions in dimension theory for finite partially ordered sets, in *Extremal Problems for Finite Sets* (eds. P. Frankl, Z. Füredi, G. Katona & D. Miklós), *Bolyai Society Mathematical Studies*, 3, (1994), 457–477.

[120] W. T. Trotter, Partially ordered sets, in *Handbook of Combinatorics* (eds. R. L. Graham, M. Grötschel & L. Lovász), Elsevier, Amsterdam (1995), 433–480.

[121] W. T. Trotter, Graphs and partially ordered sets: recent results and new directions, in *Surveys in Graph Theory* (eds. G. Chartrand & M. Jacobson), *Congressus Numerantium*, **116** (1996), 253–278.

[122] W. T. Trotter & F. Harary, On double and multiple interval graphs, *Journal of Graph Theory*, **3** (1979), 205–211.

[123] W. T. Trotter & J. I. Moore, Characterization problems for graphs, partially ordered sets, lattices, and families of sets, *Discrete Mathematics*, **16** (1976), 361–381.

[124] W. T. Trotter & J. I. Moore, The dimension of planar posets, *Journal of Combinatorial Theory, Series B*, **21** (1977), 51–67.

[125] W. T. Trotter & D. West, Poset boxicity of graphs, *Discrete Mathematics*, **64** (1987), 105–107.

[126] W. T. Trotter & P. Winkler, Ramsey theory and sequences of random variables, *Combinatorics, Probability and Computing*, in press.

[127] H. S. Wittenshausen, On intersections of interval graphs, *Discrete Mathematics*, **31** (1980), 211–216.

Department of Mathematics
Arizona State University
Tempe, Arizona 85287
U. S. A.
`trotter@ASU.edu`

Approximate Counting

Dominic Welsh

Summary I shall survey a range of very basic but provably hard counting problems which have a unifying geometrical theme. For each of them exact evaluation is $\#P$-hard but there is no obvious obstruction to the existence of a fast randomised approximation algorithm. I shall outline the techniques which are available and describe what is currently known. In most cases the results obtained only apply to a part of the input so there remain many open questions.

1 Introduction

Consider the following problem.

> How many different 4×4 matrices having nonnegative integer entries have row sums
>
> $$1000, \quad 9000, \quad 3000, \quad 440$$
>
> and column sums
>
> $$2000, \quad 4000, \quad 200, \quad 7240?$$
>
> The answer is approximately 10^{23}, and is exactly
>
> $$1602\ 5658\ 9785\ 6815\ 3518\ 4841.$$
>
> It was found in 2 seconds real time on a typical work station.† However had I posed a similar question about 5×5 matrices it would be beyond the scope of everyday machines.

A second problem for which I have thought of at least 20 bad (= wrong or too slow) algorithms is the following.

> How does one generate a random planar graph on n vertices where by "random" I mean uniformly at random over all labelled planar graphs on n vertices?

Both of these problems are typical of the sort of questions I shall be considering here. The first is provably hard and will be discussed further in §8. The second is not provably hard and some progress, mainly empirical, is contained in [15]. Here I shall discuss a range of problems of this type which have a common theme in geometric combinatorics.

†The integers were chosen by me on March 22, 1996 to test the programme which John Mount had running on the machines at MSRI, Berkeley.

Most of the concepts used will be defined as they are encountered. However although the ideas of computational complexity are now fairly familiar concepts, it might be useful to review briefly and informally the basic notions, particularly with respect to randomised computation.

1.1 Complexity

A function f is *polynomial time computable* or *P-time computable* if there exists an algorithm which computes f in time (= number of steps) bounded by a polynomial in the size of the input to the problem. We use FP to denote the class of all polynomial time computable functions, and as usual P is the class of languages (= sets of strings) *recognisable* in polynomial time. Thus, P can be identified with the class of functions in FP with range $\{0, 1\}$. Informally, *nondeterministic polynomial time*, NP, is the class of strings (objects, or theorems) which can be recognised or checked in polynomial time.

The simplest model of randomised computation is the *probabilistic Turing machine* in which the only source of randomness is a fair coin.

In a more powerful model, the *oracle coin machine*, the unbiased coin is replaced by a device which chooses transitions with probabilities p and $1 - p$, where p is the ratio of a pair of rationals presented in a special bias tape and which can be varied by the machine according to the state of the computation. Although a probabilistic machine will have difficulty simulating an oracle machine exactly, it can approximate its behaviour very accurately in polynomial time, so that the notion of polynomial time approximation algorithm is robust with respect to choice of model. A much deeper question and one which we shall duck here is how much randomness do we really need? Thus as far as the remainder of this article is concerned we will adopt the 'black box mentality' and assume that when we are working with a probabilistic Turing machine it has the capabilities of an oracle coin machine.

Random polynomial time, RP, is the subset of NP consisting of those strings which can be recognised in polynomial time by a probabilistic Turing machine. Obviously

$$P \subseteq RP \subseteq NP,$$

but neither of the inclusions is known to be proper. The class $\#P$ is the set of functions which enumerate structures which are recognisable in polynomial time. In other words there is a polynomial time algorithm which will verify whether a given object has the structure needed to be included in the count. For example, it is easy to check in polynomial time whether a given assignment of colours to the vertices of a graph is in fact a proper 3-colouring in that no two adjacent vertices are the same colour. Like NP, $\#P$ has hardest problems known as *$\#P$-complete (hard) functions.* A polynomial time algorithm for solving any of these would produce an amazing collapse in the complexity hierarchy, much more surprising than the collapse of NP to P.

It is therefore extremely strong evidence of intractability when a function is shown to be #P-hard. By the same token it is somewhat surprising that exact counting of some of the very easily recognisable objects to be discussed below turns out to be #P-hard. For more on this see Garey and Johnson [32] or Welsh [77].

2 Counting to within a ratio

Computing the number of 3-colourings of a graph G is #P-hard. It is natural therefore to ask how well can we approximate it?

For positive numbers a and $r \geq 1$, we say that a third quantity $\hat{a}$ *approximates a within ratio r* or is an *r-approximation* to a, if

$$r^{-1}a \leq \hat{a} \leq ra.$$

In other words the ratio $\hat{a}/a$ lies in $[r^{-1}, r]$.

Now consider what it would mean to be able to find a polynomial time algorithm which gave an approximation within r to the number of 3-colourings of a graph. We would clearly have a polynomial time algorithm which would decide whether or not a graph is 3-colourable. But this is NP-hard. Thus no such algorithm can exist unless $NP = P$.

But we have just used 3-colouring as a typical example and the same argument can be applied to any function which counts objects whose existence is NP-hard to decide. In other words:

Proposition 2.1 *If $f: \Sigma^* \to \mathbb{N}$ is such that it is NP-hard to decide whether $f(x)$ is non-zero, then for any constant r there cannot exist a polynomial time r-approximation to f unless $NP = P$.*

We now turn to consider a randomised approach to counting problems and make the following definition.

Let f be a function from input strings to the natural numbers. A *randomised approximation scheme* for f is a probabilistic algorithm that takes as an input a string x and a rational number ϵ, $0 < \epsilon < 1$, and produces as output a random variable Y, such that Y approximates $f(x)$ within ratio $1 + \epsilon$ with probability $\geq 3/4$.

In other words,

$$\Pr\left\{\frac{1}{1+\epsilon} \leq \frac{Y}{f(x)} \leq 1+\epsilon\right\} \geq \frac{3}{4}.$$

A *fully polynomial randomised approximation scheme* (fpras) for a function $f: \Sigma^* \to \mathbb{N}$ is a randomised approximation scheme which runs in time which is a polynomial function of n *and* ϵ^{-1}.

It is important to note that there is no significance in the constant 3/4 appearing in this definition. Any success probability *strictly greater* than 1/2 will suffice, for the following reason.

Suppose we have such an approximation scheme and suppose further that it works in polynomial time. Then we can boost the success probability up to $1-\delta$ for any desired $\delta > 0$, by using the following trick of Jerrum, Valiant and Vazirani [42]. This consists of running the algorithm $O(\log \delta^{-1})$ times and taking the median of the results.

We make this precise as follows:

An *ϵ-δ-approximation scheme* for a counting problem f is a randomised algorithm which on every input $\langle x, \epsilon, \delta \rangle$, $\epsilon > 0$, $\delta > 0$, outputs a number $\tilde{Y}$ such that

$$\Pr\{(1-\epsilon)f(x) \leq \tilde{Y} \leq (1+\epsilon)f(x)\} \geq 1-\delta.$$

Proposition 2.2 *If there exists an fpras for computing f then there exists an ϵ-δ approximation scheme for f which on input $\langle x, \epsilon, \delta \rangle$ runs in time which is bounded by $O(\log \delta^{-1})\,\mathrm{poly}(x, \epsilon^{-1})$.*

It is worth emphasising here that the existence of an fpras for a counting problem is a very strong result, it is the analogue of an RP algorithm for a decision problem and corresponds to the notion of tractability. However we should also note, that by an analogous argument to that used in proving Proposition 2.1 we have:

Proposition 2.3 *If $f\colon \Sigma^* \to \mathbb{N}$ is such that deciding if f is nonzero is NP-hard then there cannot exist an fpras for f unless NP is equal to random polynomial time RP.*

Since this is thought to be most unlikely, it makes sense only to seek out an fpras when counting objects for which the decision problem is not NP-hard.

Hence we will now concentrate on counting problems which have the following properties:

(a) Counting exactly is provably hard, usually $\#P$-hard.

(b) Deciding if the exact count is nonzero is certainly P-time, and indeed often trivial.

As typical examples consider the following. In each case condition (b) is trivially satisfied and it is mildly surprising that exact counting is in each case $\#P$-hard.

#BASES

Input: A nonsingular $m \times n$ matrix with 0/1 entries, with $m < n$.
Task: Determine the number of nonsingular $m \times m$ submatrices.

#FORESTS

Input: A graph G.
Task: Determine the number of subsets $A \subseteq E(G)$ which are forests.

3 Randomised approximation schemes

It is something of a curiosity that there seems to be a greater range of techniques for obtaining randomised algorithms which give good approximation than there are deterministic.

Most of these schemes are based on developing algorithms which will generate objects uniformly at random and then use what Lovász [52] calls the *product estimator.*

This works as follows. Suppose F is the set of objects we wish to count and that we can find a chain of subsets $F_0 \subset F_1 \subset \cdots \subset F_r = F$ such that

(a) F_0 is known

(b) $|F_{i+1}|/|F_i|$ is bounded by a polynomial in i, for each i

(c) r is polynomially bounded in the size of input

(d) for each i, $1 \leq i \leq r$, we can generate a member of F_i uniformly (or almost uniformly) at random in time polynomial in n.

Then we use (d) a polynomial number of times to estimate the ratio

$$|F_{i-1}|/|F_i|.$$

Since products preserve good approximations, we then use

$$\prod_{i=1}^{r} |F_i|/|F_{i-1}|$$

and our knowledge of F_0 to estimate $|F|$.

As an example, suppose we try to estimate the number of forests in an n-vertex graph G by this method. We would need to be able to generate a random forest of size k, for $1 \leq k \leq n-1$. As yet no one knows how to do this in polynomial time.

3.1 Almost uniform generation

Examination of the above approximation method shows that it is not necessary for the generation procedure to be exactly uniform. All that is really necessary is that it be close to uniform in some well defined sense. We now make this precise.

Throughout Σ will denote a finite alphabet which is usually taken to be $\{0,1\}$. Σ^* denotes the set of all finite strings or words of symbols from Σ. If x is a word of Σ^*, then $|x|$ denotes the length (= number of symbols) of x.

A relation $R \subseteq \Sigma^* \times \Sigma^*$ is a *p-relation* if

(i) there exists a polynomial p such that

$$\langle x, y\rangle \in R \Rightarrow |y| \leq p(|x|),$$

(ii) the predicate $\langle x, y\rangle \in R$ can be tested in deterministic polynomial time.

Typically, x may be a graph G and y might be the set of edges of a forest of G, both encoded in the alphabet Σ.

A *uniform generator* for a relation R is a probabilistic Turing machine M such that the following conditions are satisfied.

(a) There exists a function ϕ such that for all inputs x

$$\Pr\{M \text{ outputs } y\} = \begin{cases} 0 & \text{if } \langle x, y\rangle \notin R \\ \phi(x) & \text{if } \langle x, y\rangle \in R. \end{cases}$$

(b) If $\{y : x\, R\, y\}$ is nonempty then

$$\Pr\{M \text{ accepts } x\} \geq \frac{1}{2}.$$

Thus the condition (a) demands that the machine behaves uniformly.

The machine is *polynomially time bounded* if there exists a polynomial t such that for all inputs x of length n and all positive integers n, every accepting computation halts within $t(n)$ steps.

A probabilistic machine M is an *almost uniform generator* for the relation $R \subseteq \Sigma^* \times \Sigma^*$ if and only if

(c) there exists a mapping $\phi\colon \Sigma^* \to (0, 1]$ such that for all inputs $(x, \epsilon) \in \Sigma^* \times \mathbb{R}^+$ and for all words $y \in \Sigma^*$,

$$\begin{aligned} \langle x, y\rangle \notin R &\Rightarrow \Pr\{M \text{ outputs } y\} = 0 \\ \langle x, y\rangle \in R &\Rightarrow (1+\epsilon)^{-1}\phi(x) \leq \Pr\{M \text{ outputs } y\} \leq (1+\epsilon)\phi(x); \end{aligned}$$

(d) for all inputs (x, ϵ) such that $\{y \in \Sigma^* : x\, R\, y\}$ is nonempty, the probability that M accepts (x, ϵ) is at least $\frac{1}{2}$.

An almost uniform generator is *fully polynomial* (f.p.) if its running time on input (x, ϵ) is bounded by a polynomial in x and $\log(\epsilon^{-1})$. The importance of almost uniform generation is that provided the relation R is what is known as *self reducible* (and in most natural problems this seems to be the case) the existence of a f.p. almost uniform generator implies the existence of an fpras.

The concept of self reducibility was introduced by Schnorr [66], its formal definition is rather technical, but intuitively it represents the notion that the solution set associated with a given instance of a problem can be expressed in terms of the solution sets of a number of smaller instances of the same problem.

Example 3.1 Let R denote the relation $\langle G, A\rangle$ where A is the set of edges of a connected subgraph of a graph G. Then the set of solutions is the union of those where G is replaced by G'_e and G''_e where G'_e denotes the deletion of the edge e, and G''_e the effect of identifying its endpoints. ∎

When R is self reducible it is easy to see how an exact counting oracle helps with uniform generation. Consider Example 3.1; if there are 35 connected subgraphs of G'_e and 65 of G''_e then at the first step of the generation algorithm branch to G'_e with probability 0.35, and to G''_e with probability 0.65. This reduces the size of the problem by one. Now proceed recursively.

A similar method based on approximate counting gives an almost uniform generator, and this forms the basis of the proof of the main theorem of Jerrum, Valiant and Vazirani [42], which we now state.

Theorem 3.2 *If R is a self reducible relation then there exists a fully polynomial uniform generator for R if and only if there exists a fully polynomial randomised approximation scheme for the associated counting problem.*

3.2 Impossible to approximate?

We have already seen that unless $NP = P$ there cannot exist fast approximation algorithms for quantities such as the number of 3-colourings of a graph. There is another class of problems where the objects being counted have the property that they always exist, so the decision problem is trivial, but for which approximate counting is not possible unless something very unlikely is true.

A collection of such problems is given in the following proposition.

Proposition 3.3 *Unless NP=RP there cannot exist an fpras for the following:*

(a) the number of cycles in a (directed) graph,

(b) the number of paths linking two specific vertices in a graph,

(c) the number of cliques in a graph,

(d) the number of maximal cliques.

This list is not exhaustive, and in each case the proof idea is the same.

Sketch Proof I illustrate with (a).

Blow the input up by introducing a "gadget" which transforms the input G to H so that the existence or not of a Hamilton cycle in G has such a huge effect on the number of cycles in H that it will show up in the answer obtained by applying the fpras to H. Thus we would have an RP algorithm to decide whether G was Hamiltonian. ∎

For further details see [42] and [68], and for even more striking recent results see Zuckerman [81] who shows, for example, that the following is true.

Proposition 3.4 *Unless $NP = P$ it is not possible to approximate the log of the number of cliques to within a factor of n^ε, where n is the size of the input graph.*

4 Rapidly mixing Markov chains

We turn now to sampling procedures based on Markov chain simulation. Although the mathematical analysis can be very difficult the basic idea is straightforward.

Identify the set of objects being sampled with the state space Ω of a finite Markov chain constructed in such a way that the stationary distribution of the Markov chain agrees with the probability distribution on the set of objects. Now start the Markov chain at an arbitrary state and let it converge to the stationary distribution. Provided the convergence is fast, we have a good approximation to the stationary distribution fairly quickly. However the key to the success of this approach is that the time or number of steps needed to ensure that it is sufficiently close to its stationary distribution is not too large.

The first major practical application of the theory seems to have been in the work of Broder [10] on approximating the permanent. The paper of Jerrum and Sinclair [40] which built on Broder's work was a major advance, particularly in their idea of relating rapid mixing with conductance. They used this to show that there was an fpras for evaluating a dense permanent, see §7.2.

To be more specific, suppose that P is the transition matrix of an irreducible aperiodic finite Markov chain $(X_t;\ 0 \leq t < \infty)$ with state space Ω. Suppose also that the chain is *reversible* in that it satisfies the detailed balance condition

$$\pi(i)P(i,j) = \pi(j)P(j,i) \quad \text{for all} \quad i,j \in \Omega.$$

This condition implies that π is a stationary or equilibrium distribution for P and that

$$\Pr\{\lim_{t\to\infty} X_t = i\} = \pi_i$$

for all states $i \in \Omega$.

Moreover, P has eigenvalues $1 = \lambda_1 \geq \lambda_2 \geq \cdots \geq \lambda_N > -1$ where N is the number of states, and all these eigenvalues are real. The rate of convergence to the distribution π is determined by the quantity

$$\lambda_{\max} = \max\{\lambda_2, |\lambda_N|\}.$$

More precisely we have the following proposition. Let $P_{ij}(t)$ denote

$$\Pr\{X_{t+h} = j \mid X_h = i\}$$

and let

$$\Delta_i(t) = \frac{1}{2}\sum_j |P_{ij}(t) - \pi(j)|,$$

denote what is called the *variation distance* at time t. It is clearly a measure of the separation from the stationary distribution at time t.

Now define, for $\epsilon > 0$, the *mixing time* function τ_i by

$$\tau_i(\epsilon) = \min\{t : \Delta_i(s) \leq \epsilon \quad \forall s \geq t\}.$$

The following result shows the relationship between mixing times and the maximum eigenvalues. It is an extension by Sinclair [69] of a key result from [70].

Proposition 4.1 *For $\epsilon > 0$, $\tau_i(\epsilon)$ satisfies*

(i) $\tau_i(\epsilon) \leq \dfrac{1}{1-\lambda_{\max}}(\ln\dfrac{1}{\pi(i)} + \ln\dfrac{1}{\epsilon})$;

(ii) $\max\limits_i \tau_i(\epsilon) \geq \dfrac{\lambda_{\max}}{2(1-\lambda_{\max})}\ln\left(\dfrac{1}{2\epsilon}\right)$.

In order to achieve rapid convergence we need $\tau_i(\epsilon)$ be small, for all i.

It is useful to note the following trick which concentrates the interest on λ_2. Replace P by $P' = \frac{1}{2}(I + P)$ where I is the identity matrix. This only affects rates of convergence by a polynomial factor. All eigenvalues of P' are nonnegative, and the quantity $\lambda'_{\max}$ of P' is $\frac{1}{2}\lambda_{\max}$, so that henceforth we need only consider the second eigenvalue λ_2. Ideally we want $1 - \lambda_2$ to be large so λ_2 must be small.

The key idea of Sinclair and Jerrum [70] was to relate this to the very aptly named *conductance* Φ of the chain. This is defined by

$$\Phi = \max_{S \subseteq \Omega}\left\{\sum_{i\in S, j\in\Omega\setminus S} P(i,j)\pi(i) \Big/ \sum_{i\in S}\pi(i)\right\}.$$

But the bracketed term is just the conditional probability of leaving a set S in the equilibrium state. In other words Φ is a measure of the ability of the chain to escape from any subset of the state space Ω.

Theorem 4.2 *The second eigenvalue λ_2 of a reversible ergodic Markov chain satisfies*

$$1 - 2\Phi \leq \lambda_2 \leq 1 - \Phi^2/2.$$

In other words, for fast approximation we need large conductance, namely

$$\Phi \geq \frac{1}{\text{poly}(n)},$$

where poly(n) denotes some polynomial function of the input size.

4.1 Canonical paths

As we have already indicated, the fundamental objective is to get good lower bounds on the conductance Φ. To do this, the key step is to construct a *canonical path* γ_{xy} between each ordered pair of states x, y in the graph Γ having vertex set Ω (which represents the state space of the Markov chain). Provided these canonical paths can be chosen in such a way that no edge (transition) is overloaded by paths, then the chain cannot contain a bottleneck (or constriction). This implies that Φ cannot be too small, for if there were a bottleneck between S and $\Omega \setminus S$ then any choice of paths would overload the edges in the bottleneck. Thus we have shifted the problem from bounding eigenvalues, through conductances, to finding a "good" set of canonical paths having the property that the maximum loading of an edge $e = (u, v)$ of Γ, measured by

$$\rho = \max_e \frac{1}{Q(e)} \sum_{\gamma_{xy}} \pi(x)\pi(y)$$

is not too large.

Here

$$Q(e) = Q(u, v) = \pi(u)P(u, v)$$

and the sum is over all canonical paths from x to y which use e.

The relationship between ρ and conductance is that it can be shown that

$$\Phi \geq 1/2\rho.$$

The difficult part is to find a good set of canonical paths.

This was first done by Jerrum and Sinclair [40] to count matchings (see §7.2). A good account of this can be found in Motwani and Raghavan [61]. Other recent examples of the use of the method can be found in [46] and [73].

4.2 Coupling and convergence—the basic idea

Recall that we need a good upper bound on the *mixing time*

$$\tau(\varepsilon) = \max_i \tau_i(\varepsilon).$$

A *coupling* for the chain X_t is a stochastic process (Y_t, Z_t) on $\Omega \times \Omega$ with the following properties.

(i) Each of the processes Y_t and Z_t is a faithful copy of X_t for specified initial states $Y_0 = y$, $Z_0 = z$.

(ii) If $Y_t = Z_t$ then $Y_{t+1} = Z_{t+1}$.

Thus, viewed in isolation, each of Y_t and Z_t behaves like X_t, however Y_t and Z_t need not be independent. The basic idea is to construct a joint distribution which brings them close together. This is because of the crucial result that the expected time for the processes to couple is an upper bound on the mixing time. More precisely, if the *coupling time* is defined by

$$T^{y,z} = \min\{t : Y_t = Z_t \mid Y_0 = y,\ Z_0 = z\}$$

then we have

Theorem 4.3 *The mixing and coupling times are related by*

$$\tau(\varepsilon) \le 8\log_2(\varepsilon^{-1}) \max_{y,z\in\Omega} E(T^{y,z}).$$

An equivalent way of looking at coupling is to note that the variation distance of (X_t) from the stationary distribution is bounded above by the inequality

$$\Delta(t) \le \max_{y,z\in\Omega} \Pr\{Y_t \ne Z_t \mid Y_0 = y,\ Z_0 = z\}.$$

For a detailed account of the probability background we refer to Aldous [1].

Applications of this coupling technique to combinatorial problems can be found in [12, 38]. We illustrate its use in a colouring problem in §7.1.

5 Computing in a convex body

Before we consider specific computational questions, we should emphasize that there can be severe difficulties in even describing a convex body to a computer. A good account of the sort of problems which can arise is given in the monograph of Grötschel, Lovász and Schrijver [35]. In the following very brief discussion of the volume question we shall restrict attention to convex bodies with algorithmically good descriptions.

This means that the convex body $K \subseteq R^n$ is given by a *separation oracle* which for a given point x tells us whether the point is in K, and if not, it gives a hyperplane separating the point from K.

It is also assumed that K contains the unit ball $B = \{x : |x| \le 1\}$ and is contained in some ball, centred at the origin and of radius n^α for some constant α. It is known (see [45]) that every convex body can be transformed into such a ball (of radius $n^{3/2}$) by the ellipsoid method in $O^*(n^4)$ steps, where we use O^* to indicate that we are ignoring the additional terms in $(\log n)(\log \varepsilon^{-1})$ and the like.

We are mainly concerned here with algorithmic problems about convex polytopes. Here the survey of Gritzmann and Klee [33] is an invaluable source. Using their terminology, a *$\mathcal{V}$-presentation* of a polytope P consists of positive integers n and m, and m points $v_1, \ldots, v_m$ in $\mathbb{R}^n$ such that P is the convex hull of $v_1, \ldots, v_m$, written $P = \mathrm{conv}\{x_1, \ldots, x_m\}$. An *$\mathcal{H}$-presentation* of a polytope P consists of integers n, m, with $m > n \ge 1$, a rational $m \times n$ matrix A, and a vector $b \in \mathbb{R}^m$ such that $P = \{x \in \mathbb{R}^n : Ax \le b\}$.

5.1 Volume computation

Although computing volumes is not a counting problem it is intimately connected with approximate counting and can be rightly regarded as one of the success stories of the subject. First a brief history.

Elekes [29] showed it is hard to approximate closely volumes of convex bodies defined by certain types of oracles. Dyer and Frieze [18] considered the problem of computing the volume of the polytope

$$P(A, b) = \{x \in \mathbb{R}^n : Ax \leq b\}$$

and proved

Proposition 5.1 *(a) Computing* $\operatorname{vol} P(A, b)$ *is #P-hard even when A is totally unimodular. In other words, computing the volume of an* $\mathcal{H}$*-polytope even when the defining hyperplanes have a very special (totally unimodular) representation is #P-hard.*

(b) Computing the volume of a $\mathcal{V}$*-polytope is #P-hard.*

Sketch proof of (a) The key idea is to reduce computing volume to the #P-hard counting problem,

#KNAPSACK

Input: Positive integers $a_1, \ldots, a_n, b$

Task: Find the number of $(0,1)$-vectors $x = (x_1, \ldots, x_n)$ satisfying the condition

$$a^T x = \sum_{i=1}^{n} a_i x_i \leq b.$$

The particular class of polytopes used in the proof is the set

$$P = \{x \in \mathbb{R}^n : a^T x \leq b, \ 0 \leq x_i \leq 1, \ 1 \leq i \leq n\}. \tag{1}$$

It is shown that computing volumes of even this class is #P-hard, and then the total unimodularity condition is satisfied by making the substitution $y_i = a_i x_i$, $1 \leq i \leq n$. ∎

However, even though exact computation of volume of seemingly very simple polytopes is #P-hard, there are now a succession of improved randomised approximation schemes which will compute not just the volume of polytopes but of a much wider class of convex bodies.

There are already several excellent surveys on the subject, in particular [44] and [50]. We give the bare outlines.

In 1989, Dyer, Frieze and Kannan [21] designed a fully polynomial time randomised algorithm to approximate the volume of a convex body $K \subseteq \mathbb{R}^n$.

The original algorithm to find an ϵ-approximation to vol(K) with $\epsilon < 1$ and error probability less than δ demanded $O^*(n^{23})$ convex programmes.

Since then there have been improvements by Lovász [51], Lovász and Simonovits [53, 54, 55], Dyer and Frieze [20] and Applegate and Kannan [4]. The best currently known algorithm is due to Kannan, Lovász and Simonovits [45] and has running time $O^*(n^5)$.

Proof idea Let K be a convex body in $\mathbb{R}^n$. Surround K by a ball $B \subseteq \mathbb{R}^n$ whose volume we know, and ideally with B, as close as possible to K. Then, even though the volume of K may be much smaller than B, it can be shown that there exists a sequence $K = K_0 \subseteq K_1 \subseteq \cdots \subseteq K_m = B$ of convex bodies where $\mathrm{vol}(K_{i+1}) \leq 2\,\mathrm{vol}(K_i)$ and m is only a polynomial in n. Thus provided each of the ratios $\mathrm{vol}(K_i)/\mathrm{vol}(K_{i+1})$ can be estimated, these estimates can be combined to give an estimate of $\mathrm{vol}(K) = \mathrm{vol}(K_0)$. In order to make this work we need to have a method of generating a point randomly with uniform distribution in a convex body. The rough idea of an algorithm for doing this is to construct a graph G, most of whose vertices are the lattice points of a grid or mesh of $\mathbb{R}^n$ which are contained in K. (Recall that a *lattice point* is a point in $\mathbb{R}^n$ having integral coordinates and a *cube* means a unit cube whose centre is a lattice point and whose edges are parallel to the axes.) Let V consist of those lattice points which are declared to be in K by a weak separation oracle using error tolerance $\frac{1}{2}$. The graph G has edges formed by joining two lattice points of V if and only if their distance apart is say d. Then G will have exponentially many nodes (about vol(K)) and maximum degree $2n$.

The idea now is to carry out a random walk on G and to use the conductance results described earlier to show that the associated Markov chain is rapid mixing. ∎

5.2 Lattice point enumeration

Determining the number of lattice points of $\mathbb{R}^n$ contained in an n-dimensional convex polytope P is a problem which contains many key combinatorial problems as special cases. Formally we state it as follows:

#LATTICE POINTS

Input: Convex polytope P in either $\mathcal{H}$- or $\mathcal{V}$-presentation.

Task: Determine the number of lattice points contained in P.

Not only is this $\#P$-hard, it is depressingly so. Here is a particularly simple class of polytopes to illustrate this.

Take the polytope P defined by (1). The number $i(P)$ of lattice points in P is exactly the number of solutions to #KNAPSACK and this we know to be $\#P$-complete.

What is even more depressing is that even for this class of extremely well structured polytopes the simple Markov chain on lattice points is not known to be rapidly mixing. The best current result [22] is the following:

Theorem 5.2 *There exists a $2^{O(r\sqrt{n}(\log n)^{5/2})}\varepsilon^{-2}$ time randomised algorithm for estimating the number of feasible solutions of a multidimensional knapsack problem to within a ratio $1 \pm \varepsilon$ where r is the number of constraints and n is the dimension.*

Further evidence of the difficulty of lattice point counting is the following theorem of [14].

Theorem 5.3 *Given polynomial $p\colon \mathbb{Z} \to \mathbb{Z}$, and $\mathcal{H}$-polytope P in $\mathbb{R}^d$ finding positive integers α, β such that*

$$\alpha \leq i(P) + 1 \leq \beta \leq 2^{p(d)}\alpha$$

is NP-hard.

In the words of Gritzmann and Wills [34] "either the gap between the lower and upper bounds grows superexponentially with dimension *or* at least one of the two functionals is hard to compute."

On the positive side, Barvinok [8] has a P-time algorithm for exact enumeration when the polytope is of fixed dimension.

Another important and very natural question is:

#POLYTOPE VERTICES

Input: $\mathcal{H}$-presentation of convex polytope P.
Task: Determine the number of vertices of P.

This is certainly a $\#P$-hard problem. It is also the case that as far as I am aware there is no inherent obstacle to there being an fpras.

6 Partial orders

A problem which aroused considerable interest in the 1980s was to decide the difficulty of the question

#LINEAR EXTENSIONS

Input: Finite partial order P.
Task: Find the number $e(P)$ of linear extensions of P.

It was not until 1990 that Brightwell and Winkler [9] showed that #LINEAR EXTENSIONS was $\#P$-complete and a curiosity is that it was previously known to have an fpras,

Theorem 6.1 *There exists an $O^*(n^5)$ fpras for #LINEAR EXTENSIONS.*

Let $\prec$ be a partial order on the set $[n] = \{1, 2, \ldots, n\}$. Define the *order polytope* $P(\prec)$ to be the convex polyhedron in $\mathbb{R}^n$ consisting of the intersection of the unit hypercube C^n with

$$\{x : x_i \le x_j \text{ if } i \prec j\}.$$

Now let $E(\prec)$ be the set of linear extensions of $\prec$.

Theorem 6.2 $\text{vol}(P(\prec)) = \#E(\prec)/n!$.

Proof For any permutation π of $[n]$ let $S_\pi = \{x \in C^n : x_{\pi(1)} \le x_{\pi(2)} \le \cdots x_{\pi(n)}\}$. Then use the following two, easily checked, results.

- The S_π are pairwise disjoint and have equal volumes.
- The convex polytope $P(\prec) = \bigcup_{\pi \in E(\prec)} S_\pi$.

Hence $\text{vol}(S_\pi) = 1/n!$

In other words, by computing the volume of the order polytope of a partial order $\prec$ we get the number of linear extensions of $\prec$. The result follows by the volume approximation algorithm of [45]. ∎

An alternative approach due to Karzanov and Khachiyan [48] was to use a very natural Markov chain on the set of linear extensions.

Sketch Proof Let Ω be the set of all linear extensions of P and consider simple random walk on the graph with vertex set Ω and in which two vertices (states) are adjacent if the corresponding extensions differ by an adjacent transposition.

The transition probabilities $p(w, w')$ are given by

$$p(w, w') = \begin{cases} \dfrac{1}{2n-2} & w \sim w' \\ 1 - \dfrac{d(w)}{2n-2} & w = w' \\ 0 & \text{otherwise.} \end{cases}$$

Thus the diagonal entries of the transition matrix are at least $\frac{1}{2}$ and it is easy to check ergodicity with uniform stationary distribution. The hard part of the proof is showing that the conductance Φ of this chain satisfies

$$\Phi \ge 2^{-3/2} n^{-5/2},$$

which results in an $O^*(n^6)$ algorithm to produce a random linear extension.

We now use this to estimate the probability $\Pr\{x \prec y \text{ in } P\}$. If this probability is exponentially small the method is doomed, however by a result of Kahn and Saks [43] we know that if ρ_{ij} = fraction of linear extensions π with $\pi^{-1}(i) \prec \pi^{-1}(j)$ then for some i, j one or other of ρ_{ij}, ρ_{ji} is at least 3/11. Hence for some i, j, we will be able to determine a good approximation to the proportion of linear extensions with $\pi^{-1}(i) \prec \pi^{-1}(j)$. Choose the i, j which maximises $\min(\rho_{ij}, \rho_{ji})$ and add $i \prec j$ to the partial order. Proceed inductively until the order is a permutation and then the product of the inverses of these successive proportions is our required estimate of the number of linear extensions. ■

7 Graph problems

7.1 Colouring

I start with a series of problems about colourings which I first posed in [77]. We have immediately from the NP-hardness of k-colouring that:

- Unless $NP = RP$ there cannot exist an fpras for counting k-colourings for any integer $k > 2$.

This does not rule out the possibility of there being an fpras for special classes of graphs. For example: there is no inherent obstacle to there being an fpras for estimating the number of 4-colourings of a planar graph.

I do not believe such a scheme exists but cannot see how to prove it. It certainly is not ruled out by any known results and I therefore pose the specific question:

Problem 7.1 *Is there a fully polynomial randomised approximation scheme for counting the number of k-colourings of a planar graph for any fixed $k \geq 4$?*

I conjecture that the answer is negative.

Another question about colourings where decision is trivial but counting is hard is the following, considered by Bartels and Welsh [7].

#$|V|$-COLOURINGS

Input: Graph $G = (V, E)$
Task: Find the number of colourings of G using $|V|$ or fewer colours.

There is considerable evidence in [7] to suggest that a natural Markov chain which starts off in a good $|V|$-colouring and moves through the set of all good colourings using $|V|$ or fewer colours is in fact rapidly mixing. Further more substantial evidence in support of this is given by the following result of Jerrum [38].

Theorem 7.2 *For graphs of maximum degree Δ there exists an fpras for counting k-colourings for all $k \geq 2\Delta + 1$.*

Because of Brooks' theorem there is no obstacle to there existing an fpras for all $k \geq \Delta$.

Apart from the appeal of the result in Theorem 7.2 its proof is one of the nicest applications of the coupling technique described in §4.2.

Proof idea Let C be a set of k colours. Since $k > 2\Delta(G)$ there is no problem about starting the chain (Y_t, Z_t) in a pair of k-colourings. The transitions $(Y_t, Z_t) \to (Y_{t+1}, Z_{t+1})$ are defined as follows:

(a) choose a vertex v of G, uniformly at random (u.a.r.)

(b) choose a colour $c \in C$ u.a.r.

(c) compute a permutation σ of C according to procedure CHOOSE below,

(d) in colouring Y_t recolour vertex v with colour $\sigma(c)$ to obtain Y_t',

(e) in colouring Z_t recolour vertex v with colour $\sigma(c)$ to obtain Z_t',

(f) if Y_t' (respectively Z_t') is a good colouring then $Y_{t+1} = Y_t'$ (respectively $Z_{t+1} = Z_t'$) otherwise let $Y_{t+1} = Y_t$ (respectively $Z_{t+1} = Z_t$).

First note that whatever method is used to select σ, since c is u.a.r. from C, so will $\sigma(c)$ be. Thus (Y_t) and (Z_t) are both faithful copies of (X_t).

The key step is to design the procedure CHOOSE in such a way that Y_{t+1} and Z_{t+1} are stochastically closer together than Y_t and Z_t. ■

Procedure CHOOSE If v has a different colour in Y_t and Z_t then take σ as the identity permutation so that v stays different.

If v has the same colour proceed as follows.

Consider the set of neighbours $\partial(v)$ in G. Let C_Y be the set of colours c such that $\partial(v)$ uses c in Y_t but not in Z_t. Define C_Z analogously. Thus $C_Y \cap C_Z = \emptyset$. Suppose without loss of generality that $|C_Y| \leq |C_Z|$. Choose any $C' \subseteq C_Z$ such that $|C'| = |C_Y|$. Let $C_Y = \{c_1, \ldots, c_r\}$ and $C' = \{c_1', \ldots, c_r'\}$ be arbitrary orderings and take g to be the permutation $(c_1, c_1'), \ldots, (c_r, c_r')$ which interchanges the colour sets C_Y and C' and leaves the remainder of C fixed.

Now let D_t be the set of vertices on which the colourings Y_t and Z_t *disagree*. Because of the way the processes Y_t, Z_t have been chosen, $|D_t|$ is a nonnegative random variable and a not too difficult argument gives

$$\Pr\{|D_t| > 0\} \leq n e^{-t(k-2\Delta)}.$$

Thus $\Pr\{D_t \neq \emptyset\} \leq \varepsilon$ provided $t \geq a^{-1} \ln(n\varepsilon^{-1})$ and this gives mixing time

$$\tau(\varepsilon) \leq \frac{k}{k-2\Delta} n \ln\left(\frac{n}{\varepsilon}\right). \quad ■$$

7.2 Matchings

Counting matchings and perfect matchings has a prominence in this area which is arguably greater than it warrants. This is due principally to it being the success story of the seminal paper of Jerrum and Sinclair [40] and partly to its inherent attraction in (a) evaluating the permanent and (b) counting monomer-dimer coverings in statistical physics. It is therefore very well covered in all existing surveys on the Markov chain method, such as [44, 52]. Here we just highlight the basics.

The main results of [40] can be stated as follows

Theorem 7.3 *(i) There exists an fpras for the number of matchings in a graph.*

(ii) There exists an fpras for the number of perfect matchings in all $2n$-vertex graphs G which satisfy the condition

$$\alpha(G) = \frac{m_{n-1}(G)}{m_n(G)} \leq \text{poly}(n), \tag{2}$$

and where $m_k(G)$ is the number of matchings of size k.

A consequence of (ii) is:

Corollary 7.4 *There exists an fpras for the permanent of $n \times n$ $(0,1)$-matrices provided each row and column sum is at least $n/2$.*

This follows from the following two facts

- If A is an $n \times n$ $(0,1)$-matrix its permanent is the number of perfect matchings in the bipartite $2n$-vertex graph determined by A.
- The "density condition" on the row and column sum is sufficient to guarantee that condition (ii) of Theorem 7.3 is satisfied.

Since the announcement of the results of [40] there has been considerable effort to remove the denseness conditions in Corollary 7.4. We state this as an outstanding and seemingly difficult problem in

Problem 7.5 *Does there exist an fpras for the number of perfect matchings of a general graph?*

An easier and purely combinatorial problem is to extend the class of graphs for which the "slow growth condition" (2) is satisfied. This is the approach taken by Kenyon, Randall and Sinclair [49] who prove that the ratio is small for hypercubical lattices and arbitrary Cayley graphs. More specifically they prove

Theorem 7.6 *(a) If G is the d-dimensional rectangular lattice $[1,\ldots,n]^d$ with periodic boundary conditions, then $\alpha(G) \leq n^{2d}/4$, hence the running time of an fpras is polynomial in n for any fixed dimension.*

(b) Let G be an $2m$-vertex bipartite graph which is the Cayley graph of a finite group. Then $\alpha(G) \leq m^2$.

More generally we have a result attributed in [49] to Jerrum that: if G is a $2m$-vertex Cayley graph then $\alpha(G) \leq m^3$.

7.3 Trees, forests and connected subgraphs

Counting spanning trees is one of the few enumerative problems in graph theory which has a P-time algorithm, namely take the determinant of a cofactor of the Kirchhoff matrix.

Counting all trees or all forests on the other hand are both $\#P$-complete problems, see [39] and [76] respectively, even when restricted to planar graphs and their approximation versions have proved frustratingly difficult. As far as I am aware, absolutely no progress has been made on approximating the number of subtrees in a graph.

For #FORESTS, Annan [3] has shown that the following is true.

Theorem 7.7 *If $\mathcal{G}_\alpha$, for $0 < \alpha \leq 1$, denotes the classes of graphs defined by $G \in \mathcal{G}_\alpha$ if and only if each vertex in G has at least $\alpha|V(G)|$ neighbours then there exists a polynomial time randomised algorithm which for $G \in \mathcal{G}_\alpha$ generates a forest of G uniformly at random from the set of all forests in G. Moreover the running time of this algorithm is $O(n^{1/\alpha})$.*

As a consequence, if we call a class $\mathcal{C}$ of graphs *dense* if $G \in \mathcal{C} \Rightarrow G \in \mathcal{G}_\alpha$ for some $\alpha > 0$, then we have

Corollary 7.8 *There exists an fpras for #FORESTS in any class of dense graphs.*

What seems to be hard is deciding:

Question 7.9 *Can the condition of denseness be removed in Corollary 7.8?*

An immediate consequence of matroid duality and also easy to prove directly is that the number of forests in a planar graph G equals the number of connected (spanning) subgraphs in any planar dual G^*.

It follows from [37] and [76] that the following problem is $\#P$-complete.

#CONNECTED SUBGRAPHS

Input: Bipartite planar graph G
Task: Find the number of connected subgraphs of G.

Prompted by Annan's Theorem 7.7 for forests we show in [2] that:

Proposition 7.10 *For dense graphs there is an fpras for #CONNECTED SUBGRAPHS.*

This also leads to an fpras for the *reliability probability*

$$R(G;p) = \sum_A p^{|A|}(1-p)^{|E \setminus A|}$$

where the sum is over all subsets A of $E(G)$ such that $(V(G), A)$ is a connected graph, again in the case where the underlying graph G is dense.

Even more recently, Karger [47] has obtained a strengthening of this to graphs having edge connectivity $O(\log n)$.

7.4 Orientations

Counting the number of orientations of a particular type in a graph has a surprising diversity of applications. Consider first a success story.

Take a 4-regular graph, such as the 2-dimensional toroidal square lattice. In statistical physics the number of *ice configurations* is the number of distinct ways of orienting the edges in such a way that at each vertex exactly two are directed in.

More generally, on any Eulerian graph we can define:

#EULERIAN CONFIGURATIONS

Input: Eulerian graph G

Task: Find the number of orientations such that at each vertex exactly half the edges are directed in.

Mihail and Winkler [60] proved:

Theorem 7.11 *(i) Exact counting of Eulerian configurations is #P-hard.*

(ii) There exists an fpras for counting Eulerian configurations.

Proof idea The proof of (i) is a reduction from counting perfect matchings. The proof of (ii) is also based on matchings. They reduce the problem to approximately counting perfect matchings in a class of graphs for which the ratio α (defined in (2)) is bounded by a polynomial. ∎

Mihail and Winkler [60] also raised the question of counting Euler tours in a graph. No progress had been made on this until very recently when Tetali and Vempala [73] announced an fpras for the case where the inputs are regular of degree 4 or 6.

At the other end of the success/failure spectrum stands:

#ACYCLIC ORIENTATIONS

Input: Bipartite planar graph G
Task: Find the number $a(G)$ of orientations of G so that there is no directed cycle.

By Vertigan and Welsh [76] this is $\#P$-hard even for planar bipartite graphs. However despite quite a lot of effort, so far there has been absolutely no progress on finding an fpras or producing evidence for nonexistence. Apart from its key position in the Tutte plane (see [78, 79]), counting acyclic orientations is a fairly special and presumably easier case of two fundamental geometric problems, namely counting chambers in arrangements of hyperplanes (see the book by Orlik and Terao [63] or [80]) and counting vertices of totally unimodular zonotopes (see §10).

In the same way that the dual of a colouring is a flow, the concept dual to that of an acyclic orientation is a *totally cyclic* orientation. This is an orientation in which each edge belongs to a directed cycle. Again [76] gives that the following problem is $\#P$-complete:

#TOTALLY CYCLIC ORIENTATIONS

Input: Planar bipartite graph G
Task: Count the number $c(G)$ of totally cyclic orientations.

We know by duality that for planar G, $c(G) = a(G^*)$ and one of the results of Alon, Frieze and Welsh [2] is

Proposition 7.12 *For any class of strongly dense graphs ($\alpha > \frac{1}{2}$) there is an fpras for $c(G)$.*

However the inability of being able to use denseness in combination with duality means that this gives us nothing but guarded optimism in relation to one of our most pressing problems.

Problem 7.13 *Does there exist an fpras for estimating the number of acyclic orientations in any nontrivial class of graphs?*

The tenuously related problem of counting sink-free orientations is the only case where an fpras is known, see Bubley and Dyer [11].

Note All counting problems of this section can be solved *exactly* in P-time for graphs of bounded tree width, see [62].

8 Contingency tables

Given positive integers $r_1, \ldots, r_m$ and $c_1, \ldots, c_n$, let $S(r,c)$ be the set of $m \times n$ arrays with nonnegative integer entries and row (column) sums r_i (c_j) respectively. The members of $S(r,c)$ are called *contingency tables* with these given row and column sums.

Counting the number of contingency tables with given r, c is a problem with many applications, see for example the survey of Diaconis and Gangoli [16]. As might be expected, it is $\#P$-hard. In particular, we have the following recent result [24].

Proposition 8.1 *Determining the cardinality of $S(r,c)$ is $\#P$-hard even if $m = 2$ or $n = 2$.*

As another example of the inherent difficulty of this problem consider the following special case. Let $H_n(r)$ denote the number of such tables when $m = n$ and each $r_i = c_j = r$. The problem now contains the classical problem of counting magic squares and goes back to MacMahon [56] who showed

$$H_3(r) = \binom{r+2}{2} + 3\binom{r+3}{4}.$$

Stanley [72] and Ehrhart [27] show that $H_n(r)$ is a polynomial in r of degree $(n-1)^2$ but to date not even $H_7(r)$ is known.

8.1 An approximation scheme

We now sketch the ideas underlying an approximation scheme for this problem proposed recently by Dyer, Kannan and Mount [24]. As will be seen it is in polynomial time provided certain "density type conditions" are satisfied by the input.

Define the *contingency polytope* $P(r,c)$ by

$$\begin{aligned} P(r,c) &= \{x \in \mathbb{R}^{mn} : \sum_j x_{ij} = r_i \text{ for } i = 1, \ldots, m; \\ &\qquad \sum_i x_{ij} = c_j \text{ for } j = 1, \ldots, n \text{ and } x_{ij} \geq 0\}. \end{aligned}$$

Then $S(r,c)$ is the set of lattice points in $P(r,c)$.

Because of Barvinok's algorithm for counting lattice points in fixed dimension in polynomial time, we have:

- If m, n are fixed then $|S(r,c)|$ can be found in P-time.

The idea of [24] is to associate with each lattice point i in $P(r,c)$ a parallelepiped B_i so that the following conditions hold.

(i) Each box B_i has the same volume

(ii) If $i \neq j$, $B_i \cap B_j = \emptyset$

(iii) The union $B = \bigcup B_i$ over all lattice points in $P(r, c)$ is such that if P^0 is its convex hull then $\text{vol}(B) \setminus \text{vol}(P^0)$ is big enough for the volume of B to be found by using random walk to pick an "almost uniform point" in P^0 using methods described in §5. Using this we can now obtain a good enough estimate of $|S(r, c)|$.

The main result of [24] is that the above randomised algorithm runs in time $t(r, c)$ where

$$t(r, c) \leq \alpha(r, c) \, \text{poly}(n, m, \max_{ij}(\log r_i, \log c_j))$$

and where

$$\alpha(r, c) = \max_{i,j} \left(\frac{r_i + 2n}{r_i - 2n}, \frac{c_j + 2m}{c_j - 2m} \right)^{(n-1)(m-1)}.$$

Provided r_i is $\Omega(n^2 m)$ for all i and c_j is $\Omega(nm^2)$ for all j then $\alpha(r, c)$ is $O(1)$. Thus the algorithm is certainly P-time provided the row and column sums are not too small compared with the dimensions of the array. For example, Diaconis and Holmes [17] report that with $m = n = 4$ the row/column sums need only be ≥ 36.

Related questions in combinatorics are the problems of counting and random generation of $(0, 1)$-matrices with prescribed row and column sums, generating (or counting) labelled graphs with a given degree sequence, and generating a Latin square of order n uniformly at random.

In [41] Jerrum and Sinclair used the Markov chain method to solve the problem of generating regular graphs and various other classes of degree sequence. Very recently, Kannan, Tetali and Vempala [46] have proposed a new algorithm which may work for all degree sequences.

9 Matroid problems

Consider the following open problems.

> Does there exist an fpras for counting (a) forests, or (b) connected subgraphs of a given size, in a graph?

These problems are particular instances of the problem of estimating the number of bases in a matroid of some class. For example the question about forests is just counting the number of bases in truncations of graphic matroids.

Similarly, the problem #BASES described in §2 is a particular instance of the same underlying problem of counting bases in vector matroids.

A *matroid* M is just a pair (E, r) where E is a finite set and r is a *rank function* mapping $2^E \to \mathbb{Z}$ and satisfying the conditions

$$0 \leq r(A) \leq |A| \quad \text{for } A \subseteq E,$$

$$A \subseteq B \Rightarrow r(A) \leq r(B),$$

$$r(A \cup B) + r(A \cap B) \leq r(A) + r(B) \quad \text{for } A, B \subseteq E.$$

The edge set of any graph G with its associated rank function as defined by

$$r(A) = |V(G)| - k(A),$$

where $k(A)$ is the number of connected components of the graph $G : A$ having vertex set $V = V(G)$ and edge set A, is a matroid. This defines a very small subclass of matroids:—known as *graphic matroids.*

Given $M = (E, r)$ the *dual matroid* $M^* = (E, r^*)$ where r^* is defined by

$$r^*(E \setminus A) = |E| - r(E) - |A| + r(A).$$

A set X is *independent* if $r(X) = |X|$, it is a *base* if it is a maximal independent subset of E. An easy way to work with the dual matroid M^* is not via the rank function but by the following definition.

- M^* has as its bases all sets of the form $E \setminus B$, where B is a base of M.

M is *representable* over a field F if it is isomorphic with the matroid induced on the columns of some matrix over F by linear independence. It is *binary* if F can be taken to be GF(2) and is *unimodular* or *regular* if and only if it is representable over GF(2) and the reals. This is equivalent to having a representation over $\mathbb{R}$ by a totally unimodular matrix. For more details see Oxley [64].

In its most general form, and avoiding niceties of how input is to be described, the crucial underlying question may be described as:

Problem 9.1 *How difficult is it to count the number of bases of a matroid?*

In this form, and in terms of standard complexity this is easily answered for the following rather trivial reason. On a set of n elements there are $O(2^{2^n})$ distinct matroids and hence the input size to describe a general matroid is $O(2^n)$. Exhaustive search to count bases can be carried out in time polynomially bounded in the size of this huge input. Hence in this form the problem is in P.

However this is not interesting. Most interesting classes of matroids have what we might call "succinct representations", either as graphs, matrices, or the like. Hence the specific forms of Problem 9.1 that we wish to consider are those when the class of matroids has such a representation.

To see what is known, consider some of our earlier examples.

- Graphic matroids have as bases their spanning trees (or forests in the disconnected case) and these can be counted *exactly* by Kirchhoff's determinantal formula.

- Binary matroids: counting bases is known to be #P-hard by Vertigan [75]. No nontrivial approximation algorithm is known and this is one of the key open problems in approximate counting.

Proposition 9.2 *The number of bases of regular (unimodular) matroids can be counted exactly in polynomial time.*

Proof Given, a totally unimodular matrix A coordinatising the regular matroid M, then $\det(AA^T)$ is the number of bases of M. This follows from the Binet-Cauchy expansion and the fact that all maximal minors have determinant ± 1 or zero. ∎

For general matroids having some reasonable presentation, Azar, Broder and Frieze [5] prove the following negative result. Suppose that a matroid is presented by an *independence oracle* (see [65]) which when presented by a set A determines if it is independent and in such a case provides a base of the matroid containing A. Such a call to the oracle is called a *probe*. The main result of [5] is that any deterministic algorithm which relies on probes must make an exponential number in order to approximate the number of bases to within some constant ratio. However, on the positive side, it is shown in [13] that for what is believed to be a majority of matroids, a randomised algorithm based on probes gives a $(1+\varepsilon)$-approximation in polynomial time.

9.1 The matroid polyhedron

The *matroid polyhedron* $P(M)$ is defined to be the convex polytope in $\mathbb{R}^E$ consisting of the intersection of half spaces

$$\sum_{e\in A} x_e \le r(A), \quad A \subseteq E, \quad x_e \ge 0.$$

The fundamental property of this polytope proved in [25] is

Theorem 9.3 *$P(M)$ has all its vertices integral, moreover they are the* $(0,1)$ *incidence vectors of the independent sets of M.*

Suppose now that we wish to find f_k, the number of forests of size k, where k may be a function of n. In general this is a #P-hard problem.

Geometrically it means we want to count the integer points in the slice of $\mathbb{R}^E$ given by

$$S_G^{(k)} \equiv P(M(G)) \cap \left\{ \sum x_i = k \right\}.$$

Thus one approach to the problem of generating a random forest in G, would be to carry out random walk on the 1-skeleton of this polytope $P(M(G))$ and hope that the convergence to stationarity is rapid.

The polytope is well understood. We know the vertices and adjacency in the 1-skeleton is completely characterised in the following theorem from Hausmann and Korte [36] who prove

Proposition 9.4 *Two distinct independent sets F_1, F_2 are adjacent if and only if $|F_1 \Delta F_2| = 1$ or $|F_1 \Delta F_2| = 2$ and $F_1 \cup F_2$ is dependent.*

However showing random walk on this polytope is rapidly mixing is a special case of the much wider conjecture about general $(0,1)$-polytopes made by Mihail and Vazirani [59] see [74]. This can be stated as follows.

The polytope conjecture: *For any bipartition of the vertices of a $(0,1)$-polytope, the number of edges in the cut set is at least as large as the number of vertices in the smaller block of the partition.*

The importance of this algorithmically is as follows.

Consider a class $\mathcal{C}$ of polytopes satisfying the additional conditions:

(i) There is a polynomial p such that the maximum degree of a vertex in an n-dimensional member of $\mathcal{C}$ is bounded by $p(n)$.

(ii) There is a P-time algorithm for enumerating the neighbours of a vertex in the polytope.

(iii) There is a P-time algorithm which outputs a vertex of the polytope.

Then there is a P-time algorithm which will generate almost uniformly at random a vertex of a polytope in the class $\mathcal{C}$.

Two important classes of polytopes which satisfy the conditions (i)—(iii) are: (a) matroid polyhedra, and (b) basis polytopes of matroids, provided they are suitably presented.

In case (b) the 1-skeleton is the usual *basis exchange graph*, in which two bases B_1, B_2 are adjacent vertices if and only if $|B_1 \oplus B_2| = 2$.

We close this section by stating a special case of the Polytope Conjecture which is purely combinatorial.

The cut set expansion conjecture: *The basis exchange graph $G(M)$ of any matroid has the property that for any bipartition (V_1, V_2) of the vertices of $G(M)$, the number of edges in the cut set (V_1, V_2) is at least as large as* $\min(|V_1|, |V_2|)$.

Notice that this is demanding more than is needed for random walk on basis graphs to be rapidly mixing.

An interesting new light on these problems is outlined in the paper of Feder and Mihail [30] who show that a sufficient condition for the natural random walk on the set of bases of M to be rapidly mixing is that M and its minors are *negatively correlated.* That is, they satisfy the constraint, that if B_R denotes a random base of M and e, f are distinct elements of $E(M)$, then

$$\Pr\{e \in B_R \mid f \in B_R\} \leq \Pr\{e \in B_R\}. \tag{3}$$

This concept goes back to 1975 when Seymour and Welsh [67] related it with the log-concavity of various matroid quantities. It is mildly intriguing that this concept has reappeared now in an entirely new context.

Call matroids all of whose minors satisfy (3) *balanced.* Balanced binary matroids include regular matroids but "not all that much more" since any binary matroid containing a specific 8-element S_8 (see Oxley [64] or [67]) as a minor is not balanced. Since we can count exactly bases in regular matroids the practical improvement produced by balance is not that significant. However the proof techniques are.

10 Zonotopes

Zonotopes form a particularly well structured class of convex polytopes. They can be viewed as projections of cubes or as Minkowski sums of line segments. Formally, a d-dimensional *zonotope* $Z = Z[a_1, \ldots, a_n]$ consists of a subset Z of $\mathbb{R}^d$ which can be expressed in the form

$$Z = \{x \in \mathbb{R}^d : x = c + \sum_{i=1}^{n} \lambda_i a_i,\ 0 \leq \lambda_i \leq 1\}$$

In other words, forgetting about the translation vector c, if A is the matrix $[a_1, \ldots, a_n]$ then $Z = Z[A]$ is the *Minkowski sum* of the line segments $S_i = [0, a_i]$, $1 \leq i \leq n$.

The zonotope is *unimodular* if A is a totally unimodular matrix. Unimodular zonotopes have particularly nice features, see for example the following theorem of McMullen [58].

Theorem 10.1 *The zonotope $Z[A]$ tiles (or packs) $\mathbb{R}^d$ if and only if A is unimodular.*

Here we shall be concerned with some of the combinatorial features of zonotopes, particularly those related to problems discussed earlier.

Since the zonotopes we consider will usually be described by giving a list of line segments we call this an *S-presentation* as in [33]. Now given an $r \times n$ matrix A whose columns determine $Z[A]$, a classical formula gives

$$\operatorname{vol} Z[A] = \sum_{1 \leq i_1 < i_2 < \cdots < i_r \leq n} |\det(a_{i_1}, \ldots, a_{i_r})| \tag{4}$$

Computing this in general is #P-hard, [23] but, in contrast we have:

Proposition 10.2 *There is a P-time algorithm for computing the volume of a unimodular zonotope from its $\mathcal{S}$-presentation.*

Proof The sum on the right hand side of (4) is just the number of bases of the matroid represented by the totally unimodular matrix A and we know that this is in P by Proposition 9.4. ∎

10.1 The Ehrhart polynomial

Classical results of Ehrhart [26] (see also [28]) show that any convex polytope P whose vertices are points of a lattice in $\mathbb{R}^d$ determines a polynomial which counts the number of lattice points of L contained in the integer expansions of P. First recall that a *lattice* L in $\mathbb{R}^d$ is a discrete subset of $\mathbb{R}^d$ of the following form: there is a basis $\{v_1, \ldots, v_d\}$ such that

$$L = \left\{ \sum_{i=1}^{d} m_i v_i : m_1, \ldots, m_d \in \mathbb{Z} \right\}$$

An *L-polytope* is one which has its vertices members of L. Ehrhart's results may be stated in the following form:

Theorem 10.3 *For each lattice L and L-polytope P of $\mathbb{R}^d$ there exists a polynomial*

$$I(P;\lambda) = \sum_{j=0}^{d} c_j(P)\lambda^j$$

with the property that for each positive integer k, the number of points of L contained in kP is given by $I(P;k)$.

Furthermore the number of points of L contained in the interior of P is given by

$$\hat{I}(P;k) = (-1)^{\dim(P)} I(P;-k). \tag{5}$$

The polynomial I is called the *Ehrhart polynomial* of P with respect to L and (5) is often referred to as the reciprocity law.

The coefficients $c_j(P)$ obviously also depend on the lattice L. In the case where L is the hypercubical lattice $\mathbb{Z}^d$ the leading term $c_d(P) = \mathrm{vol}(P)$.

For further details see the survey of Gritzmann and Wills [34].

We now relate the Ehrhart polynomial of a unimodular zonotope with the more familiar Tutte polynomial of the underlying matroid.

Recall that for any matroid M, the *Tutte polynomial* is defined by

$$T(M;x,y) = \sum_{A \subseteq E(M)} (x-1)^{r(E)-r(A)}(y-1)^{|A|-r(A)}.$$

Routine checking shows that

$$T(M;x,y) = T(M^*;y,x).$$

In particular, when G is a planar graph and G^* is any plane dual of G,

$$T(G;x,y) = T(G^*;y,x).$$

Of the problems we have already discussed many are special cases of evaluating T at appropriate points in the (x,y)-plane. For example, the number of bases is just $T(M;1,1)$. Other evaluations are

- $T(M;2,1)$ is the number of independent sets (forests in a graph).
- $T(M;2,0)$ is the number of acyclic orientations when $M = M(G)$.
- $T(G;1-k,0)$ is the number of k-colourings.
- When G is connected the reliability probability $R(G;p)$ is given by

$$R(G;p) = p^{|V|-1}(1-p)^{|E|-|V|+1}T(M(G);1,\frac{1}{1-p}).$$

Many more applications of T to topics as diverse as knots, statistical physics and codes are given in Welsh [77]. A recent evaluation which is not given there is the following which can be deduced by combining work of McMullen [57], Stanley [71] with known interpretations of T from [77].

Theorem 10.4 *Let M be a rank r regular matroid on n elements and let A be an $r \times n$ totally unimodular representation of M. Then the unimodular zonotope $Z[A]$ has Ehrhart polynomial given by*

$$I(Z[A];\lambda) = \lambda^r T(M;1+\frac{1}{\lambda},1).$$

Turning now to counting lattice points in a zonotope, Gritzmann and Klee [33] give a simple argument showing that the existence of a P-time algorithm for this would allow the recovery of the volume using Ehrhart polynomials and Lagrange interpolation. However, since computing volumes of unimodular zonotopes is in P, this argument does not apply to the unimodular case. Nevertheless, we still have:

Proposition 10.5 *Consider the problems of counting lattice points and vertices of a unimodular zonotope from an S-presentation. Then the following statements are true.*

(a) Both problems are $\#P$-hard.

(b) *If counting lattice points has an fpras then so does #FORESTS.*

(c) *If counting either interior lattice points or vertices has an fpras then so does #ACYCLIC ORIENTATIONS.*

A proof of these results is obtained by considering the polytope P_G of $\mathbb{R}^{|V|}$ defined by

$$\begin{aligned} \sum_{i \in U} x_i &\leq e(U) \quad \text{for } U \subseteq V(G), \\ x_i &\geq 0, \end{aligned}$$

where $e(U)$ is the number of edges incident with U.

We prove in [6] what can be regarded as a polytopal analogue of Theorem 7.7 namely,

Theorem 10.6 *There exists an fpras for the number of lattice points in P_G for all dense graphs G.*

It also turns out that the intersection of P_G with $\sum x_i = |E|$ is a unimodular zonotope S_G whose Ehrhart polynomial is given by

$$I(S_G; \lambda) = \lambda^{r(E)} T(G; 1 + \frac{1}{\lambda}, 1).$$

Since the method of proof of Theorem 10.6 is completely different from the earlier comparable results described in §7 which also demanded some form of "denseness" in the input this is slight evidence that denseness is indeed necessary for an fpras in these cases. It adds further interest to the recent attempt by Frieze and Kannan [31] to understand the role of denseness in many of these problems.

Finally, although all $\#P$-complete problems "are equal" as regards exact computation it does seem to be true that there are different levels of difficulty when it comes to approximation. Of those not infeasible by virtue of links with NP-complete problems, the two questions of approximating the number of lattice points and vertices of a unimodular zonotope seem the hardest considered here. Producing an algorithm or showing it is not possible would be a major advance.

Acknowledgements

I am grateful to John Mount for allowing me to use the example in §1 and also for helpful discussions with him. Further discussions with Eric Bartels, Ravi Kannan and Prasad Tetali have also been very helpful.

Research partially supported by RAND-REC EC US030.

References

[1] D. Aldous, Random walks on finite groups and rapidly mixing Markov chains, in *Séminaire de Probabilités XVII 1981/1982*, *Lecture Notes in Mathematics*, 986, Springer-Verlag, Berlin (1983), pp. 243–297.

[2] N. Alon, A. M. Frieze & D. J. A. Welsh, Polynomial time randomised approximation schemes for Tutte-Grothendieck invariants: the dense case, *Random Structures and Algorithms*, **6** (1995), 459–478.

[3] J. D. Annan, A randomised approximation algorithm for counting the number of forests in dense graphs, *Combinatorics, Probability and Computing*, **3** (1994), 273–283.

[4] D. Applegate & R. Kannan, Sampling and integration of near log-concave functions, in *Proceedings of the 23rd Annual Association for Computing Machinery Symposium on Theory of Computing*, ACM Press, New York (1990), pp. 156–163.

[5] Y. Azar, A. Z. Broder & A. M. Frieze, On the problem of approximating the number of bases of a matroid, *Information Processing Letters*, **50** (1993), 9–11.

[6] E. Bartels, J. Mount & D. J. A. Welsh, The win polytope of a graph, *Annals of Combinatorics*, (1997), in press.

[7] E. Bartels & D. J. A. Welsh, The Markov chain of colourings, in *Proceedings of the Fourth Conference on Integer Programming and Combinatorial Optimisation*, *Lecture Notes in Computer Science*, 920, Springer-Verlag, Berlin (1995), pp. 373–387.

[8] A. I. Barvinok, A polynomial time algorithm for counting integral points in polyhedra when the dimension is fixed, in *Proceedings of the 34th Annual IEEE Symposium on Foundations of Computer Science* (1993), pp. 566–572.

[9] G. Brightwell & P. Winkler, Counting linear extensions, *Order*, **8** (1991), 225–242.

[10] A. Z. Broder, How hard is it to marry at random?, in *Proceedings of the Eighteenth Annual Association for Computing Machinery Symposium on Theory of Computing, (Berkeley, California)*, ACM Press, New York (1986), pp. 50–58.

[11] R. Bubley & M. Dyer, Graph orientations with no sink and an approximation for a hard case of #SAT, preprint, 1996.

[12] R. Bubley, M. Dyer and M. Jerrum, A new approach to polynomial-time generation of random points in convex bodies, preprint, 1996.

[13] L. Chávez Lomelí & D. J. A. Welsh, Randomised approximation of the number of bases, in *Matroid Theory* (eds. J. E. Bonin, J. G. Oxley & B. Servatius), *Contemporary Mathematics*, 197, American Mathematical Society, Providence, Rhode Island (1996), pp. 371–376.

[14] W. J. Cook, M. Hartmann, R. Kannan & C. J. H McDiarmid, On integer points in polyhedra, *Combinatorica*, **12** (1992), 27–37.

[15] A. Denise, M. Vasconcellos & D. J. A. Welsh, The random planar graph, *Congressus Numerantium*, **13** (1996), 61–79.

[16] P. Diaconis & A. Gangoli, Rectangular arrays with fixed margins, in *Discrete Probability and Algorithms* (eds. D. Aldous, P. Diaconis, J. Spencer & J. M. Steele), *IMA Volumes in Mathematics and its Applications*, 72, (1995), pp. 15–42.

[17] P. Diaconis & S. Holmes, Three examples of Monte-Carlo Markov chains: at the interface between statistical computing, computer science, and statistical mechanics, in *Discrete Probability and Algorithms* (eds. D. Aldous, P. Diaconis, J. Spencer & J. M. Steele), *IMA Volumes in Mathematics and its Applications*, 72, (1995), pp. 43–56.

[18] M. E. Dyer & A. M. Frieze, The complexity of computing the volume of a polyhedron, *SIAM Journal on Computing*, **17** (1988), 967–974.

[19] M. E. Dyer & A. M. Frieze, Computing the volume of convex bodies: A case where randomness provably helps, in *Probabilistic Combinatorics and its Applications* (ed. B. Bollobás), *Symposia in Applied Mathematics*, 44, American Mathematical Society, Providence, Rhode Island (1991), pp. 123–169.

[20] M. E. Dyer & A. M. Frieze, Random walks, totally unimodular matrices and a randomised dual simplex algorithm, in *Proceedings of the Second Conference on Integer Programming and Combinatorial Optimisation, (Carnegie Mellon University, 1992)*, (1992), pp. 72–84.

[21] M. E. Dyer, A. M. Frieze & R. Kannan, A random polynomial-time algorithm for approximating the volume of convex bodies, *Journal of the Association for Computing Machinery*, **38** (1991), 1–17.

[22] M. Dyer, A. Frieze, R. Kannan, A. Kapoor, L. Perkovic & U. Vazirani, Approximating the number of solutions to a multidimensional knapsack problem, *Combinatorics, Probability and Computing*, **2** (1993), 271–284.

[23] M. E. Dyer, P. Gritzmann & A. Hufnagel, On the complexity of computing (mixed) volumes, preprint, Universität Trier, 1994.

[24] M. Dyer, R. Kannan & J. Mount, Sampling contingency tables, preprint, 1996.

[25] J. Edmonds, Submodular functions, matroids, and certain polyhedra, in *Combinatorial Structures and their Applications* (eds. R. Guy, H. Hanani, N. Sauer & J. Schönheim), Gordon and Breach, New York (1970), pp. 69–87.

[26] E. Ehrhart, Sur un problème de géometrie diophantienne linéaire I, II, *Journal für die Reine und Angewandte Mathematik*, **226** (1967), 1–29, and **227** (1967), 25–49. Correction **231** (1968), 220.

[27] E. Ehrhart, Sur les carrés magiques, *Comptes Rendus de l'Académie des Sciences de Paris*, **227 A** (1973), 575–577.

[28] E. Ehrhart, *Polynômes Arithmétiques et Méthodes des Polyédres en Combinatoire*, Birkhäuser, Basel (1977).

[29] G. Elekes, A geometric inequality and the complexity of computing volume, *Discrete and Computational Geometry*, **1** (1986), 289–292.

[30] T. Feder & M. Mihail, Balanced matroids, in *Proceedings of the 24th Annual Association for Computing Machinery Symposium on the Theory of Computing*, ACM Press, New York (1992), pp. 26–38.

[31] A. Frieze & R. Kannan, The regularity lemma and approximation schemes for dense problems, *Foundations of Computer Science*, (1996), in press.

[32] M. R. Garey & D. S. Johnson, *Computers and Intractability—A guide to the theory of NP-completeness*, W. H. Freeman, San Francisco (1979).

[33] P. Gritzmann & V. Klee, Some basic problems in computational convexity II. Volume and mixed volumes, in *Polytopes: Abstract, Convex and Computational* (eds. T. Bisztriczky, P. McMullen, R. Schneider & A. Ivic Weiss), Kluwer, Boston (1994), pp. 373–466.

[34] P. Gritzmann & J. M. Wills, Lattice points, in *Handbook of Convex Geometry B* (eds. P. M. Gruber & J. M. Wills), North Holland, Amsterdam (1993), pp. 765–798.

[35] M. Grötschel, L. Lovász & A. Schrijver, *Geometric Algorithms and Combinatorial Optimization*, Springer Verlag, Berlin (1988).

[36] D. Hausmann & B. Korte, Colouring criteria for adjacency on 0-1 polyhedra, in *Mathematical Programming Study*, 8, (eds. M. L. Balinski & A. J. Hoffman), North Holland, New York (1978), pp. 106–127.

[37] F. Jaeger, D. L. Vertigan & D. J. A. Welsh, On the computational complexity of the Jones and Tutte polynomials, *Mathematical Proceedings of the Cambridge Philosophical Society*, **108** (1990), 35–53.

[38] M. R. Jerrum, A very simple algorithm for estimating the number of k-colourings of a low-degree graph, *Random Structures and Algorithms*, **7** (1995), 157–165.

[39] M. R. Jerrum, Counting trees in a graph is $\#P$-complete, *Information Processing Letters*, **51** (1994), 111-116.

[40] M. R. Jerrum & A. Sinclair, Approximating the permanent, *SIAM Journal on Computing*, **18** (1989), 149–178.

[41] M. Jerrum & A. J. Sinclair, Fast uniform generation of regular graphs, *Theoretical Computer Science*, **73** (1990), 91–100.

[42] M. R. Jerrum, L. G. Valiant & V. V. Vazirani, Random generation of combinatorial structures from a uniform distribution, *Theoretical Computer Science*, **43** (1986), 169–188.

[43] J. Kahn & M. Saks, Every poset has a good comparison, in *Proceedings of the 25th Annual IEEE Symposium on Foundations of Computer Science* (1984), pp. 299–301.

[44] R. Kannan, Markov chains and polynomial time algorithms, in *Proceedings of the 35th Annual IEEE Symposium on Foundations of Computer Science (Santa Fe, New Mexico)*, (1994), pp. 656–671.

[45] R. Kannan, L. Lovász & M. Simonovits, Random walks and an $O^*(n^5)$ volume algorithm for convex bodies, preprint, January 1996.

[46] R. Kannan, P. Tetali & S. Vempala, Simple Markov chain algorithms for generating bipartite graphs and tournaments, in *Proceedings of the 7th Annual ACM-SIAM Symposium on Discrete Algorithms*, ACM Press, New York (1997), in press.

[47] D. R. Karger, A randomised fully polynomial time approximation scheme for the all terminal network reliability problem, in *Proceedings of the 36th Annual IEEE Symposium on Foundations of Computer Science*, (1995), pp. 328–337.

[48] A. Karzanov & L. G. Khachiyan, On the conductance of order Markov chains, Technical Report DCS TR 268, Rutgers University, 1990.

[49] C. Kenyon, D. Randall & A. J. Sinclair, Matchings in lattice graphs, in *Proceedings of the 25th Symposium on Theoretical Computer Science* (1993), pp. 738–746.

[50] L. Lovász, Geometric algorithms and algorithmic geometry, in *Proceedings of the International Congress of Mathematicians (Kyoto, Japan)* (1990), pp. 139–154.

[51] L. Lovász, How to compute the volume?, in *Jahresberichte der Deutschen Mathematiker-Vereiningung, Jubiläumstagung 1990*, B. G. Teubner, Stuttgart (1992), pp. 138–151.

[52] L. Lovász, Random walks on graphs: a survey, in *Combinatorics, Paul Erdős is Eighty* (Volume 2) (eds. D. Miklós, V. T. Sós & T. Szőnyi), *Bolyai Society Mathematical Studies*, 2, Bolyai Society, Budapest (1996), 353–398.

[53] L. Lovász & M. Simonovits, Mixing rate of Markov chains, an isoperimetric inequality, and computing the volume, in *Proceedings of the 31st Annual IEEE Symposium on Foundations of Computer Science*, (1990), pp. 346–355.

[54] L. Lovász & M. Simonovits, On the randomized complexity of volume and diameter, in *Proceedings of the 33rd Annual IEEE Symposium on Foundations of Computer Science*, (1992), pp. 482–491.

[55] L. Lovász & M. Simonovits, Random walks in a convex body and an improved volume algorithm, *Random Structures and Algorithms*, **4** (1993), 359–412.

[56] P. MacMahon, *Combinatorial Analysis*, Cambridge University Press, Cambridge (1916).

[57] P. McMullen, On zonotopes, *Transactions of the American Mathematical Society*, **159** (1971), 91–109.

[58] P. McMullen, Space-tiling zonotopes, *Mathematika*, **22** (1975), 202–211.

[59] M. Mihail & U. Vazirani, On the magnification of 0-1 polytopes, Technical Report TR 05-89, Harvard University, 1989.

[60] M. Mihail & P. Winkler, On the number of Eulerian orientations of a graph, Technical Memorandum TM-ARH-018829, Bellcore, 1991.

[61] R. Motwani & P. Raghavan, *Randomised Algorithms*, Cambridge University Press, Canbridge (1995).

[62] S. D. Noble, Evaluating the Tutte polynomial for graphs of bounded tree-width, preprint, 1995.

[63] P. Orlik & H. Terao, *Arrangements of Hyperplanes*, Springer Verlag, Berlin (1991).

[64] J. G. Oxley, *Matroid Theory*, Oxford University Press, Oxford (1992).

[65] G. C. Robinson & D. J. A. Welsh, The computational complexity of matroid properties, *Mathematical Proceedings of the Cambridge Philosophical Society*, **87** (1980), 29–45.

[66] C. P. Schnorr, Optimal algorithms for self-reducible problems, in *Proceedings of the 3rd International Colloquium on Automata, Languages and Programming* (1976), pp. 322–337.

[67] P. D. Seymour & D. J. A. Welsh, Combinatorial applications of an inequality from statistical mechanics, *Mathematical Proceedings of the Cambridge Philosophical Society*, **77** (1975), 485–497.

[68] A. Sinclair, Randomised Algorithms for Counting and Generating Combinatorial Structures, Ph.D. thesis, University of Edinburgh, 1988.

[69] A. J. Sinclair, Improved bounds for mixing rates of Markov chains and multicommodity flow, *Combinatorics, Probability and Computing*, **1** (1992), 351–370.

[70] A. J. Sinclair & M. R. Jerrum, Approximate counting, uniform generating and rapidly mixing Markov chains, *Information and Computing*, **82** (1989), 93–133.

[71] R. Stanley, *Enumerative Combinatorics*, Wadsworth, Monterey (1973).

[72] R. Stanley, Linear homogeneous diophantine equations and magic labelings of graphs, *Duke Mathematical Journal*, **40** (1973), 607–632.

[73] P. Tetali & S. Vempala, Generating Euler tours uniformly at random, preprint, 1996.

[74] U. Vazirani, Rapidly mixing Markov chains, in *Probabilistic Combinatorics and its Applications* (ed. B. Bollobás), *Symposia in Applied Mathematics*, 44, American Mathematical Society, Providence, Rhode Island (1991), 99–121.

[75] D. L. Vertigan, private communication, 1992.

[76] D. L. Vertigan & D. J. A. Welsh, The computational complexity of the Tutte plane: the bipartite case, *Probability, Combinatorics and Computer Science*, **1** (1992), 181–187.

[77] D. J. A. Welsh, *Complexity: Knots, Colourings and Counting*, *London Mathematical Society Lecture Note Series*, 186, Cambridge University Press, Cambridge (1993).

[78] D. J. A. Welsh, Randomised approximation in the Tutte plane, *Combinatorics, Probability and Computing*, **3** (1993), 137–143.

[79] D. J. A. Welsh, Randomised approximation schemes for Tutte-Grothendieck invariants, in *Discrete Probability and Algorithms* (eds. D. Aldous, P. Diaconis, J. Spencer & J. M. Steele), *IMA Volumes in Mathematics and its Applications*, 72, (1995), pp. 133–148.

[80] T. Zaslavsky, *Facing up to arrangements: face count formulas for partitions of spaces by hyperplanes*, *Memoirs of the American Mathematical Society*, 154, (1975).

[81] D. Zuckerman, NP-complete problems have a version that's hard to approximate, in *Proceedings of the 8th Annual IEEE Conference on Structure in Complexity Theory* (1993), pp. 305–312.

Mathematical Institute
University of Oxford
24–29 St. Giles
Oxford OX1 3LB
dwelsh@maths.ox.ac.uk

Author Index

Subject Index

Printed in the United States
By Bookmasters